Graduate Texts in Mathematics 6

Managing Editor: P. R. Halmos

D. R. Hughes F. C. Piper

Projective Planes

Springer-Verlag New York Heidelberg Berlin

Daniel R. Hughes

Professor of Mathematics, Westfield College, University of London

Fred C. Piper

Reader in Pure Mathematics, Westfield College, University of London

AMS Subject Classification (1970)
Primary: 05 B 25, 17 E 05, 50 D 35; Secondary: 94 A 10, 62 K 99

© 1973 by Springer-Verlag New York Inc.
Library of Congress Catalog Card Number 72-75719.

Printed in the United States of America.

ISBN 0-387-90044-6 Springer-Verlag New York Heidelberg Berlin (soft cover)
ISBN 0-387-90043-8 Springer-Verlag New York Heidelberg Berlin (hard cover)
ISBN 3-540-90044-6 Springer-Verlag Berlin Heidelberg New York (soft cover)

Preface

This book has grown out of notes on lecture courses which we have given on projective planes and related topics. It has two main aims: to provide an elementary treatment of the subject, suitable for advanced undergraduates, and yet to give enough insight into the research areas that it may also be used for postgraduate courses. The subject matter is such that there are virtually no mathematical prerequisites for studying many of the earlier chapters, although a maturity in the basic concepts of modern algebra is often an advantage.

We would like to thank Michigan State University for providing us with a home during the summer of 1970, when we finished most of the manuscript. We are also indebted to the Consiglio Nazionale delle Ricerche and the universities of Rome and Perugia for the assistance they gave the first author in the academic year 1970-71.

Many people have helped us with the material and with the preparation of the manuscript. Our original version was read, criticized and improved by Dr. Marion Kimberley, and the finished manuscript was studied in similar fashion by Professor Heinz Lüneburg. Both of them corrected so many errors and suggested so many improvements that it is impossible for us to thank them adequately. Similarly we are indebted to Dr. C. W. Norman for reading various sections and providing us with many clarifications, examples and problems. Finally we must thank Dr. P. D. Chawathe and Professor C. W. Garner for the care and precision with which they read the proofs.

London
November 8, 1972

D. R. Hughes
F. C. Piper

v

Table of Contents

Projective Planes

Introduction

The modern study of projective planes, which has largely but not entirely meant non-desarguesian and finite projective planes, has grown enormously in the past twenty years. In 1951 it was mostly a collection of ad hoc and somewhat unrelated results; the fundamental work of Reinhold Baer and of Marshall Hall had taken place, the Bruck-Ryser Theorem had been proved, and the Bruck-Kleinfeld Theorem was to appear that year. But one could not speak of a coherent theory in any sense. Today the situation is completely different, the subject has acquired the shape and outline of a developed area and it is possible to think of the theory of projective planes. This book is meant to be a text-book in the field, an introduction for both undergraduates and postgraduates to a fascinating branch of mathematics which has many fruitful connections with other fields and many interesting unsolved problems; a book which can also prepare the student with a little knowledge of algebra for reading the literature in the field and for research.

Chapter I reviews the main ideas required from algebra; here we must assume a certain basic knowledge of group theory and of linear algebra, but much of the theory of skewfields and fields which will be needed can be at least referred to. However it is important to note that many parts of the book can be studied without any understanding of Chapter I; in particular, it is easy to construct an elementary course in projective planes which avoids the necessity for Chapter I. Chapter II is the longest in the book, and deals with classical projective planes (to a small extent, also with classical projective geometry); many students today may not have had any course whatsoever in classical geometry and this chapter is meant to provide a basic understanding of the subject, especially of classical projective planes. Despite its length this chapter can be skipped if the student knows classical projective geometry (though reference to particular results might have to be made later), or also if the student is willing to forego understanding completely the background or the motivation of certain results in the later chapters. In fact Chapter II contains much of the material which might be found in a one term undergraduate geometry course; very little needs to be added (and that mostly about higher dimensions) to make this chapter alone into a suitable course.

The rest of the book is concerned with projective planes from the more general and modern point of view mentioned above. Chapters III to VI are basic. They should be studied in that order and may even be taken on their own for an elementary course. (Perhaps omitting the last four sections of Chapter VI in the case of very elementary courses.) Chapters III and IV are introductory, with the basic combinatorial results in Chapter III and the elementary development of automorphisms or collineations in Chapter IV. In Chapter V coordinates are introduced and some of the simplest properties of planar ternary rings are worked out; latin squares are introduced briefly at the end of this chapter. Chapter VI develops many of the most important equivalences between algebraic properties of planar ternary rings and geometric properties of projective planes; the latter parts of this chapter become more difficult and include the fundamental results on inverse property division rings (Moufang planes) which are necessary for the rest of the book. Chapter VII contains a detailed study of quasifields and is followed, in Chapter VIII, by a similar discussion of division rings. The results of these two chapters are then exploited in Chapter IX to give examples of non-desarguesian projective planes. These three chapters, which should be studied in the order given, are fairly important and should be included in any complete course on the subject. Chapter XI studies the generation of projective planes, especially "free" generation, and gives further examples of planes. This chapter is completely coordinate free and could be read immediately after Chapter VI if desired. (In fact it could even be studied after Chapter IV.) Chapter X gives yet another method of constructing planes, but cannot be read before Chapter IX. Polarities are studied in Chapter XII and the problems concerned with the sets of absolute points are discussed in detail. This chapter contains the very important work of Baer together with Segre's Theorem on ovals in finite desarguesian planes. Examples of polarities of non-desarguesian planes are given and so this chapter cannot be read before Chapter IX. Chapter XIII is possibly the most advanced chapter in the book. It collects together a number of more advanced theorems about collineations or collineation groups: the extremely important orbit theorem is in this chapter, and also some results about groups containing perspectivities (leading to more characterizations of desarguesian projective planes). In addition, Singer groups and Hall's multiplier theorem are in this chapter (in fact Singer groups are also studied in Chapter II). Finally Chapter XIV proves Wagner's theorem and includes its important corollary, the Ostrom-Wagner Theorem.

As we have already indicated there are now many possibilities for constructing a course from this book. A fairly elementary course can be made up of Chapters III, IV, V, and part of VI; a slightly more advanced course might include Chapters VII, VIII, and IX, and perhaps X or XI or both. Probably some attention should be paid to Chapter II in any advanced course. Finally, the last three chapters offer the kind of advanced material that would round off a one-year postgraduate course. Chapter II requires familiarity with linear algebra, and the last three chapters will be hard going for a student without a decent background in group theory. For those whose interests are purely combinatorial, these three chapters, with their emphasis on groups, might seem redundant: but in fact both Chapters XII and XIII contain much material of combinatorial interst (e.g., difference sets, unitals).

We have included a minimum of references, always grouped together at the end of the chapters. One reference is indispensable, but will hardly ever be mentioned: "Finite Geometries", by Peter Dembowski. It is probably too difficult a book to be used as a text, and it is concerned with all sorts of finite combinatorial structures, not just projective planes (and also it is not concerned at all with infinite structures, hence not with infinite projective planes); but no serious student of the area can be without it. It has a bibliography so complete that we have felt no guilt at having one so sparse.

Notationally, we have adopted some conventions, but we have felt free to adapt or modify these whenever it is convenient or comfortable. Such systems as skewfields, fields, planar ternary rings and so on are in upper case Latin letters, and their elements are lower case Latin; also vector spaces are upper case Latin, but vectors are (at least in Chapter II) lower case bold face (hence scalars and vectors are easier to tell apart). Mappings are usually lower case Greek, written as superscripts, but linear and semilinear transformations are bold face lower case Greek, to distinguish them from the collineations they induce; some mappings (e.g., the ternary mapping) are upper case Latin, and not superscripts (and usually mappings appear on the right, but again this is violated by the ternary mapping). Groups are upper case Greek and their elements are lower case Greek (since most groups encountered here are groups of mappings), but again at least once we use Latin letters for groups and their elements (in a context where we want Greek for the elements of a group algebra). Sets in general are upper case script. The points and lines of a projective plane are upper case and lower case Latin, respectively (at least after Chapter II), the plane itself is upper case script.

In Chapter II we refer to both points and lines by capital letters, since we are thinking of them as subspaces. But this need cause no confusion, as the point of view changes so drastically anyway after Chapter II. For instance in Chapter II we speak of "projective plane" as if the only examples there were those constructed over skewfields; in fact they are the *desarguesian* planes of the rest of the book. But if a student does only Chapter II he hardly need concern himself with this, and if he starts with Chapter III then "desarguesian planes" are simply those whose coordinatizing planar ternary ring is a skewfield; the proof that the two attitudes come to the same thing is in Chapter VI. When dealing with projective planes over skewfields it is possible to construct them using either a right or left vector space; the difference will be only notational over a field of course. We have used left spaces more than right in Chapter II, since this is convenient when studying collineations and so forth. But for the rest of the book, whenever we refer to the desarguesian projective plane $\mathscr{P}_2(K)$ over the skewfield K, we mean the one constructed by means of the *right* vector space.

We deal with higher dimensional projective geometries only casually, but the reader may be curious about the possibility of similar "non-desarguesian" higher dimensional geometries: using the usual *synthetic* definitions of a projective geometry, it can be shown that in the case of g-dimension > 2 any projective geometry is isomorphic to one constructed over a skewfield. This has the concise corollary: a projective plane is desarguesian if and only if it can be embedded in a projective geometry of g-dimension three.

Most chapters contain a large number of exercises and they should form an integral part of any course. Those marked with an asterisk are considered to be difficult. The standard of the others varies considerably and the reader should not assume that he is wrong if a few appear to be trivial; occasionally we include a trivial exercise merely to bring the conclusion to the reader's notice or, alternatively, so that we may refer back to it.

I. Review of Basic Algebra

1. Introduction

In this chapter we give a brief review of the algebraic definitions and results which will be needed later in the book. For the most part these results are ones which the student will have met in his introductory courses, but we state them so that we can easily refer to them in the later chapters. Occasionally, particularly in the section on permutation groups, we include proofs of results. This is not an indication that the result is difficult, but merely that the result, in the form that we want it, is possibly not taught in the elementary courses. Any reader with a strong algebraic background can skip this chapter and refer back to it if, in the later chapters, we make references to results unknown to him. However, since all the results given here will be used in the book, it will be helpful to look through the chapter.

The results on fields, skewfields, vector spaces and linear algebra are very important for Chapter II but, apart from some of the simplest properties of finite fields, are not really necessary for the other chapters. The section on group theory, on the other hand, is of maximal importance in a number of the later chapters. However the fundamental introductory chapters on projective planes, (i.e. Chapters III, IV, V) may be understood by a reader with virtually no knowledge of this chapter.

At the end of the chapter we list a few books which, between them, contain proofs of all the results listed in this chapter. The linear algebra is contained in [1, 3] the abstract group theory in the early chapters of [2] and the permutation group theory is in the beginning of [5]. The results on skewfields, including a proof of Wedderburn's theorem (Result 1.3), are in [3, 4]. The fundamental concepts of all three sections can be found in [4].

2. Skewfields and Fields

A *skewfield* is a non-empty set K with two binary operations called *addition* and *multiplication*, so that if a and b are in K, then $a + b$ and ab (or $a \cdot b$) are elements in K all satisfying:

(1) Both operations are associative; i.e., $a + (b + c) = (a + b) + c$, and $a(bc) = (ab) c$ for all a, b, c in K.

(2) Addition is commutative; i.e., $a + b = b + a$ for all a, b in K.

(3) Each operation has an identity; so there is an element 0 in K such that $0 + b = b + 0 = b$ for all b in K and there is an element 1 in K such that $1b = b1 = b$ for all b in K. In addition $0 \neq 1$.

(4) For each b in K there is an element $-b$ in K such that $b + (-b) = 0$, and for each $b \neq 0$ in K there is an element b^{-1} in K such that $bb^{-1} = b^{-1}b = 1$. The element $-b$ is called the additive inverse of b while b^{-1} is its multiplicative inverse.

(5) $a(b + c) = ab + ac$, and $(a + b)c = ac + bc$ for all a, b, c in K.

From these axioms we can prove most of the expected rules for addition and multiplication. For instance $0x = x0 = 0$ for all x in K (thus 0 cannot have a multiplicative inverse), and $x(-y) = (-x)y = -(xy)$ for all x, y in K; so we can write for instance $x - y = x + (-y)$. Given an equation of the sort $a + x = b$, or $y + a = b$, we can solve uniquely for x and y; similarly $ax = b$ or $ya = b$ will have unique solutions for x and y if $a \neq 0$. Thus the inverses and the identities are unique. We refer to the system K with its addition as the *additive group of* K and to the system of non-zero elements of K (for which we often write K^*) with its multiplication as the *multiplicative group of* K, or of K^*.

The intersection of any family of subskewfields of K is also a subskewfield, and so we can speak of the minimal skewfield of K, called the *prime field of* K. If a skewfield has commutative multiplication it is called a *field*, and a field with no non-trivial subfields is called a *prime field*. It is easy to see that, for any skewfield K, the prime field of K is a prime field. For any prime p the set Z_p of integers modulo p (with operations modulo p) is a prime field. The set of rationals, under the ordinary addition and multiplication, is also a prime field. These, in fact, are the only prime fields.

Result 1.1. *Any prime field is either Z_p, for some prime p, or is the field of rationals.*

The *centre* of a skewfield K is the set of all elements in K which commute under multiplication with every other element of K. For any skewfield the centre is a subfield which contains the prime field.

If the prime field of the skewfield K is Z_p, then we say that K has *characteristic* p; if the prime field is the field of rationals, then K has *characteristic* 0. In a field of characteristic n, if x is any element, $x \neq 0$, then the sum of m x's, which we write as mx, is zero if and only if m is a multiple of n. We shall be much more interested in fields than in general skewfields, but we mention the classical example of a skewfield which is not a field:

Let F be any subfield of the real numbers (e.g., the rationals, or the reals themselves), and let i, j, k be three new symbols; suppose K is the set of all elements of the form $a + bi + cj + dk$, with addition prescribed by:

$$(a + bi + cj + dk) + (e + fi + gj + hk)$$
$$= (a + e) + (b + f)i + (c + g)j + (d + h)k.$$

and with multiplication given by the rule:

$$(a + bi + cj + dk)(a_1 + b_1 i + c_1 j + d_1 k)$$
$$= (aa_1 - bb_1 - cc_1 - dd_1) + (ab_1 + ba_1 + cd_1 - dc_1) i$$
$$+ (ac_1 - bd_1 + ca_1 + db_1) j + (ad_1 + bc_1 - cb_1 + da_1) k.$$

Then K is a skewfield which we call the *F-quaternions* or, if F is the reals, merely the *quaternions*. The quaternions are often useful as counter examples to theorems which are true for fields but false for skewfields.

If K is a field, we may speak of polynomials over K: that is, expressions of the form $f(x) = a_n x^n + a_{n-1} x^{n-1} + \cdots + a_1 x + a_0$, where the a_i are elements of K and x is some new symbol, called an *indeterminate*. These polynomials add and multiply in the familiar manner. For any b in K we may speak of the value of $f(x)$ at b; that is $f(b) = a_n b^n + a_{n-1} b^{n-1} + \cdots + a_0$, which is an element of K. We may consider the possible factors of $f(x)$, and we say that $f(x)$ is *irreducible* if there are no polynomials $g(x)$ and $h(x)$, both of lower degree, such that $f(x) = g(x) h(x)$. If K is a field and $f(x)$ is any polynomial over K, than a *splitting field* for $f(x)$ over K is a field F which contains K and in which $f(x)$ can be factored completely into linear factors of the form $bx + c$, and such that no subfield has the same property.

Result 1.2. *Given a field K and an irreducible polynomial $f(x)$, a splitting field for $f(x)$ over K always exists, and any two splitting fields are isomorphic.*

The special case that interests us the most is that of an irreducible quadratic $f(x) = ax^2 + bx + c$. If $d = b^2 - 4ac$, and if the characteristic of K is not two, then the splitting field of $f(x)$ over K consists of all elements of the form $r + s\sqrt{d}$, where r and s are arbitrary elements of K, and where addition and multiplication are given by:

$$\left(r_1 + s_1\sqrt{d}\right) + \left(r_2 + s_2\sqrt{d}\right) = (r_1 + r_2) + (s_1 + s_2)\sqrt{d}$$

and

$$\left(r_1 + s_1\sqrt{d}\right)\left(r_2 + s_2\sqrt{d}\right) = r_1 r_2 + s_1 s_2 d + (r_1 s_2 + s_1 r_2)\sqrt{d}.$$

We shall be particularly interested in finite fields. However before discussing finite fields we note a very difficult but very important theorem.

Result 1.3 (Wedderburn's theorem). *A finite skewfield is a field.*

Finite fields have all been classified and their structure is well known.

Result 1.4. *Let p be a prime and $q = p^n$. Then there is (up to isomorphism) a unique field, called the Galois Field of order q and written $GF(q)$, with q elements, and any finite field is isomorphic to some $GF(q)$. In addition:*

(i) *$GF(q)$ has characteristic p.*

(ii) *The multiplicative group of $GF(q)$ is cyclic.*

(iii) *If $p = 2$ then every element of $GF(q)$ is a square, while if $p \neq 2$, then exactly half of the non-zero elements of $GF(q)$ are squares.*

(iv) *Every element of $GF(q)$ is a sum of two squares.*

We give a proof of (iv), since all of the other results are well known. Let $K = GF(q)$. From (iii) there is nothing to prove if $p = 2$, so suppose $p \neq 2$. Then the squares make up a set S which includes half the non-zero elements plus zero. If there is some element which is not a sum of two squares, then it must be an element z which is also not a square. Now the non-zero squares N are a subgroup of index two of the multiplicative group and hence, since N has just two cosets in K^*, every non-square is equal to z multiplied by a square. If one of these, zx^2 say, were a sum $a^2 + b^2$, then $z = (a/x)^2 + (b/x)^2$ would be a sum of two squares as well. So *no* non-square is a sum of two squares. Thus every sum of two squares is again a square, and the set S is closed under addition, as well as multiplication. It is now an easy exercise to show that this means that S is a subfield. But S must have $(q-1)/2 + 1 = (q+1)/2$ elements, and clearly has the same characteristic as $GF(q)$. Thus, by the earlier parts of the result, S must also be isomorphic to $GF(p^m)$ for some m. But this is impossible since, if $q = p^n$, then $(p^n + 1)/2$ is never a power of p. $\square$

Result 1.5. *If p is a prime, then the field $GF(p^n)$ has a subfield isomorphic to $GF(p^m)$ if and only if m divides n; in that case the subfield is unique.*

For any skewfield K, an *automorphism* of K is a one-to-one mapping α of K onto K such that $(x + y)^\alpha = x^\alpha + y^\alpha$ and $(xy)^\alpha = x^\alpha y^\alpha$, for all x and y in K. The set of all such automorphisms is a group, written $\mathrm{Aut}(K)$. An *anti-automorphism* of K is a one-to-one mapping β of K onto K such that $(x + y)^\beta = x^\beta + y^\beta$ and $(xy)^\beta = y^\beta x^\beta$ for all x and y in K. Clearly if K is a field, then any anti-automorphism is an automorphism. But the student should verify that the mapping $(x + iy + jz + kw)^\beta = x - iy - jz - kw$ of the skewfield of quaternions is an anti-automorphism (and therefore not an automorphism).

Result 1.6. (a) *If p is a prime, then $\mathrm{Aut}(GF(p^n))$ is a cyclic group of order n, generated by $\alpha : x \to x^\alpha = x^p$.*

(b) *If K is the rational or real field, $\mathrm{Aut}(K) = 1$.*

(c) *If K is the field of complex numbers then $\mathrm{Aut}(K)$ contains the element $\alpha : x + iy \to x - iy$.*

(d) *If K is a field and Γ is an automorphism group of order n, then the set $F = K_\Gamma$ of all elements fixed by Γ is a subfield, and K is an n-dimensional vector space over F.*

(e) *If d is a non-square in a field F and $K = F(\sqrt{d})$ is the field of all elements of the form $x + y\sqrt{d}, x, y \in F$, then $x + y\sqrt{d} \to x - y\sqrt{d}$ is an automorphism of K whose fixed elements are exactly the elements of F.*

If α is an automorphism of order n of the field K, then writing $x^{\beta + \gamma}$ for $x^\beta x^\gamma$, the elements of the form $x^{1 + \alpha + \alpha^2 + \cdots \alpha^{n-1}}$ are called the *norms* of

α in K. Clearly any norm of α is fixed by α. We shall be interested in a particular instance of a norm.

Result 1.7. *Let $K = GF(q^2)$ and let $F = GF(q)$ be contained in K. Then the automorphism $\alpha : x \to x^q$ has order 2 and the norms of α are exactly the elements of F.*

To prove this result we note that the norms of α are merely the $(q + 1)$-st powers of the elements of K^* plus, of course, 0. But since K^* is a cyclic group of order $q^2 - 1$ and F^* is a cyclic subgroup of order $q - 1$ (see Result 1.4 part (ii)) the result now follows from elementary group theory. $\square$

Obviously the concept of norm is extendable to other fields, and the reader will find it an easy exercise to show that in the field of complex numbers over the reals, the set of norms of the automorphism mapping every element to its conjugate is exactly the set of positive real numbers, while in the case of the complex numbers over the rationals, it is an even more restricted set of rationals.

3. Group Theory

We shall assume that the basic results and definitions of abstract group theory are already familiar to the reader. In particular we shall make repeated use of the isomorphism theorems and, to a lesser extent, the Sylow theorems. In this book we shall be mainly concerned with permutation groups and so in this section we give a brief review of those permutation group theorems which will be useful to us.

If $\mathscr{S}$ is a set then a *permutation* of $\mathscr{S}$ is a one-to-one mapping of $\mathscr{S}$ onto itself. The set of all these mappings forms a group under the usual composition of mappings: if α and β are permutations of $\mathscr{S}$ then $\alpha\beta$ is the permutation defined by $x^{\alpha\beta} = (x^{\alpha})^{\beta}$. This group is the *symmetric group* on $\mathscr{S}$ and clearly depends only on the cardinality of $\mathscr{S}$. If $\mathscr{S}$ is a finite set with n elements, then the symmetric group on $\mathscr{S}$ is often denoted by Σ_n and has order $n!$. If Γ is a group, then a *permutation representation* of Γ is a homomorphism of Γ into the symmetric group of some set $\mathscr{S}$. The kernel of the given homomorphism is called the *kernel* of the representation. If this kernel is trivial, i.e. if the homomorphism is an isomorphism, then the representation is *faithful*.

A *permutation group* is a subgroup of some symmetric group, and so a permutation representation of a group is simply a homomorphism of that group onto a permutation group. Not surprisingly we are most interested in faithful representations which amounts to saying that we are interested in finding a permutation group which is abstractly the same as our given group. But there may be many such permutation groups and so, in order to distinguish between them, we must define permutation iso- morphism. If Γ is a permutation group on a set $\mathscr{S}$ and if Σ is a permutation group on a set $\mathcal{T}$ then a *permutation isomorphism* between the permutation groups is an isomorphism α from Γ onto Σ together with a one-to-one mapping β from $\mathcal{S}$ onto $\mathcal{T}$ such that, for any $s \in \mathcal{S}$ and any $\gamma \in \Gamma$, $(s^{\gamma})^{\beta} = (s^{\beta})^{\gamma^{\alpha}}$.

If Γ is a permutation group on a set $\mathcal{S}$ then, for any $x \in \mathcal{S}$, the *orbit* of x under Γ is the set $\{x^{\gamma} \,|\, \gamma \in \Gamma\}$. We denote this set by $x\Gamma$. (To be completely consistent we should write x^{Γ} as, for example, Wielandt does [5]. But the notation is somewhat easier this way since we often use groups which are themselves notationally rather complicated.) Clearly $y \in x\Gamma$ if and only if $x \in y\Gamma$ so that the orbits of Γ partition $\mathcal{S}$. As an immediate consequence of the definitions we have:

Result 1.8. *Let Γ be a permutation group on $\mathcal{S}$ and let Σ be a subgroup of Γ. Then, for any $x \in \mathcal{S}$, $x\Sigma \leqq x\Gamma$ so that any orbit of Γ is a union of orbits of Σ.*

If $\mathcal{S}$ is the only orbit of Γ then we say that Γ is *transitive* on $\mathcal{S}$. More generally we say that Γ is *t-transitive* on $\mathcal{S}$ if for every pair of ordered t-tuples of distinct elements in $\mathcal{S}$ there is an element in Γ sending the first t-tuple onto the second. Thus one-transitive is the same as transitive. We shall often refer to a two-transitive group as being *doubly transitive*.

If Γ is a permutation group on a set $\mathcal{S}$ then, for any $x \in \mathcal{S}$, we denote by Γ_x the set of all elements in Γ which fix x. Then Γ_x is called the *stabilizer* of x and it is easy to see that Γ_x is a subgroup of Γ. Furthermore, if for any subgroup Δ of Γ and any α in Γ we write Δ^{α} to mean the conjugate $\alpha^{-1}\Delta\alpha$ of Δ by α, we have

Result 1.9. *If Γ is a permutation group on a set $\mathcal{S}$, then for any x in $\mathcal{S}$ and any α in Γ, $\Gamma_{x^{\alpha}} = (\Gamma_x)^{\alpha}$.*

As a corollary to Result 1.9 we see that if x and y are in the same orbit under Γ then $\Gamma_x \cong \Gamma_y$. The next result gives a test, which we shall sometimes use, to see whether a given group is t-transitive.

Result 1.10. *If $t > 1$ and $|\mathcal{S}| > 2$, the group Γ is t-transitive on the set $\mathcal{S}$ if and only if for any element x in $\mathcal{S}$ the stabilizer Γ_x is $(t - 1)$-transitive on $\mathcal{S} \setminus \{x\}$.*

If Γ is any group and Δ is a subgroup of Γ, then we can represent Γ as a permutation group on the right cosets of Δ as follows: if γ is in Γ, then the mapping $\bar{\gamma}$ sends the right coset $\Delta\alpha$ onto $\Delta\alpha\gamma$. The set of all these mappings $\bar{\gamma}$ forms a group $\bar{\Gamma}$, and, since $\bar{\gamma}\bar{\beta} = \overline{\gamma\beta}$, the mapping $\gamma \to \bar{\gamma}$ is a homomorphism of Γ onto $\bar{\Gamma}$. So we have a permutation representation of Γ.

Result 1.11. *The permutation representation of Γ onto $\bar{\Gamma}$ as given above is transitive on the set $\mathcal{S}$ of right cosets of Δ. Every transitive permutation representation of Γ is permutation isomorphic to a representation of this sort.*

This classifies, in a sense, all transitive permutation representations. But we still want to know when such a representation is faithful. We make a definition: If Δ is a subgroup of the group Γ, then the *core* of Δ is the intersection of all the conjugates of Δ.

Result 1.12. *The core of Δ is the largest subgroup of Δ which is normal in Γ. Furthermore, the core of Δ is the kernel of the representation of Γ on the right cosets of Δ, as above.*

Result 1.13. (a) *Let Γ be represented as a permutation group on a set $\mathscr{S}$, and let x be in $\mathscr{S}$. Then $|\Gamma| = |x\Gamma| \cdot |\Gamma_x|$.*

(b) *If $\mathscr{T}_1, \mathscr{T}_2$ are orbits of Γ with $(|\mathscr{T}_1|, |\mathscr{T}_2|) = 1$ then, for any x in $\mathscr{T}_1, \Gamma_x$ is transitive on $\mathscr{T}_2$.*

(Here we are using $|\mathscr{T}|$ to mean the number of elements in the set $\mathscr{T}$.)

An important result on permutation groups is the following, where we write $f(\alpha)$ to mean the number of elements in the set $\mathscr{S}$ which are fixed by the permutation α.

Result 1.14. *If Γ is a permutation group on a set $\mathscr{S}$, then*

$$t|\Gamma| = \sum_{\alpha \in \Gamma} f(\alpha)$$

where t is the number of orbits of Γ on $\mathscr{S}$.

This is the first result we have used which is not found in every elementary group theory text-book, so we sketch a proof. If x is an element of $\mathscr{S}$ and α an element of Γ, then we call the ordered pair (x, α) a *flag* if $x^\alpha = x$; we wish to count the flags. On the one hand, for each x in $\mathscr{S}$, there are $|\Gamma_x|$ elements α such that (x, α) is a flag, and so the number of flags is

$$\sum_{x \in \mathscr{S}} |\Gamma_x| \, .$$

But the number of elements fixed by α is $f(\alpha)$, and so we must have

$$\sum_{x} |\Gamma_x| = \sum_{\alpha} f(\alpha) \, .$$

Each $|\Gamma_x|$ can be changed (by Result 1.13) to $|\Gamma|/|x\Gamma|$. In each orbit $x\Gamma$, the term $|\Gamma|/|x\Gamma|$ will occur $|x\Gamma|$ times (that is, once for each element in the orbit), and so the term $|\Gamma|/|x\Gamma|$, occurring $|x\Gamma|$ times, contributes $|\Gamma|$ to the sum on the left. Thus the sum on the left will contribute $|\Gamma|$ once for each orbit, and hence the left is equal to $t|\Gamma|$. This proves the result. $\square$

A special sort of permutation group is one in which Γ_x is always the identity; such a group is called *semi-regular*. If a semi-regular group is also transitive, it is called *regular*.

We now list three results which follow from the earlier results in this section.

Result 1.15. *If Γ is a regular permutation group on a set $\mathscr{S}$, then Γ is permutation isomorphic to the representation of Γ on the right cosets of the identity subgroup, and conversely.*

Result 1.16. *If $\bar{\Gamma}$ is a permutation representation of Γ on the right cosets of the subgroup Δ, then $\bar{\Gamma}$ is regular if and only if Δ is a normal subgroup of Γ.*

Result 1.17. *Any transitive abelian permutation group is regular.*

An even more interesting class of permutation groups is given by the transitive groups such that a stabilizer (any stabilizer will do, from the observations following Result 1.9) is semi-regular. These are called *Frobenius groups*, and have been the subject of much deep group-theoretic research for many years.

Result 1.18. *Let Γ be a finite Frobenius group on a set $\mathscr{S}$. Then Γ has a normal subgroup Δ which is regular on $\mathscr{S}$, such that $\Gamma = \Delta\Gamma_x$, for any stabilizer Γ_x (and hence Γ is the semi-direct product of Δ by Γ_x).*

The normal subgroup Δ of Result 1.18 is called the *Frobenius kernel* of Γ.

A characteristic sort of Frobenius group can be given as follows. Let K be a skewfield and let G be a multiplicative subgroup of the multiplicative group K^*; for each a in G and each b in K, define the mapping $\phi = \phi(a, b)$ of K onto K by

$$x^\phi = xa + b \, .$$

Then it is not difficult to show that the set Γ of all these mappings form a group, where $\phi(a, b)\,\phi(c, d) = \phi(ac, bc + d)$. This group is Frobenius and in the finite case its Frobenius kernel is the subgroup Δ of all elements $\phi(1, b)$. (Note that from Result 1.18 we only know that finite Frobenius groups possess Frobenius kernels; but Δ has the properties of the Frobenius kernel even in the infinite case.) The student should verify these remarks and should find the stabilizer of a point, say 0.

If Γ is a permutation group on a set $\mathscr{S}$, then it may be possible to partition $\mathscr{S}$ into subsets $\mathscr{S}_1, \mathscr{S}_2, \ldots$ such that Γ induces a permutation group on the $\mathscr{S}_i$ (that is, if x in $\mathscr{S}_i$ is sent by α to x^α in $\mathscr{S}_j$, then α sends every element of $\mathscr{S}_i$ to an element of $\mathscr{S}_j$). Of course it is always possible to do this in two trivial ways: either let $\mathscr{S}$ itself be the only such subset, or let each subset consist of a single element of $\mathscr{S}$. These two ways of constructing the $\mathscr{S}_i$ will be called *trivial*. If a non-trivial partition exists we will call the $\mathscr{S}_i$ *systems of imprimitivity*, and we say that Γ is *imprimitive*. If only trivial such decompositions exist, then we say that Γ is *primitive*. If Γ is not transitive, then the orbits themselves give systems of imprimitivity, and so a primitive group must be transitive. In addition we have:

Result 1.19. *If Γ is t-transitive, $t > 1$, then Γ is primitive.*

Thus the interesting primitive groups are those which are transitive but not two-transitive. It is not hard to prove:

Result 1.20. *Let Γ be a transitive permutation group on a set $\mathscr{S}$. Then Γ is primitive if and only if any stabilizer Γ_x in Γ is a maximal subgroup of Γ (that is, there are no subgroups of Γ properly between Γ and Γ_x).*

4. Vector Spaces

The subject of vector spaces is so large and is so fundamental to all of modern mathematics that we shall assume that the student is familiar with the basic ideas of vector space, linear dependence and independence, bases, dimension, subspaces, and so on; but since sometimes we want to consider vector spaces over skewfields instead of the more usual fields, we must make a distinction between left and right vector spaces. So "V is a *left* vector space over the skewfield K" will mean that if v is a vector in V and k is an element of K, then the product kv is defined (as a vector of V), and that all the usual associativity and distributivity properties hold, but with the "scalars" always on the left of the vectors. A right vector space is defined analogously. The student will be familiar with the theorem which asserts that all vector spaces of dimension n over the field K are isomorphic; over skewfields, we can only say that all left vector spaces of dimension n over the skewfield K are isomorphic.

When we come to linear transformations and semi-linear transformations, again a little care is required in the definitions (and sometimes in the theorems). So let V and W be left vector spaces over the skewfield K, and let α be an automorphism of K. If ϕ is a mapping of V into W with the property that (i) for all v_1, v_2, in V, $(v_1 + v_2)^\phi = v_1^\phi + v_2^\phi$, and (ii) for all v in V, k in K, $(kv)^\phi = k^\alpha(v^\phi)$, then ϕ is a *semi-linear transformation* from V to W, with *associated automorphism* α. We say that ϕ is a *linear transformation* if its associated automorphism is the identity. It is clear what should be the correct definitions when V and W are right vector spaces, but there is a third possibility for which the definitions are less obvious. If V is a left vector space and W is a right vector space, then everything is as expected, except that we change the one rule to $(kv)^\phi = v^\phi k^\alpha$, where α is an anti-automorphism of K. (Otherwise we would get into trouble with things of the sort $a(bv)$, where a and b are elements of K.)

Now if we consider the set of semi-linear transformations of the left vector space V into itself we have a closed set: that is, there is a natural multiplication, where if ϕ_1 and ϕ_2 are semi-linear transformations with associated automorphisms α_1 and α_2 respectively, then $\phi_1\phi_2$ is defined by $v^{\phi_1\phi_2} = (v^{\phi_1})^{\phi_2}$, and will have the associated automorphism $\alpha_1\alpha_2$. This multiplication is associative. A semi-linear transformation of V to V is said to be *non-singular* if it is one-to-one and onto; the set of non-singular semi-linear transformations forms a group, written $\Gamma L(V)$. From the remarks above, we can map each element of $\Gamma L(V)$ onto its associated automorphism, and this mapping will be a homomorphism into (even onto, in fact) $\text{Aut}\, K$, whose kernel is the set of non-singular linear transformations; this last group is called the *general linear group* of V, written $GL(V)$.

Now in fact it is easy to show that if V is finite dimensional then $\text{Aut}\, K$ is, as stated above, a subgroup of $\Gamma L(V)$ with the following properties:

Result 1.21. *If V is a finite dimensional left or right vector space over the skewfield K, then $\Gamma L(V) = A \cdot GL(V)$, where $GL(V)$ is normal in $\Gamma L(V)$, $A \cap GL(V) = 1$, and A is isomorphic to* Aut K.

If a basis of a (left or right) vector space is chosen, then we can represent the vectors of the space in terms of their coefficients in that basis, in a well-known way. So if V is a finite dimensional left vector space over K, with a basis $e_1, e_2, \ldots, e_n$, then for each v in V there is a unique n-tuple $(x_1, x_2, \ldots, x_n)$ such that $v = x_1 e_1 + x_2 e_2 + \cdots + x_n e_n$. We may choose to represent v simply by the n-tuple $(x_1, x_2, \ldots, x_n)$, and then all the familiar laws for operating with these "row vectors" are valid, remembering that since we have a *left* vector space we only multiply the n-tuple on the *left* by elements of K. A linear transformation of V is completely determined by its effect on a basis, and hence we can represent it by a square matrix, in the usual fashion, where the effect of the square matrix (a_{ij}) on the n-tuple (the vector, in other words) $(x_1, x_2, \ldots, x_n)$ is to send it to the ordinary matrix product; that is, the row vector whose j-th term is $\sum_i x_i a_{ij}$.

Something a little awkward comes up here. Suppose (to illustrate) we have a three-dimensional left vector space V over K and we represent the elements of V by row vectors, that is, by their coordinates in some fixed basis. Consider the mapping $\eta: (x, y, z) \to (xa, ya, za)$, for some fixed a in K, $a \neq 0$; similarly, consider the mapping $\mu: (x, y, z) \to (ax, ay, az)$, for the same a. Then the student should verify the following statements: (i) η is a non-singular linear transformation; (ii) μ is a non-singular semi-linear transformation; (iii) μ is linear if and only if a is in the centre of K, in which case it equals η. Finally he should determine the automorphism associated with μ.

Result 1.22. *The group $GL(V)$ is transitive on the bases of V, and the subgroup of $GL(V)$ fixing every vector in some chosen basis is the identity.*

Now when K is a field we can compute the determinant of a matrix and it is well-known that this determinant is in fact an invariant of the linear transformation that goes with the matrix: that is, if A and B are matrices representing the same linear transformation in different bases, then they have equal determinants. (In fact the characteristic equation is an invariant of the linear transformation and hence its coefficients are invariants; one of these is the determinant, another is the trace.)

Furthermore, if $\det(A)$ means the determinant of A, then $\det(AB) = \det(A) \cdot \det(B)$, so the mapping $A \to \det(A)$ is a homomorphism. Since a linear transformation is in $GL(V)$ if and only if its determinant is not zero, this assures us that the mapping $A \to \det(A)$ is a homomorphism of $GL(V)$ into the multiplicative group of the field K. In fact it is onto (the student should be able to construct a linear transformation whose determinant has any prescribed value), and the kernel, the set of linear transformations (or perhaps matrices) of determinant one, is called the *special linear group*, written $SL(V)$.

Result 1.23. *Let V be a finite dimensional left (or right) vector space over the skewfield K. If N is the group of all semi-linear transformations of V which fix every one dimensional subspace of V, i.e. $N = \{\gamma \in \Gamma L(V) | \langle v^{\gamma} \rangle = \langle v \rangle$ for all v in $V\}$, then $N \cong K^*$ and if the dimension of V is greater than one, $N \cap GL(V) \cong Z(K^*)$.*

Result 1.24. *Let V be a finite dimensional left (or right) vector space over the field K. If a semi-linear transformation α has the property that it fixes every one dimensional subspace of V then α is linear and is in the centre of $GL(V)$. In matrix terms α is represented by a scalar multiple of the identity matrix.*

If V is a left vector space over the skewfield K then a *linear functional* on V is a mapping f' of V into K such that (i) for any v, w in V, $(v + w) f' = vf' + wf'$ and (ii) for any v in V and k in K, $(kv)f' = k(vf')$. (We are writing linear functionals as (transposed) vectors, rather than mappings, in anticipation of the following discussion.) If f' and g' are two linear functionals, then we define $f' + g'$ by: $v(f' + g') = vf' + vg'$, while if f' is a linear functional and b is an element of K, we define $f'b$ by: $v(f'b) = (vf') b$. Under these operations the set V' of linear functionals is a *right vector space* over K. If $e_1, e_2, \ldots, e_n$ is a basis for V, and if we define mappings f'_i, for $i = 1, 2, \ldots, n$ by $(x_1 e_1 + x_2 e_2 + \cdots + x_n e_n)f'_i = x_i$ then it is not difficult to show that the f'_i are elements of V', that they are linearly independent and that they span V'. We say that f'_i are the basis vectors corresponding to the e_i and that V' is the *dual space* of V. Hence:

Result 1.25. *The dimension of V' is the same as the dimension of V, when V is finite dimensional. Furthermore, V'' is naturally isomorphic to V.* (The reader may interpret "naturally" in an intuitive way; however the expert will see it has an exact meaning.)

Now if $g' = f'_1 a_1 + f'_2 a_2 + \cdots + f'_n a_n$ is an element of V', then the effect of g' on V can be expressed as follows:

$$(x_1 e_1 + x_2 e_2 + \cdots + x_n e_n) g' = x_1 a_1 + x_2 a_2 + \cdots + x_n a_n.$$

Hence if we use row vectors $(x_1, x_2, \ldots, x_n)$ for the elements of V and column vectors

$$\begin{pmatrix} a_1 \\ a_2 \\ \vdots \\ a_n \end{pmatrix}$$

for the elements of V', then we can express the effect of the column vector w' on the row vector v by vw'; that is, by the matrix product of the row vector v and the column vector w'. (We shall consistently use a symbol like v to represent a row vector, and a symbol like w' therefore will represent the transpose of a row vector, i.e., a column vector.)

Now suppose α is a non-singular semi-linear transformation of V onto V, and in some fixed basis $e_1, e_2, \ldots, e_n$ of V, α is represented by the matrix A and the automorphism σ; so if $v = (x_1, x_2, \ldots, x_n)$, then

$$v^\alpha = v^\sigma A = (x_1^\sigma, x_2^\sigma, \ldots, x_n^\sigma)\, A \ .$$

If we define a mapping α' of V' into V' by: $(y_1, y_2, \cdots, y_n)'^{\alpha'} = A^{-1}(y_1^\sigma, y_2^\sigma, \cdots, y_n^\sigma)'$, then we can say:

Result 1.26. *If α is a non-singular semi-linear transformation of V onto V, then α' is a non-singular semi-linear transformation of V' onto V', and with the same associated automorphism σ. Furthermore, $v^\alpha(w')^{\alpha'} = (vw')^\sigma$, for all v in V and w' in V'. The mapping $\alpha \to \alpha'$ is an isomorphism of $\Gamma L(V)$ onto $\Gamma L(V')$, with the property that $\alpha'' = \alpha$. The mapping also sends $GL(V)$ onto $GL(V')$.*

Note that if V is a right vector space then V' is a left vector space. Hence it will be convenient to represent the vectors of V by column vectors and those of V' by row vectors. A linear transformation of a right vector space V, in matrix form, will be a mapping $w' \to Aw'$ and so on. In Chapter II we shall slightly prefer left vector spaces, since mappings then appear on the right and the product of linear transformations with matrices A and B will have matrix AB (otherwise it would have matrix BA).

The correspondence $\alpha \to \alpha'$ above not only maps $GL(V)$ onto $GL(V')$ but does it in a way which preserves the effect of the linear functionals, a most powerful and useful property.

There is another interesting relationship between V and V'. For each subspace W of V, we define the *annihilator* W^{av} of W to be the set of all elements in V' which map every vector of W onto 0 in K.

Result 1.27. *If V is n-dimensional, and W is any subspace of V, then $W^{av\,av} = W$, and the dimension of W plus the dimension of W^{av} equals n.*

References

Naturally there are many books which cover the material discussed in this chapter. The books in the following short list contain proofs of all the results which we have listed.

1. Birkhoff, G., MacLane, S.: A survey of modern algebra. London-New York: Macmillan 1965.
2. Hall, M., Jr.: The theory of groups. London-New York: Macmillan 1959.
3. Herstein, I. N.: Topics in algebra. Waltham: Blaisdell 1964.
4. van der Waerden, B. L.: Modern algebra, Vols. I and II. New York: Frederick Ungar Publ., Co. 1949.
5. Wielandt, H.: Finite permutation groups. London-New York: Academic Press 1964.

II. Classical Projective Planes

1. Introduction

In this chapter we give a fairly concise introduction to those parts of classical geometry that are relevant to the study of projective planes in general. We include it because we feel that much of the point of the study of projective planes is in knowing what is true for the classical case and is not (or might not be) true in the general case. Most of the theorems in the later chapters take on a new aspect from this point of view and are much more meaningful when compared to the classical theorems. Similarly, the strange and pathological behaviour of free closures, even though they are in some sense the most "natural" projective planes, is only really appreciated when contrasted with the classical situation.

Although in the later chapters we occasionally make references to classical situations, the reader may always ignore these references and the material that begins in Chapter III is then completely self-contained. Consequently the student may skip this chapter without affecting his comprehension of the later chapters. *Throughout this chapter we act as if there are no geometries other than those we are discussing:* so we do not use words like "classical" or "desarguesian".

Another reason for including this chapter is the historical one. At first people studied projective planes over the reals, then over the complexes, then over arbitrary fields and skewfields; later the notion of geometric dimension higher than three crept in. Finally an attempt was made to make up a list of axioms, that is a set of synthetic incidence properties between the various kinds of objects in a projective geometry, which should characterize projective geometries. These axioms worked quite nicely for geometric dimension greater than two, in the sense that they defined exactly the class of things that everybody wanted to call projective geometries. But for geometric dimension two every attempt to show that the axioms (which become particularly simple in this case; see Theorem 2.2 and Chapter III) characterized the projective planes constructed over fields or skewfields led to failure. At last, in the latter part of the nineteenth century, it was discovered that there were projective planes that could not arise from skewfields in the standard way. These were the first "non-desarguesian" projective planes. In the early years of the twentieth century the first examples of finite non-desarguesian projective planes were discovered.

We shall study geometries in this chapter almost completely from an algebraic point of view; this has the disadvantage that the real "geometric meaning" of much of what we do is obscure sometimes. But it has the advantage of being neat and often rather easier than if we had taken a more conventional point of view. We assume that the reader is fairly comfortable with linear algebra; it seems to us that studying projective geometry with linear algebra is much the easiest and most natural way to grasp the subject. Affine geometries will not make their appearance until late in the chapter, at which point some of the "geometric meaning" will become clearer. In fact we will often phrase definitions and so on in terms of arbitrary dimension, but we only do this to illustrate that the special case of the plane is not really much different from the general case. However for any interesting results, we usually restrict ourselves to the plane.

We have included a fair amount of material on polarities, on conics, and on unitals, including some material on the transitivity properties of the orthogonal and unitary groups. A reader with more skill in group theory will not find it difficult to show that these groups are non-soluble (though the simple pieces are not so easy to identify). This material is included since in Chapter XII, where we study similar material for arbitrary projective planes, it will be useful to have considerable detailed knowledge of the classical case. Section 2.9 contains material on *affine* conics (so we can speak of ellipses, hyperbolas and parabolas), and, among other things, classifies them according to the affine group of the plane. In fact, this project can only be carried out completely in certain cases, since ellipses cause a lot of trouble and would carry us into deep problems of algebraic number theory etc. Empty conics (which can lead to metrics and angles) are not studied, but we show that they cannot arise in the finite case. Unitals are not studied in the same detail, mostly because the problems are either too difficult, or are even unsolved.

Since the projective planes over skewfields do not have the same geometric properties as those over fields (see, for example, Theorem 2.6), we often discuss vector spaces over skewfields. However there are many situations where we restrict ourselves to the field case; either because the proof in the skewfield case is too complicated or because the skewfield case is of no interest to us. Thus the reader should exercise a certain amount of caution and, when he sees $\mathscr{P}(V)$ written without any immediate reference to a field or skewfield K, should refer back to see which case is under consideration. It is safest to assume that K is a field unless told otherwise; of course, there is no distinction in the finite case.

Finally we emphasize that this chapter is only intended as an introduction to certain aspects of classical projective geometry. Many text books already exist which treat the topic in much more detail than we have room for here. We refer the interested reader to Artin [1], Baer [2] or Gruenberg and Weir [4].

2. Basic Definitions and Results

Let K be a skewfield and V a finite dimensional (left or right) vector space over K; we want to study the *lattice of subspaces* of V, instead of V itself. The collection of subspaces of V, together with the natural containment relation, will be called the *projective geometry* $\mathscr{P}(V)$, and we shall say that the subspace W is *incident with* the subspace U if either W contains U, or U contains W. Whenever it is appropriate we shall adopt the standard terminology, either algebraic or geometric, such as U lies on W (or in W), W contains U, W passes through U, etc. While it is not obvious that this definition of a projective geometry has any relation to our "natural" notions of geometry, we shall soon see that this is in fact the case.

If V is $(n + 1)$-dimensional over K, we shall say that $\mathscr{P}(V)$ has *g-dimension* n. The subspaces of V of (algebraic) dimension $i + 1$ will be said to have *g*-dimension i, and if $d(W) = i + 1$, we can write $gd(W) = i = d(W) - 1$. The subspaces of *g*-dimension zero (i.e., algebraic dimension one) will be called *points*, those of *g*-dimension one will be called *lines*, those of *g*-dimension two will be called *planes*, and those of *g*-dimension $n - 1$ will be called *hyperplanes*. If we were to think of the points of $\mathscr{P}(V)$ as its most fundamental objects, then their intersection (or indeed, the intersection of any two of them) would be the "natural" empty set: so the zero-subspace of V, whose dimension is zero and whose *g*-dimension is -1, is the *empty space* in $\mathscr{P}(V)$. If W is a subspace of V, then $\mathscr{P}(W)$ is contained in a natural way in $\mathscr{P}(V)$, and all the objects of $\mathscr{P}(W)$ are also objects of $\mathscr{P}(V)$, with the same *g*-dimension. If K is a field then V is defined uniquely by its dimension (i.e. it does not matter whether we take V to be a left vector space or a right one). In this case, if V has dimension $n + 1$ then $\mathscr{P}(V)$ has *g*-dimension n and we may sometimes write $\mathscr{P}_n(K)$ for $\mathscr{P}(V)$. If K is a finite field $GF(q)$ we may even write $\mathscr{P}_n(q)$ for $\mathscr{P}_n(K)$.

As an immediate consequence of the dimension theorem for vector spaces we have the following *g*-dimension theorem for projective geometries.

Lemma 2.1 (Grassman's identity). *Let U and W be subspaces of a given vector space V. Then $gd(U + W) + gd(U \cap W) = gd(U) + gd(W)$.*

The following three exercises give our first indication that a projective geometry has any similarity to our "natural" geometric ideas.

Exercise 2.1. Let P and Q be distinct points in $\mathscr{P}(V)$. Show that there is a unique line of $\mathscr{P}(V)$ which contains both P and Q.

Exercise 2.2. Let E be a line and W a hyperplane in $\mathscr{P}(V)$. Show that either E lies in W or E and W meet in a unique point.

Exercise 2.3. Let E and F be distinct lines in $\mathscr{P}(V)$. Show that either E and F meet in a unique point, in which case E and F lie in a unique plane of $\mathscr{P}(V)$, or E and F have no point in common, in which case E and F are contained in a unique space of g-dimension three in $\mathscr{P}(V)$.

The reader will find it very important to work through, as well as understand, the three exercises above. They all depend on Lemma 2.1 whose importance is basic. In a similar vein we have an exercise which is fundamental:

Exercise 2.4. If $\mathscr{P}(V)$ has g-dimension two show that every pair of distinct lines meet in a unique point. (This is a corollary of Exercise 2.3.)

We have already observed that if W is a subspace of a vector space V then $\mathscr{P}(W)$ is contained in $\mathscr{P}(V)$. There is another possible sense in which we could speak of one projective geometry being contained in another. Suppose that V is an n-dimensional left vector space over a skewfield K. If K has a sub-skewfield F then V is also a left vector space over F and, hence, contains a set of vectors U which form an m-dimensional vector space over F, where m may be larger than n. Clearly, since U is contained in V, $\mathscr{P}(U)$ is contained in $\mathscr{P}(V)$ but U need not be a sub-vector space of V. (Note, of course, that there is no need to insist that V, U are left vector spaces. They could equally well both be right vector spaces but we could not allow one to be left and the other right. This will often be the case and we shall in future not bother to point out this fact.) To distinguish these two types of containment we shall say that $\mathscr{P}(U)$ is a *complete sub-geometry* of $\mathscr{P}(V)$ if U is a sub-vector space of V over K. In any other case we shall call $\mathscr{P}(U)$ an *incomplete sub-geometry* of $\mathscr{P}(V)$.

Exercise 2.5. Let V be a (left) vector space over a skewfield K. If $\mathscr{P}(W)$ is a sub-projective geometry of $\mathscr{P}(V)$, show that $\mathscr{P}(W)$ is complete if and only if some line E of $\mathscr{P}(W)$ has the property that all the points on E in $\mathscr{P}(V)$ are on E in $\mathscr{P}(W)$.

Very soon we shall be interested only in projective geometries of g-dimension one or two, which we will call, respectively, *projective lines* and *projective planes*. However, some of the basic ideas are almost more easily understood for arbitrary g-dimension than for these special cases, and so we do not completely discard the general case yet. Thinking about the case of g-dimension two, we see that all "interesting" subspaces have dimension one or two, that is, are points or lines, and that the containment relation between these two classes of objects constitutes the entire geometry (for everything contains the zero-space, or the empty object, and everything is contained in the entire space). But the projective line is very peculiar, in that there are no interesting containment relations: besides the zero-space and the entire space, we have only a collection of one-dimensional subspaces. That is, the projective line might be said to look like a line with a lot of points on it, and to have no other structure. This fact will pose us certain difficulties, and until we have decided how to overcome the problem of the lack of incidence structure, we shall often have to make exceptions for g-dimension one.

We first look at the incidence structure of a projective geometry of g-dimension two.

Theorem 2.2. *Let $\mathscr{P}(V)$ have g-dimension two. Then the points and lines of $\mathscr{P}(V)$ satisfy:*

(i) *every pair of distinct points are incident with a unique common line;*

(ii) *every pair of distinct lines are incident with a unique common point;*

(iii) *$\mathscr{P}(V)$ contains a set of four points with the property that no three of them lie on a common line.*

Proof. Parts (i) and (ii) were proved in Exercises 2.1 and 2.4. The proof of (iii) depends on showing that the subspaces generated by the following four vectors have (as points) the required property: $(1, 0, 0), (0, 1, 0), (0, 0, 1),$ $(1, 1, 1)$. (Here we are using some fixed basis of V, as we are entitled to do.) Suppose, for instance, that the first three vectors generate subspaces which, as points, lie on a common line; that means that there is a two-dimensional subspace containing the three vectors. But the three vectors are linearly independent, which is a contradiction. The reader should finish the proof by considering all other subsets of three vectors taken from the given four. $\square$

(The object of Theorem 2.2 is rather obscure at this time, so we give a brief explanation. At one time it was thought, or hoped, that the three properties of Theorem 2.2 characterized projective geometries of g-dimension 2, in the sense that a collection of abstract points and lines satisfying them must be in some sense the same as a $\mathscr{P}(V)$. This was shown not to be so by Hilbert and Moulton and later in this book we shall use the three properties given as the *definition* of a projective plane. One of our chief concerns will then be to study what else must be assumed, besides (i), (ii), (iii) of Theorem 2.2, in order to give us a projective geometry of g-dimension two.)

Obviously we shall need definitions of isomorphism, homomorphism, automorphism, etc.; all of these will be completely natural. Thus an *isomorphism* from $\mathscr{P}(V)$ onto $\mathscr{P}(W)$ is a one-to-one mapping α of the subspaces of $\mathscr{P}(V)$ onto the subspaces of $\mathscr{P}(W)$ which preserves the containment relation, so $E < F$ in $\mathscr{P}(V)$ if and only if $E^{\alpha} < F^{\alpha}$ in $\mathscr{P}(W)$. If $\mathscr{P}(V) = \mathscr{P}(W)$ then the isomorphism α is called an *automorphism*, or a *collineation*, of $\mathscr{P}(V)$. It is clear what a *homomorphism* would be, but we shall not be much concerned with these excepting in Chapter XI. An *anti-isomorphism* is a one-to-one mapping β from $\mathscr{P}(V)$ to $\mathscr{P}(W)$ which reverses incidence: that is, $E < F$ in $\mathscr{P}(V)$ if and only if $F^{\beta} < E^{\beta}$ in $\mathscr{P}(W)$. If $\mathscr{P}(V) = \mathscr{P}(W)$ then an anti-isomorphism β is called a *correlation* and is a *polarity* when it has order two (that is, $\beta^2 = \beta\beta = 1$, the identity automorphism).

But all these definitions are not much good for the projective line. For instance, any one-to-one mapping of the points onto themselves would satisfy any of them, and while the set of all one-to-one mappings of a set to another set, or to itself, is a very interesting thing, we shall need more precise definitions. The trouble of course is that the projective line appears to have no structure, and so we shall have to provide some: after having done so (in a natural way) we will be able to give definitions for g-dimension one. Until then we must understand that all the definitions of isomorphism, etc., given above are only for g-dimension greater than one.

3. Dual Spaces and Homogeneous Coordinates

If V is a left vector space of dimension $n + 1$ over the skewfield K, then (see Chapter I) the dual space V' of V is a right vector space of the same dimension over K; and if V is a right vector space then V' is a left one. So we may discuss the two projective geometries $\mathscr{P}(V)$ and $\mathscr{P}(V')$, and the relationship that may exist between them. They are not necessarily isomorphic (though they certainly would be so in the important special case that K is a field), but we saw in Chapter I that there was a connection between them, which can be expressed in terms of the mapping a_V from subspaces of V to subspaces of V' given by: if $W \subset V$, then $W^{a_V} = \{v' \text{ in } V' \mid wv' = 0 \text{ for all } w \text{ in } W\}$. Since the subspaces of V are the elements of $\mathscr{P}(V)$ we see that, while a_V was not a mapping from V to V', it is a mapping from $\mathscr{P}(V)$ to $\mathscr{P}(V')$. Writing $a_{V'}$ for the dual mapping from V' to $V'' = V$ (see Result 1.25) the following theorem is a slight extension of Result 1.27.

Theorem 2.3. *The mappings a_V and $a_{V'}$ are anti-isomorphisms (respectively from $\mathscr{P}(V)$ to $\mathscr{P}(V')$ and from $\mathscr{P}(V')$ to $\mathscr{P}(V)$), and $a_V a_{V'} = 1 = a_{V'} a_V$.* (Here we are abusing notation to the extent of using "1" to mean the identity mapping of both $\mathscr{P}(V)$ and $\mathscr{P}(V')$.)

Corollary. *If V is a vector space over a field K, then $\mathscr{P}(V)$ always possesses polarities.*

Proof. Certainly $\mathscr{P}(V)$ possesses correlations, since if ϕ is any isomorphism from V' to V then $a_V \phi$ induces an anti-isomorphism of $\mathscr{P}(V)$ onto itself. But if ϕ is chosen as the isomorphism from V' to V which maps the basis $f_1', \ldots, f_n'$ of V' onto the basis $e_1, \ldots, e_n$ of V (where the e_i and f_j' are related by $e_i f_j' = 1$ or 0 according as $i = j$ or not; see Chapter I) then it is an easy exercise to see that the mapping induced by $a_V \phi$ is even a polarity. $\quad\square$

Exercise 2.6. If V and W are vector spaces over a skewfield K and β is any semi-linear transformation from V into W, show that β induces a homomorphism from $\mathscr{P}(V)$ into $\mathscr{P}(W)$. Show that the induced homomorphism is an isomorphism if and only if β is an isomorphism.

We introduce here an important convention. If β is a semi-linear transformation from V into W then β will indicate the homomorphism of $\mathscr{P}(V)$ into $\mathscr{P}(W)$ induced by β; if β is one-to-one and onto then β is an isomorphism. If β is a non-singular semi-linear transformation of V, then β is a collineation, or automorphism, of $\mathscr{P}(V)$.

Note also that the anti-isomorphism a_V of Theorem 2.3 is between a projective geometry over a left vector space and one over a right vector space. For the moment we leave aside the interesting question of the existence (or not) of isomorphisms between $\mathscr{P}(V)$ and $\mathscr{P}(V')$, and move onto another useful application of the dual space. Suppose V is a left vector space of dimension $n + 1$ over K. Then a_V maps subspaces of V of dimension i onto subspaces of V' of dimension $n + 1 - i$, and in particular sends the points of $\mathscr{P}(V)$ (i.e. the one-dimensional subspaces of V) onto the hyperplanes of $\mathscr{P}(V')$ (i.e. the subspaces of dimension n); similarly $a_{V'}$ sends points of $\mathscr{P}(V')$ onto hyperplanes of $\mathscr{P}(V)$ and, by Theorem 2.3, $a_{V'}$ sends the hyperplanes of $\mathscr{P}(V')$ back onto the points which are their pre-images under a_V.

If U is any hyperplane of $\mathscr{P}(V)$ then the point U^{a_V} in $\mathscr{P}(V')$ uniquely defines, and is uniquely defined by, U. Recalling the definition of a_V we see that the point E of $\mathscr{P}(V)$ is on the hyperplane U of $\mathscr{P}(V)$ if and only if $e u' = 0$ for every e in E and every u' in U^{a_V}. Since E and U^{a_V} both have algebraic dimension one, E consists of the multiples $k e$, as k ranges over K, of some non-zero vector e in E, and similarly U^{a_V} consists of the multiples $u' k$, as k ranges over K, of some non-zero vector u' in U^{a_V}. So we can represent E by any one of its non-zero vectors, and represent the hyperplane U by any one of the non-zero vectors u' in U^{a_V}. Then the incidence will be given by the simple rule: if E is identified by e, U by u', then E is on U if and only if $e u' = 0$. This very important rule enables us to completely identify, for instance, every element in a projective geometry of g-dimension two, and every incidence. For suppose V is a three-dimensional left vector space over K, and V' its dual space. Then $\mathscr{P}(V)$ can be thought of as the object whose points are the one-dimensional subspaces of V and whose lines are the one-dimensional subspaces of V', and where the incidence rule is given by "annihilation", i.e. $E = \langle e \rangle$ is on $L = \langle w' \rangle$ if and only if $e w' = 0$. If we have chosen a coordinate system, then $E = \langle (x, y, z) \rangle$ is on $L = \langle (a, b, c)' \rangle$ if and only if $x a + y b + z c = 0$. That is, the ordinary "inner product" of the row vector (x, y, z) and the column vector $(a, b, c)'$ is zero. (The reader should convince himself that in this simple rule it does not really matter which vector e is chosen in E, so long as $e \neq 0$, nor does it matter which $w' \neq 0$ is chosen in L.)

In fact, this rule is as effective (at least in principle) for projective geometries of higher dimension, since it can be shown that any projective geometry is completely defined by its points and hyperplanes and the incidence rules between them.

This use of the dual space to give "coordinates" for hyperplanes leads to what are called *homogeneous coordinates*: once a basis has been chosen any point may be represented by a row vector v, or the subspace $\langle v \rangle$, and any hyperplane by a column vector w', or the subspace $\langle w' \rangle$. In the case of the plane, where lines and hyperplanes are the same, every element is then represented either by a row or column vector (or the subspaces they generate). We shall often use homogeneous coordinates without further comment but, in order not to confuse points of the plane with vectors of the space, will represent points by $\langle (x, y, z) \rangle$ and not (x, y, z). Similarly we shall represent lines by $\langle (a, b, c)' \rangle$.

Now of course if V is a right vector space then everything will be reversed: the points of $\mathcal{P}(V)$ will be represented by column vectors and the lines of $\mathcal{P}(V)$, that is the points of $\mathcal{P}(V')$, are then row vectors. The point $\langle (x, y, z)' \rangle$ is on the line $\langle (a, b, c) \rangle$ if and only if $ax + by + cz = 0$.

In the study of vector spaces the concept of a basis plays a very central role. The equivalent role in the study of projective geometries is taken by a "frame". If $\mathcal{P}(V)$ has g-dimension n, then a *frame* is an ordered set of $n + 2$ points in $\mathcal{P}(V)$ such that no $n + 1$ are contained in a hyperplane. Such things exist:

Lemma 2.4. *Let V be an $(n + 1)$-dimensional left vector space over a skewfield K and let $e_1, e_2, \ldots, e_{n+1}$ be a basis for V. Then $E_1 = \langle e_1 \rangle$, $E_2 = \langle e_2 \rangle, \ldots, E_{n+1} = \langle e_{n+1} \rangle$, $E_{n+2} = \langle e_1 + e_2 + \cdots + e_{n+1} \rangle$ is a frame for $\mathcal{P}(V)$.*

Proof. It is easy to see that the $n + 2$ vectors which are given as the generators of the E_i have the property that any $n + 1$ of them are linearly independent, and so they could not possibly be contained in a subspace of dimension n. □

In fact as the following lemma shows, Lemma 2.4 gives the most general possible example of a frame.

Lemma 2.5. *Let $\mathcal{P}(V)$ have g-dimension n, and let $E_1, E_2, \ldots, E_{n+2}$ be a frame in $\mathcal{P}(V)$. Then there exists a basis $e_1, e_2, \ldots, e_{n+1}$ of V such that $E_i = \langle e_i \rangle$ for $i = 1, 2 \ldots, n + 1$ and $E_{n+2} = \langle e_1 + e_2 + \cdots + e_{n+1} \rangle$.*

Proof. Without any loss of generality we may take V to be a left vector space over a skewfield K. Let $E_i = \langle f_i \rangle$, for $i = 1, 2, \ldots, n + 1$. If the f_i did not form a basis for V, then one of them would be a linear combination of the others. But if f_i, say, were a linear combination of the other f_j, then E_i would be contained in the space generated by all the others, which could not have dimension greater than n (since it has n generators), and so there would be a hyperplane containing $n + 1$ of the E_i. Thus the f_i are a basis. Now if $E_{n+2} = \left\langle \sum_{i=1}^{n+1} x_i f_i \right\rangle$ and one of the x_i were zero, then E_{n+2} would be contained in the subspace generated by n of the f_i, and there would be a set of $n+1$ E_j contained in a hyperplane. As this is not so each x_i is non-zero and so if we let $e_i = x_i f_i$, it is clear that the e_i form a basis which is of the kind called for in the lemma. $\square$

Exercise 2.7. If $E_1, E_2, \ldots, E_{n+2}$ is a frame in $\mathscr{P}(V)$, then no set of $r+1$ points of the frame, $r < n+1$, lies in a space of g-dimension $r-1$.

It is probably worthwhile to pause and emphasize the relevance of the preceeding discussion on frames. Given a basis of V we will have determined $n+1$ points in $\mathscr{P}(V)$ but these $n+1$ points do not determine the given basis. Any multiples of the given basis vectors could be used instead to give the same $n+1$ points. But once the $(n+2)$-nd point of the frame is given, then the particular basis *(up to a common multiple)* is fixed. In the next section we shall see more clearly how the role played by the frame is analogous to that of the basis. Meanwhile we give two exercises which determine all frames for a $\mathscr{P}(V)$ of g-dimension one or two.

Exercise 2.8. Let $\mathscr{P}(V)$ have g-dimension one. Then show that the three points E_1, E_2, E_3 are a frame if and only if they are distinct.

Exercise 2.9. Let $\mathscr{P}(V)$ have g-dimension two. Then show that a set of four points is a frame if and only if no three of the points are collinear.

We now prove an interesting classical theorem which gives us a chance to use the ideas introduced so far.

Theorem 2.6. *Let A_1, B_1, C_1, be three distinct points on a line L_1 and A_2, B_2, C_2, three distinct points on a line $L_2 \neq L_1$, all in $\mathscr{P}(V)$ of g-dimension two. Define three points A_3, B_3, C_3 as follows: $A_3 = (B_1 + C_2) \cap (B_2 + C_1)$, $B_3 = (A_1 + C_2) \cap (A_2 + C_1)$, $C_3 = (A_1 + B_2) \cap (A_2 + B_1)$. If no three of A_1, B_1, A_2, B_2 are collinear and C_1, C_2 are arbitrary then the points A_3, B_3, C_3 are collinear if and only if the skewfield K is commutative.*

Proof. We first point out that, as a consequence of Exercise 2.9, the condition on A_1, B_1, A_2, B_2 merely says that they are a frame. The special cases, by the way, when one of C_1 or C_2 coincides with a point of the given frame are trivial and are left for the reader to verify.

Suppose that V is a left vector space over K and that a basis has been chosen so that $A_1 = \langle (1,0,0) \rangle$, $B_1 = \langle (0,1,0) \rangle$, $A_2 = \langle (0,0,1) \rangle$, $B_2 = \langle (1,1,1) \rangle$. An arbitrary point $C_1 \neq A_1$ on the line L_1 will have the form $\langle (x,1,0) \rangle$, while the arbitrary point $C_2 \neq A_2$ on L_2 will be $\langle (1,1,y) \rangle$. Since $\mathscr{P}(V)$ has g-dimension 2 a line of $\mathscr{P}(V)$ is also a hyperplane of $\mathscr{P}(V)$. Thus, using the rule for incidence between a point and a hyperplane we see that $A_1 + B_2$, (the line joining A_1 to B_2), is $\langle (0,1,-1)' \rangle$, $A_2 + B_1$ is $\langle (1,0,0)' \rangle$, and hence C_3 is $\langle (0,1,1) \rangle$. The line

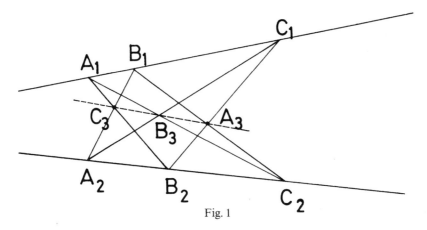

Fig. 1

$A_1 + C_2$ will be $\langle (0, -y, 1)' \rangle$, and $A_2 + C_1$ is $\langle (1, -x, 0)' \rangle$, so B_3 must be $\langle (x, 1, y) \rangle$. Finally, $B_1 + C_2$ is $\langle (-y, 0, 1)' \rangle$ and $B_2 + C_1$ is $\langle (1, -x, -1 + x)' \rangle$, and so A_3 must be $\langle (1, x^{-1} + y(1 - x^{-1}), y) \rangle$. Note that in each case the proof that the appropriate elements are incident is simply a matter of taking the inner products; we already know that two lines meet exactly once and two points have just one line in common, so it is never necessary to verify uniqueness.

Now the line joining B_3 and C_3 will be $\langle (x^{-1}(1 - y), -1, 1)' \rangle$, and so A_3 is on that line if and only if $x^{-1}(1 - y) - x^{-1} - y(1 - x^{-1}) + y = 0$. This last equation is valid if and only if $x^{-1}y = yx^{-1}$. Since x and y are arbitrary, we have shown that the appropriate collinearity occurs if and only if K is commutative. □

The theorem above is very famous and is known as *Pappus' Theorem*; the configuration of nine points and nine lines that it involves is *Pappus' Configuration* (see Fig. 1).

Another application of the ideas developed here can be made by attempting to "coordinatize" a projective plane. Let V be a three-dimensional left vector space over K, and let E_1, E_2, E_3, E_4 be a frame for $\mathscr{P}(V)$; choose a basis for V, as in Lemma 2.5, so $E_1 = \langle e_1 \rangle$, $E_2 = \langle e_2 \rangle$, $E_3 = \langle e_3 \rangle$, $E_4 = \langle e_1 + e_2 + e_3 \rangle$. The line $E_1 + E_2$ in $\mathscr{P}(V)$ is the two dimensional subspace of V spanned by e_1 and e_2. Thus every point of $E_1 + E_2$, other than E_1, has the form $\langle (y_1 e_1 + y_2 e_2) \rangle$ where y_1, y_2 are in K such that $y_2 \neq 0$. Furthermore the point $\langle (z_1 e_1 + z_2 e_2) \rangle$ is the point $\langle (y_1 e_1 + y_2 e_2) \rangle$ if and only if there is an element k in K with $y_1 = k z_1$, $y_2 = k z_2$. Hence every point of $E_1 + E_2$ other than E_1 has the form $\langle (x e_1 + e_2) \rangle$ for a unique x in K. The line $E_3 + E_4$ meets the line $E_1 + E_2$ in the point $\langle e_1 + e_2 \rangle$; for certainly the vector $e_1 + e_2$ is in the subspace $E_1 + E_2$, and since $e_1 + e_2 = (e_1 + e_2 + e_3) - e_3$, it lies in $E_3 + E_4$. (Again we note that we never need verify uniqueness. In order to find the point common to two lines it is always sufficient to find a single non-zero vector which lies in both subspaces.) Every other point on $E_3 + E_4$ (that is, not on $E_1 + E_2$) can be uniquely represented in the form $\langle x e_1 + x e_2 + e_3 \rangle$.

Now the line from E_2 to $\langle x e_1 + e_3 \rangle$ meets $E_3 + E_4$ in $\langle x e_1 + x e_2 + e_3 \rangle$; similarly the line from E_1 to $\langle y e_2 + e_3 \rangle$ meets $E_3 + E_4$ in $\langle y e_1 + y e_2 + e_3 \rangle$. An arbitrary point P not on the line $E_1 + E_2$ has a unique representation in the form $\langle x e_1 + y e_2 + e_3 \rangle$, and using the arguments as above, we can see that the line $E_2 + P$ meets $E_1 + E_3$ in $\langle x e_1 + e_3 \rangle$, while the line $E_1 + P$ meets $E_2 + E_3$ in $\langle y e_2 + e_3 \rangle$. We have now set up a kind of co-ordinatization of the points of $\mathscr{P}(V)$ not on $E_1 + E_2$. By this we mean there is a one-to-one correspondence between such points and the ordered pairs (x, y) from K. If we could imagine $E_1 + E_3$ as an x-axis and $E_2 + E_3$ as a y-axis (somehow ignoring the points E_1 and E_2 themselves), then the point $\langle x e_1 + y e_2 + e_3 \rangle$, that is, the point corresponding to the ordered pair (x, y) "projects" from E_2 onto the x-axis in the point $(x, 0)$, and from E_1 onto the y-axis in the point $(0, y)$. These ideas are only suggestive and inexact at this stage but will be clarified later. However we proceed with our "coordinatization".

The points on the line $E_2 + E_1$, again excepting one of them which will be E_2 this time, all have the form $\langle m e_2 + e_1 \rangle$. If we fix such a point (i.e. fix the element m in K), and choose a point $\langle b e_2 + e_3 \rangle$ on the line $E_2 + E_3$ (that is, our unofficial y-axis above), then the line which joins them will contain, besides $\langle m e_2 + e_1 \rangle$, points of the form $\langle x e_1 + y e_2 + e_3 \rangle$.

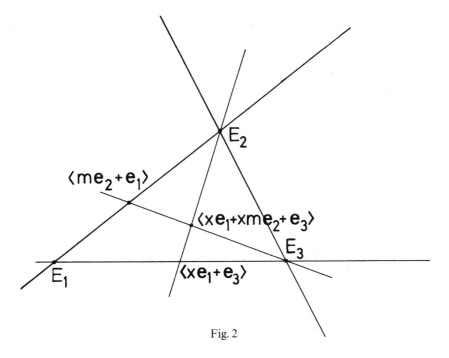

Fig. 2

Lemma 2.7. *A point* $\langle x e_1 + y e_2 + e_3 \rangle$ *is on the line joining* $\langle m e_2 + e_1 \rangle$ *to* $\langle b e_2 + e_3 \rangle$ *if and only if* $y = xm + b$.

Proof. The point $\langle x e_1 + y e_2 + e_3 \rangle$ is on the line joining $\langle m e_2 + e_1 \rangle$ to $\langle b e_2 + e_3 \rangle$ if and only if the vector $x e_1 + y e_2 + e_3$ is a linear combination of $m e_2 + e_1$ and $b e_2 + e_3$. But if $x e_1 + y e_2 + e_3 = c(m e_2 + e_1) + d(b e_2 + e_3)$ then, by first equating the coefficients of e_3 and then those of e_1, we have $d = 1$ and $c = x$. Equating the coefficients of e_2 now gives $y = xm + b$ as required. Conversely it is easily seen that if $y = xm + b$ then $x e_1 + y e_2 + e_3 = x(m e_2 + e_1) + b e_2 + e_3$. □

Lemma 2.7 gives us a geometrical interpretation of the skewfield operations. For example, if $b = 0$ then the line joining $\langle m e_2 + e_1 \rangle$ to E_3 meets the line joining $\langle x e_1 + e_3 \rangle$ to E_2 in the point $\langle x e_1 + x m e_2 + e_3 \rangle$. Thus we have a geometric construction which, given x and m, determines xm for us (see Fig. 2).

Exercise 2.10. By using Lemma 2.7 with $m = 1$ give a construction (similar to that for finding xm) to find $x + b$. (Draw a diagram!)

Lemma 2.7 may be regarded as giving an equation for the line joining $\langle m e_2 + e_1 \rangle$ to $\langle b e_2 + e_3 \rangle$ in the sense that it gives an equation which must be satisfied by the "coordinates" of any point on the given line provided that the point is not on $E_1 + E_2$.

We shall return to this subject, and that of relating geometrical constructions to the operations of K, in later sections.

Exercise 2.11. If $K = GF(q)$ show that every line of $\mathscr{P}_2(K)$ has $q + 1$ points on it and that every point is on $q + 1$ lines. Show also that $\mathscr{P}_2(K)$ has $q^2 + q + 1$ points and $q^2 + q + 1$ lines.

Exercise 2.12. Show that if the frame $E_1 = \langle e_1 \rangle$, $E_2 = \langle e_2 \rangle, \ldots,$ $E_{n+1} = \langle e_{n+1} \rangle$, $E_{n+2} = \langle e_1 + e_2 + \cdots + e_{n+1} \rangle$ is replaced by $E_1^* = \langle a_1 e_1 \rangle$, $E_2^* = \langle a_2 e_2 \rangle, \ldots, E_{n+1}^* = \langle a_{n+1} e_{n+1} \rangle, E_{n+2}^* = \langle a_1 e_1 + a_2 e_2 + \cdots + a_{n+1} e_{n+1} \rangle$ then the point $\langle (x_1, x_2, \ldots, x_{n+1}) \rangle$ becomes $\langle (x_1 a_1^{-1}, x_2 a_2^{-1}, \ldots, x_{n+1} a_{n+1}^{-1}) \rangle$.

(Note that the two frames differ only by a single point.)

4. Isomorphisms, the Fundamental Theorem and Related Topics

In this section we discuss the problem of determining all isomorphisms, anti-isomorphisms, automorphisms, and correlations of projective planes. In fact, our results are equally valid (properly interpreted) for higher g-dimension, and the proofs are not basically different; but they are sufficiently more complicated to make it worth our while to restrict ourselves to the plane.

Suppose V and W are left vector spaces over K, and β is a semi-linear transformation from V onto W. Then, since β preserves the relation of incidence among subspaces, it is clear that β induces a homomorphism β of $\mathscr{P}(V)$ onto $\mathscr{P}(W)$, as pointed out in the previous section.

If V and W have the same dimension then β induces an isomorphism from $\mathscr{P}(V)$ onto $\mathscr{P}(W)$. The remarkable fact is that all isomorphisms from $\mathscr{P}(V)$ onto $\mathscr{P}(W)$ are of this form, a theorem which we prove for the projective plane.

Theorem 2.8. *Let V be a left vector space over the skewfield K and W a left vector space over the skewfield F, and suppose each has dimension three. If there is an isomorphism from $\mathscr{P}(V)$ onto $\mathscr{P}(W)$, then F and K are isomorphic and the isomorphism from $\mathscr{P}(V)$ onto $\mathscr{P}(W)$ is induced by a semi-linear transformation from V onto W.*

Proof. Let β be the isomorphism from $\mathscr{P}(V)$ onto $\mathscr{P}(W)$ and choose a frame E_1, E_2, E_3, E_4, for $\mathscr{P}(W)$, given by a basis e_1, e_2, e_3 (so $E_i = \langle e_i \rangle$ for $i = 1, 2, 3$ and $E_4 = \langle e_1 + e_2 + e_3 \rangle$). The pre-images of the E_i are four points G_i in $\mathscr{P}(V)$, and clearly they are a frame for $\mathscr{P}(V)$; so choose a basis g_1, g_2, g_3 for $\mathscr{P}(V)$ such that $G_i = \langle g_i \rangle$ for $i = 1, 2, 3$ and $G_4 = \langle g_1 + g_2 + g_3 \rangle$. (Such a choice is possible by Exercise 2.9.)

Clearly all lines of the sort $G_i + G_j$ must be sent by β onto lines $E_i + E_j$, and intersections of such lines must similarly be sent onto the appropriate intersection in $\mathscr{P}(W)$. So a point $\langle x g_1 + g_3 \rangle$ on $G_1 + G_3$ goes to a point $\langle x^\phi e_1 + e_3 \rangle$ in $\mathscr{P}(W)$, where ϕ is some one-to-one mapping of F onto K. The point $\langle x g_1 + x g_2 + g_3 \rangle$ is the intersection of the line $G_3 + G_4$ with the line joining G_2 to the point $\langle x g_1 + g_3 \rangle$ on $G_1 + G_3$. Thus, since β sends $G_3 + G_4$ onto $E_3 + E_4$, G_2 onto E_2 and $\langle x g_1 + g_3 \rangle$ onto $\langle x^\phi e_1 + e_3 \rangle$, the image of $\langle x g_1 + x g_2 + g_3 \rangle$ must be the point $\langle x^\phi e_1 + x^\phi e_2 + e_3 \rangle$. Similarly by "projecting" from G_1 onto the line $G_2 + G_3$ (the unofficial y-axis), we see that $\langle y g_2 + g_3 \rangle^\beta = \langle y^\phi e_2 + e_3 \rangle$. From this it is easy to see that any point of the form $\langle x g_1 + y g_2 + g_3 \rangle$ has for its image under β the point $\langle x^\phi e_1 + y^\phi e_2 + e_3 \rangle$. Thus we have found the image of every point not on the line $G_1 + G_2$ in terms of the mapping ϕ which, since $G_4^\beta = E_4$, maps the "1" of K onto the "1" of F. In order to completely determine the action of β we must find the image of an arbitrary point $\langle m g_2 + g_1 \rangle$ on $G_1 + G_2$. The point $\langle m g_2 + g_1 \rangle$ is the intersection of $G_1 + G_2$ with the line joining G_3 to $\langle g_1 + m g_2 + g_3 \rangle$ and hence $\langle m g_2 + g_1 \rangle^\beta$ is the intersection of $E_1 + E_2$ with the join of E_3 and $\langle g_1 + m g_2 + g_3 \rangle^\beta = \langle e_1 + m^\phi e_2 + e_3 \rangle$. Thus $\langle m g_2 + g_1 \rangle^\beta = \langle m^\phi e_2 + e_1 \rangle$.

In order to establish that ϕ is a multiplicative isomorphism we utilize the geometric structure given in Lemma 2.7 to determine first the product xm. The point $\langle x g_1 + x m g_2 + g_3 \rangle$ is the intersection of the line joining G_3 to $\langle m g_2 + g_1 \rangle$ with the line joining G_2 to $\langle x g_1 + g_3 \rangle$. Thus $\langle x g_1 + x m g_2 + g_3 \rangle^\beta$, which we already know is $\langle x^\phi e_1 + (xm)^\phi e_2 + e_3 \rangle$, is the intersection of the line joining E_3 to $\langle m^\phi e_2 + e_1 \rangle$ with the join of E_2 with $\langle x^\phi e_1 + e_3 \rangle$. Thus, since this point is $\langle x^\phi e_1 + x^\phi m^\phi e_2 + e_3 \rangle$, we have $(xm)^\phi = x^\phi m^\phi$ for all x, m in K. Diagrams are obviously helpful in arguments of this type and the reader is urged to draw them for each proof.

In order to establish the isomorphism between K and F we have to prove $(x + b)^\phi = x^\phi + b^\phi$ for all x, b in K. But the proof of this follows similar reasoning to that used for the multiplicative property of ϕ and is left as an important exercise. To conclude the proof of Theorem 2.8 we simply observe that β is induced by the semi-linear transformation $\pmb{\beta}$ of V onto W which is the linear transformation sending the $\pmb{g}_i$ onto the $\pmb{e}_i$ followed by the field isomorphism ϕ. $\square$

Corollary. (The Fundamental Theorem of Projective Geometry – for the case of g-dimension two.) *The group of all automorphisms of a projective plane $\mathcal{P}(V)$ onto itself is induced by the group of all non-singular semi-linear transformations of V onto V.*

Now in fact the method of proof of Theorem 2.8 can be used with the obvious and simple modifications to give us:

Theorem 2.9. *Let V be a left vector space over the skewfield K and W a right vector space over the skewfield F, and suppose each has dimension three. If there is an isomorphism from $\mathcal{P}(V)$ onto $\mathcal{P}(W)$, then K and F are anti-isomorphic and the isomorphism from $\mathcal{P}(V)$ onto $\mathcal{P}(W)$ is induced by a semi-linear transformation from V onto W.*

(N.B. Recall that a semi-linear transformation from a left to a right vector space involves an anti-isomorphism, not an isomorphism; see Chapter I.)

Corollary. *Let V and W be left vector spaces of dimension three over the skewfields K and F respectively and suppose there is an anti-isomorphism β from $\mathcal{P}(V)$ onto $\mathcal{P}(W)$. Then K and F are anti-isomorphic. In addition, β has form $\beta = \theta_1 a_{W'} = a_V \theta_2$, where a_V is the annihilator map from $\mathcal{P}(V)$ onto $\mathcal{P}(V')$, $a_{W'}$ is the annihilator map from $\mathcal{P}(W')$ onto $\mathcal{P}(W)$, θ_1 is an isomorphism induced by a semi-linear transformation θ_1 from V onto W' and θ_2 is an isomorphism induced by a semi-linear transformation θ_2 from V' onto W.*

Proof of Corollary. Since the mapping βa_W is the product of two anti-isomorphisms it is an isomorphism, and goes from $\mathcal{P}(V)$ to $\mathcal{P}(W')$; call it θ_1. Then it has the required form by Theorem 2.9, and so $\beta = \beta a_W a_{W'} = \theta_1 a_{W'}$. The other half is similar (and of course the anti-isomorphism between K and F is a consequence of Theorem 2.9 as well). $\square$

Exercise 2.13. Analyze the situation of an anti-isomorphism from a $\mathcal{P}(V)$ to a $\mathcal{P}(W)$, where V is a left and W is a right vector space over skewfields K and F. In particular, are K and F isomorphic or anti-isomorphic?

All the above theorems and corollaries are equally true for arbitrary $\mathscr{P}(V)$, so long as the g-dimension is greater than one (and when we have made the correct definitions, they will be true for g-dimension one as well); the interested student will not find it difficult to extend the proofs to higher dimension. Hence we have a motive for analyzing the group of all non-singular semi-linear transformations of a vector space V onto itself, and the group induced by this first group on $\mathscr{P}(V)$. With this in mind we make some definitions and introduce some notation. Let V be a left vector space over the skewfield K (arbitrary finite dimension). Let $\Gamma L(V)$ be the group of all non-singular semi-linear transformations of V onto V and let $GL(V)$ be the subgroup consisting of all non-singular linear transformations (see Chapter I). In case the underlying skewfield is a field, we define $SL(V)$ to be the subgroup of $GL(V)$ consisting of elements of determinant one. Each of these groups induces a group of automorphisms of $\mathscr{P}(V)$ onto itself, which will be indicated by writing the letter "P" in front of the group: so $P\Gamma L(V)$, $PGL(V)$ and $PSL(V)$ (when this last can be defined) are the groups induced by $\Gamma L(V)$, $GL(V)$ and $SL(V)$ on $\mathscr{P}(V)$.

Naturally we wish to determine these new groups:

Theorem 2.10. *Let V be a left vector space over a skewfield K and let $N = \{\gamma \in \Gamma L(V) \,|\, \langle v^\gamma \rangle = \langle v \rangle$ for all v in $V\}$. Then $P\Gamma L(V) \cong \Gamma L(V)/N$, $PGL(V) \cong GL(V)/N \cap GL(V)$ and, if K is a field, $PSL(V) \cong SL(V)/N \cap SL(V)$.*

Proof. Since N is the set of semi-linear transformations which fix every one-dimensional subspace of V, it is the set of all these semi-linear transformations which induce the identity automorphism on $\mathscr{P}(V)$. Thus N is the kernel of the "natural" homomorphism which sends an element of $\Gamma L(V)$ onto the element it induces in $P\Gamma L(V)$. Similarly $N \cap GL(V)$ and $N \cap SL(V)$ are the kernels of the other natural homomorphisms.

Theorem 2.11. *Let V be a left vector space over a skewfield K, let $\operatorname{Aut} K$ be the automorphism group of K, let K^* be the multiplicative group of non-zero elements of K and let $\operatorname{In} K$ be the group of automorphisms of K given by the inner automorphisms of K^*. Then $GL(V)$ is normal in $\Gamma L(V)$ and, when K is a field, $SL(V)$ is normal in $GL(V)$; hence $PGL(V)$ is normal in $P\Gamma L(V)$ and $PSL(V)$ is normal in $PGL(V)$. Furthermore*
  (a) $\Gamma L(V)/GL(V) \cong \operatorname{Aut} K$,
  (b) $P\Gamma L(V)/PGL(V) \cong \operatorname{Aut} K/\operatorname{In} K$,
  (c) $GL(V)/SL(V) \cong K^*$.

Proof. Consider the mapping ϕ that sends a semi-linear transformation α onto its associated automorphism $\phi(\alpha)$, i.e. if k is in K and v in V then

$(k\,v)^{\alpha} = k^{\phi(\alpha)}\,v^{\alpha}$. Then, since $k^{\phi(\alpha\beta)}\,v^{\alpha\beta} = (k\,v)^{\alpha\beta} = (k^{\phi(\alpha)}v^{\alpha})^{\beta} = k^{\phi(\alpha)\,\phi(\beta)}\,v^{\alpha\beta}$, ϕ is a homomorphism. Clearly the kernel of ϕ is $GL(V)$ and, in order to see that the image is $\operatorname{Aut} K$, the student should convince himself that every automorphism of K can occur in some semi-linear transformation. This proves (a).

From Theorem 2.10 and the isomorphism theorems, $P\Gamma L(V)/PGL(V)$ $= (\Gamma L(V)/N)/(N \cdot GL(V)/N) \cong \Gamma L(V)/N \cdot GL(V)$. If we write $A = \operatorname{Aut} K$, then $\Gamma L(V) = A \cdot GL(V)$, and since N is contained in $\Gamma L(V)$, we have $\Gamma L(V) = A \cdot N \cdot GL(V)$. So $\Gamma L(V)/N \cdot GL(V) = A \cdot N \cdot GL(V)/N \cdot GL(V)$ $\cong A/(A \cap N \cdot GL(V))$. But suppose an element α in A is also in $N \cdot GL(V)$; then if β is the map $\beta : e \to ke$, for an appropriate k in K^{*}, it follows that $\alpha\beta$ is a linear transformation. So if a is in K, then $(av)^{\alpha\beta} = a(v^{\alpha\beta})$, and this forces $a^{\alpha} = k^{-1}ak$. So $A \cap N \cdot GL(V) = \operatorname{In} K$, the inner automorphism group of K, proving (b).

The last part follows from considering the mapping $A \to \det A$, where A is a linear transformation over a field. This is well known to be a homomorphism, and its kernel is exactly $SL(V)$. Its image is all of K^{*}, since the mapping whose matrix (in some basis) has x in the upper left corner and 1's on the rest of the main diagonal, 0's elsewhere, has determinant x. ☐

Exercise 2.14. Determine $PGL(V)/PSL(V)$.

Now we can ask for the group-theoretic properties of the various groups above. In particular, questions of transitivity are of importance. The first result is easy:

Theorem 2.12. *The group $PGL(V)$ (and hence $P\Gamma L(V)$) is transitive on the frames of $\mathscr{P}(V)$. The subgroup of $PGL(V)$ fixing a frame pointwise is isomorphic to $\operatorname{In} K$ and the subgroup of $P\Gamma L(V)$ fixing a frame pointwise is isomorphic to $\operatorname{Aut} K$.*

Proof. From Result 1.22 the group $GL(V)$ is transitive on the bases of V and hence, since by Lemma 2.5 each frame is completely determined by a basis, $PGL(V)$ is transitive on frames.

If an element of $P\Gamma L(V)$ fixes a frame pointwise then it is induced by an element of $\Gamma L(V)$ leaving invariant the one-dimensional subspace spanned by each basis vector associated with the frame. Let $E_1 = \langle e_1 \rangle, \ldots,$ $E_{n+1} = \langle e_{n+1} \rangle$ $E_{n+2} = \langle e_1 + \cdots + e_{n+1} \rangle$ be a frame. Any element of $P\Gamma L(V)$ fixing $E_1, \ldots, E_{n+1}$ must be of the form $\Sigma x_i e_i \to \Sigma x_i^{\alpha} k_i e_i$ where each k_i is in K^{*} and $\alpha \in \operatorname{Aut} K$, and this element fixes E_{n+2} if and only if $k_i = k_j$ for all i, j. Let $\theta(\alpha, k)$ be the element of $P\Gamma L(V)$ which maps $\Sigma x_i e_i$ onto $\Sigma x_i^{\alpha} k e_i$. Then $\theta(\alpha, k) = 1$ if and only if there is an element h in K^{*} such that $\Sigma x_i^{\alpha} k e_i = \Sigma h x_i e_i$ for all x_i in K. Clearly the only possible value for h is k and thus $\theta(\alpha, k) = 1$ if and only if $\alpha = \omega_{k^{-1}}$, where ω_k is conjugation by k, i.e. $\omega_k : x \to k^{-1}xk$.

Let G denote the subgroup of $P\Gamma L(V)$ fixing the given frame pointwise, let $H = \{\theta(\alpha, k) \mid \alpha \in \operatorname{Aut} K, k \in K^*\}$ and let $N = \{\theta(\alpha, k) \mid \alpha = \omega_{k^{-1}}\}$.

Since N is the subgroup of H which induces the identity on $\mathscr{P}(V)$, $G \cong H/N$. Let ϕ be the mapping from H onto $\operatorname{Aut} K$ given by $\phi: \theta(\alpha, k) \to \alpha \omega_k$. Straightforward computation gives $\theta(\alpha, k)\,\theta(\beta, h) = \theta(\alpha\beta, k^\beta h)$ and thus, since $\alpha\beta\,\omega_{k^\beta h} = \alpha\omega_k \beta\omega_h$, ϕ is a homomorphism of H onto $\operatorname{Aut} K$. The kernel of ϕ is $\{\theta(\alpha, k) \mid \alpha\omega_k = 1\} = N$ and so $\operatorname{Aut} K \cong H/N \cong G$.

Clearly $\theta(\alpha, k)$ is linear if and only if $\alpha = 1$. Repeating the above argument the mapping $\theta(1, k) \to \omega_k$ gives a homomorphism from $GL(V) \cap H$ onto $\operatorname{In} K$ with kernel $GL(V) \cap N$, and completes the proof.   □

We are now in a position to give a definition of the automorphism group of a projective geometry of g-dimension one, or more generally, of isomorphism, anti-isomorphism, correlation etc. We want the theorems above to be true for g-dimension one, and so we simply *define* an isomorphism from a projective line to another to be a mapping induced by a semi-linear transformation of the given vector space, and so on. This will guarantee that the group $\operatorname{Aut}\mathscr{P}(V)$ is always $P\Gamma L(V)$, for all g-dimensions. For the moment we shall be content with this approach, but later when we consider the structures induced on a projective line by its embedding in a projective geometry of higher g-dimension, we will offer another justification for it.

This justification will be given in the next section. Meanwhile we study the projective line over a field in more detail. Let V be a two-dimensional vector space over the field K, so $\mathscr{P} = \mathscr{P}(V)$ is the line.

A frame (see Exercise 2.9) is any set of three distinct points in $\mathscr{P}$; let them be E_1, E_2, E_3, and let V have a basis e_1 and e_2 such that $E_1 = \langle e_1 \rangle$, $E_2 = \langle e_2 \rangle$, $E_3 = \langle e_1 + e_2 \rangle$. A point in $\mathscr{P}$ is a subspace $\langle xe_1 + ye_2 \rangle$, where x and y are arbitrary elements of K, not both zero. If $y \neq 0$, then we may write $\langle xe_1 + ye_2 \rangle = \langle y^{-1} xe_1 + e_2 \rangle$, and so the points of $\mathscr{P}$ can be put in one-to-one correspondence with the elements of K, excepting for the "extra" point E_1, for which we will use the symbol ∞. This is the representation of $\mathscr{P}$ in *parametric coordinates*: the points of $\mathscr{P}$ are coordinatized by the marks of K, plus the one new symbol ∞. Notice that if we had used another basis for the same frame, then it would have been $f_i = be_i$ for some b in K, and so the point $\langle xe_1 + ye_2 \rangle$ would have been $\langle xb^{-1} f_1 + yb^{-1} f_2 \rangle$, whence the parametric coordinate for the point would be $(yb^{-1})^{-1} xb^{-1}$; this last is equal to $y^{-1} x$, since we are only considering the case where K is a field. This is the chief reason that we have restricted ourselves to fields in this part: we want parametric coordinates to be well-defined, and while it is quite proper that they should depend on the frame, it is not so useful if they also depend upon the basis that gives the frame.

Using our definition of $\operatorname{Aut}\mathscr{P}$ above, we naturally ask: what is the effect of an element of $\operatorname{Aut}\mathscr{P}$ on the elements of $\mathscr{P}$ written in parametric form? An element of $\text{Aut}\,\mathscr{P}$ is induced by a semi-linear transformation, that is by a mapping that sends a vector $x e_1 + y e_2$ onto $(x^\alpha a + y^\alpha b) e_1 + (x^\alpha c + y^\alpha d) e_2$, where the determinant $ad - bc \neq 0$ and α is in $\text{Aut}\,K$. The mapping of $\mathscr{P}$ induced by this semi-linear transformation will be given as follows: the point k in K corresponds to $\langle k e_1 + e_2 \rangle$, which is sent to $\langle (k^\alpha a + b) e_1 + (k^\alpha c + d) e_2 \rangle$, and the parametric coordinate of this last point is $(k^\alpha c + d)^{-1}(k^\alpha a + b)$, if $k^\alpha c + d \neq 0$; if $k^\alpha c + d = 0$, then its parametric coordinate is ∞. The point ∞ is mapped to $c^{-1} a$ if $c \neq 0$, and is fixed if $c = 0$. If we make the simple assumption that the inverse of ∞ is 0, and the inverse of 0 is ∞, then we can introduce certain other conventions, abuses of mathematical notation really, to give us a useful form for the group $\text{Aut}\,\mathscr{P}$: for instance, if α is an automorphism, then $\infty^\alpha = \infty$, $c \cdot \infty + d = c \cdot \infty$, $c \cdot \infty = \infty$ but $(a \cdot \infty)(c \cdot \infty)^{-1} = ac^{-1}$, and so on. So we have:

Lemma 2.13. *If K is a field and $\mathscr{P} = \mathscr{P}_2(K)$ is the projective line over K, then the group $\text{Aut}\,\mathscr{P}$ can be represented on the points of $\mathscr{P}$ in parametric form by the set of mappings $\theta(a, b, c, d, \alpha)$ below: for each α in $\text{Aut}\,K$ and for each choice of a, b, c, d in K such that $ad - bc \neq 0$, $\theta(a, b, c, d, \alpha)$ is defined by*

$$k \to \frac{k^\alpha a + b}{k^\alpha c + d}.$$

The reader should verify that this formula works in all the various "degenerate" cases.

Lemma 2.14. *Using the terminology of Lemma 2.13, the group $PGL(V)$ of the projective line is given by the mappings of that lemma in which $\alpha = 1$, and the group $PSL(V)$ of the projective line is given by the mappings of that lemma in which $\alpha = 1$ and $ad - bc$ is an arbitrary non-zero square.*

Proof. The first part is obvious, since $\alpha = 1$ for a linear transformation. For the second, we must ask what "parametric" form is induced by a linear transformation of determinant one. The matrix

$$\begin{bmatrix} a & c \\ b & d \end{bmatrix}$$

of the mapping $x \to (ax + b)/(cx + d)$ has determinant one if and only if $ad - bc = 1$. But replacing each of a, b, c, d by a non-zero (but constant) multiple ta, tb, tc, td will not change the mapping, in its parametric form (hence in its geometric form), but will change the determinant by t^2. Since t is arbitrary, any (non-zero) square can result. $\square$

Theorem 2.15. *Let V be a two dimensional vector space over a field K. Then the group $PSL(V)$ is two-transitive on the points of $\mathscr{P}(V)$ but not in general three-transitive, while $PGL(V)$ is sharply three-transitive, in every case.*

Proof. Since K is a field, $\text{In}\, K = 1$ and thus, by Theorem 2.12, $PGL(V)$ is sharply transitive on frames which is exactly equivalent to saying that it is sharply three-transitive on points (see Exercise 2.8). To show $PSL(V)$ is two-transitive we show first that it is transitive: the mapping $x \to (ax + b)/(cx + d)$ sends ∞ onto a/c, and clearly we can choose a/c to be an arbitrary element of K while keeping $ad - bc$ a square. (If $a/c = t \neq 0$ put $b = 0$ and $d = ct$ so that $ad - bc = (ct)^2$, if $a = 0$ put $d = 0, c = 1, b = -1$.)

The subgroup of $PSL(V)$ fixing the point ∞ will be exactly the subgroup with $c = 0$; i.e., the group $x \to (a/d)\, x + b/d$, where the determinant $ad - bc = ad =$ square. But $ad = (a/d)\, d^2$ which is a square if and only if a/d is a square. So the subgroup we are after is the group $x \to ax + b$, where a is a non-zero square. In this group 0 is sent onto elements $0 + b = b$, where b is arbitrary: that is, this group is transitive on K, i.e. on the elements of $\mathscr{P}(V) \backslash \infty$. Thus $PSL(V)$ is two-transitive.

The stabilizer of the two points ∞ and 0 is now the subgroup of all mappings $x \to ax$, where a is a non-zero square. The point 1 is sent onto a, and so, since in general not every element of K^* is a square, $PSL(V)$ is not generally three-transitive. In fact:

Lemma 2.16. *Let V be a two dimensional vector space over a field K. Then $PSL(V)$ is three-transitive if and only if $PSL(V) = PGL(V)$, which is equivalent to demanding that every non-zero element of the field K is a square.*

Exercise 2.15. If $K = GF(q)$ is a finite field and V is a two-dimensional vector space over K, then show that $PSL(V)$ is three-transitive if and only if q is even. (Here q is an arbitrary prime-power.)

Exercise 2.16. If p is a prime, $K = GF(p^t)$ and V is two-dimensional over K, then show that the orders of $P\Gamma L(V)$, $PGL(V)$ and $PSL(V)$ are respectively $t p^t (p^{2t} - 1)$, $p^t (p^{2t} - 1)$, and $p^t (p^{2t} - 1)/j$, where $j = 1$ or 2 according as p is even or odd.

Exercise 2.17. Suppose K is a field and V is two-dimensional over K. Show that if Σ is the subgroup of $PSL(V)$ fixing two points of $\mathscr{P}(V)$, then there is a one-to-one correspondence between the number of orbits of Σ on the remaining points of $\mathscr{P}(V)$ and the index of the subgroup of squares in the multiplicative group of K. Furthermore Σ is semi-regular on each of these orbits (that is, the orbits other than the two fixed points).

Exercise 2.18. Let V be a vector space of any dimension greater than one over a field K. Show that if a collineation in $PGL(V)$ fixes three collinear points in $\mathscr{P}(V)$ then it fixes all the points of that line.

Exercise 2.19. If V has dimension three over a field K show that any element of $PGL(V)$ fixing three concurrent lines must fix all the lines through the common point.

Exercise 2.20. Show that Exercises 2.18 and 2.19 are both false if we write $P\Gamma L(V)$ instead of $PGL(V)$.

For completeness, we ask what of all this treatment of the line is preserved if K is only a skewfield? In the first place, Aut $\mathscr{P}$ will still be three-transitive, since the group of semi-linear (even linear) transformations is always transitive on bases, hence the group induced by it is transitive on frames. What of the stabilizer of three points, that is, the subgroup of Aut $\mathscr{P}$ fixing some three points?

Exercise 2.21. If K is a skewfield and $\mathscr{P}$ is a projective line over K (right or left does not matter) then find the stabilizer of any three points of $\mathscr{P}$ in Aut $\mathscr{P}$.

5. The Line in the Plane

We now return for a while to the case where K is an arbitrary skewfield. Since any $\mathscr{P}(V)$ of g-dimension one can also be thought of a sub-geometry of a geometry of higher dimension, we may examine the effect of this "embedding". For instance, a projective line in a projective plane will have some structure induced upon it by the surrounding plane. Also the full group of collineations of the plane has a subgroup consisting of all those elements which fix the given line, and we can compare this subgroup with the group $P\Gamma L(V)$ defined above. From now on whenever we wish to stress the g-dimension n of $\mathscr{P}(V)$ we shall write $\mathscr{P}_n(V)$.

First we speak about a notational difficulty, one which could cause delicate problems in some situations, but which we can ignore once we have recognized it. If $\mathscr{P}_2(V)$ is a projective plane a given line W in $\mathscr{P}_2(V)$ can be thought of as the collection of its points, or as an object in $\mathscr{P}_2(V)$ which happens to be incident with certain other objects (the points, or one-dimensional subspaces). From the first point of view we are perhaps thinking of $\mathscr{P}_1(W)$, which strictly speaking is not the same as the line W in $\mathscr{P}_2(V)$. But it is clear that they both "contain" or are "incident" with the same set of points: that is, the one-dimensional subspaces of V which happen to lie in W. We shall not concern ourselves about this difference, at least not notationally.

If V is a three-dimensional (left or right) vector space over K then $\mathscr{P}(V)$ is a projective plane and each two-dimensional subspace of V gives rise to a line in $\mathscr{P}(V)$. Since all two-dimensional left (say) vector spaces over K are isomorphic, all these lines are isomorphic. Furthermore there is an isomorphism induced by a non-singular semi-linear transformation from any one of them to any other. But we would like more: is there an automorphism of $\mathscr{P}(V)$ which maps one line onto another, and which has the same effect as the mapping induced by a semi-linear transformation? More precisely: given a semi-linear transformation α from one two-dimensional subspace to another, is there an element of $P\Gamma L(V)$ which has the same effect as the mapping induced by α? And conversely, given an element β of $P\Gamma L(V)$ which sends one given line to another, is there a semi-linear transformation $\boldsymbol{\beta}$ which induces the given element β?

Both questions are easy to answer:

Lemma 2.17. *Let W_1 and W_2 be two-dimensional subspaces of a three-dimensional vector space V over K, and let α be an arbitrary non-singular semi-linear transformation from W_1 to W_2. Then there exists a non-singular semi-linear transformation α^* of V which sends W_1 to W_2 such that if w is a vector in W_1, then $w^\alpha = w^{\alpha^*}$. Hence in particular there is an element of $P\Gamma L(V)$ which agrees with the mapping α induced by α from the projective line W_1 (or $\mathscr{P}(W_1)$) to the projective line W_2 (or $\mathscr{P}(W_2)$).*

Proof. We shall show that α can be extended to a non-singular semi-linear transformation of V in a number of ways. In fact for any e not in W_1 and any f not in W_2, α can be extended to a non-singular semi-linear transformation α^* sending e onto f. Since V has dimension three and W_1, W_2 both have dimension two, any vector of V has unique representations of the form $w_1 + xe$ and $w_2 + yf$ where $w_1 \in W_1$, $w_2 \in W_2$. It is now easy to see that the mapping α^* defined by $(w_1 + xe)^* = w_1^\alpha + x^\phi f$, where ϕ is the automorphism of K associated with α, is a non-singular semi-linear transformation of V which maps W_1 onto W_2 and that $w_1^{\alpha^*} = w_1^\alpha$ for all $w_1 \in W_1$. (Why must we insist that ϕ be the automorphism associated with α?) $\square$

A little later we shall give a more precise answer to the first question by showing that α can always be extended in a certain special way.

Lemma 2.18. *Let W_1 and W_2 be two-dimensional subspaces of V and let β be an element of $P\Gamma L(V)$ which sends the line W_1 to the line W_2. Then β is induced by a semi-linear transformation $\boldsymbol{\beta}$ of V which sends W_1 onto W_2.*

Proof. Obvious. $\square$

Corollary. *Let W be a line in the projective plane $\mathscr{P}(V)$. Then the subgroup Λ of $P\Gamma L(V)$ which fixes W induces $P\Gamma L(W)$ on W.*

Proof. By the two lemmas above every semi-linear transformation of W (or of $\mathscr{P}(W)$, depending on the point of view) is induced by an element of $P\Gamma L(V)$ fixing W, and conversely. $\square$

The kernel of the representation of Λ on W is the set of elements of $P\Gamma L(V)$ which fix every point on the line W. In Chapter IV we shall study such collineations in great detail, and we want to use (without proof) one of the simplest of their properties. If a collineation fixes all the points of a line in a projective plane $\mathscr{P}(V)$, then the collineation is called a *perspectivity*, and the line of fixed points is the *axis*. It is shown (Theorem 4.9) that for a non-identity perspectivity there must always be a special point, called the *centre*, such that all lines passing through this point are fixed. In addition, there are no fixed objects in $\mathscr{P}(V)$ other than those already listed; for convenience the identity automorphism is considered to be a perspectivity with any line as axis and any point as centre. So the kernel Φ of the representation of Λ on W is the group of perspectivities of $\mathscr{P}(V)$ which have W as axis and, by the corollary above, the factor group Λ/Φ is isomorphic to $P\Gamma L(W)$. In fact in many ways it might have been better to have *defined* $\operatorname{Aut}\mathscr{P}(W)$ as the factor group Λ/Φ; certainly it would have been more natural geometrically. The kernel Φ of the representation of Λ on W is never the identity and we shall now show that perspectivities are abundant in $\mathscr{P}_2(V)$.

Lemma 2.19. *Let W_1 and W_2 be two lines in a projective plane $\mathscr{P}(V)$ and X a point not on either line. Then there is a perspectivity with centre X sending W_1 to W_2.*

Proof. Choose a frame for $\mathscr{P}(V)$ as follows: $E_1 = X = \langle e_1 \rangle$, $E_2 = W_1 \cap W_2 = \langle e_2 \rangle$, $E_3 = \langle e_3 \rangle =$ any point on W_1 (not equal to E_2 naturally), and $E_4 = \langle e_1 + e_2 + e_3 \rangle =$ any point on W_2 (not equal to E_2 or to $W_2 \cap (E_1 + E_3)$).

Let β be the linear transformation of V defined as follows:

$$\beta : x_1 e_1 + x_2 e_2 + x_3 e_3 \to (x_1 + x_3) e_1 + x_2 e_2 + x_3 e_3 .$$

It is easy to verify that β is non-singular. The line W_1 is the two-dimensional subspace of V spanned by e_2 and e_3, so that any arbitrary point of W_1 is $\langle x_2 e_2 + x_3 e_3 \rangle$. Thus any point of W_1 other than E_2 is of the form $\langle x_2 e_2 + e_3 \rangle$. Similarly any point of W_2 other than E_2 is $\langle e_1 + y_2 e_2 + e_3 \rangle$. If $\bar{\beta}$ is the element of $PGL(V)$ induced by β then $E_2^{\bar{\beta}} = E_2$ and $\langle x_2 e_2 + e_3 \rangle^{\bar{\beta}} = \langle e_1 + x_2 e_2 + e_3 \rangle$, i.e. $W_1^{\bar{\beta}} = W_2$.

Similar considerations show that $\bar{\beta}$ fixes every point on the line $E_1 + E_2$; but we want to show that it fixes every line on the point $X = E_1$. Since any line is uniquely determined by two of its points, any line U on X has the form $E_1 + Y$, where Y is a point on $E_2 + E_3$, and hence has the form $Y = \langle x e_2 + y e_3 \rangle$. But $(x e_2 + y e_3)^\beta = y e_1 + x e_2 + y e_3$ and consequently, since it is a linear combination of e_1 and $x e_2 + y e_3$, $(x e_2 + y e_3)^\beta$ lies in the subspace $E_1 + Y = U$. Thus, since it sends Y to a point on the line U, and sends E_1 to E_1, $\bar{\beta}$ fixes the line U. []

The perspectivity exhibited in Lemma 2.19 has the property that its centre is incident with its axis. This does not follow from the definition and, in fact, need not have been the case.

Exercise 2.22. With the notation of Lemma 2.19, if $K \neq GF(2)$ find a perspectivity δ with centre X sending W_1 onto W_2 such that X is not incident with the axis of δ.

Exercise 2.23. With the notation of Lemma 2.19 show that there is a unique involutory perspectivity with centre X which interchanges W_1 and W_2. (Hint: the characteristic of the underlying skewfield K will determine whether the centre is incident with the axis.)

The duals of the preceeding lemma and exercises are also true.

Exercise 2.24. Given two points X_1 and X_2 and a line W passing through neither, show that there is a perspectivity with axis W sending X_1 to X_2 and show that there is a unique involutory perspectivity with this property.

So far we have been concerned with projective planes $\mathscr{P}(V)$, where V is a vector space over a skewfield K. In this case we have had to worry about whether V was left or right and so could not write $\mathscr{P}_2(K)$ for $\mathscr{P}(V)$. We shall now look at the situation when K is a field and establish a stronger property of $PSL(V)$.

Theorem 2.20. *If K is a field, then the group $PSL(V)$ is two-transitive on the points of $\mathscr{P}(V) = \mathscr{P}_2(K)$.*

Proof. The linear transformation β of the proof of Lemma 2.19 has the following matrix with respect to the basis e_1, e_2, e_3 of V:

$$\begin{bmatrix} 1 & 0 & 0 \\ 0 & 1 & 0 \\ 1 & 0 & 1 \end{bmatrix}.$$

Since this has determinant one, β is in $PSL(V)$. Given any two lines of $\mathscr{P}(V)$, there is always a point on neither (why?) and so, by Lemma 2.19, there is an element of $PSL(V)$ which sends the first to the second. Hence $PSL(V)$ is transitive on lines.

Now let W be a line of $\mathscr{P}(V)$, and consider the group $PSL(W)$, which by Theorem 2.15 is two-transitive on the points of W. We now show that every element of $PSL(W)$ can be thought of as an element of $PSL(V)$ which fixes the line W. If we choose a frame where $\langle e_1 \rangle$ and $\langle e_2 \rangle$ are points of W, then a matrix

$$\begin{bmatrix} a & b \\ c & d \end{bmatrix}$$

which has determinant one acting on the line W, can be thought of as the restriction of the transformation α which has matrix

$$\begin{bmatrix} a & b & 0 \\ c & d & 0 \\ 0 & 0 & 1 \end{bmatrix}.$$

But this matrix has determinant one and, hence, the transformation α (which clearly fixes W) belongs to $PSL(V)$.

Let X and Y be distinct points, and U and Z be another pair of distinct points. Then we must show that $PSL(V)$ contains an element sending X onto U and Y onto Z. Since we know that $PSL(V)$ is transitive on the lines of $\mathscr{P}(V)$ there is a mapping in $PSL(V)$ which sends the line $X + Y$ onto the line $U + Z$. Under this mapping, X and Y go to points X_1 and Y_1, respectively, both of which are on $U + Z$. But now there is an element of $PSL(U + Z) < PSL(V)$ which sends X_1 to U and Y_1 to Z, and so the product of the two automorphisms sends X and Y onto U and Z, respectively. Thus $PSL(V)$ is two-transitive on points. □

Exercise 2.25. Let V be a three dimensional vector space over a field K. If an element α of $PGL(V)$ induces a perspectivity of $\mathscr{P}_2(K)$ whose centre is on its axis show that α is in $PSL(V)$.

There is an alternate approach to the set of problems, lemmas, exercises and so on given above. Suppose $\mathscr{P}$ is a projective plane over a skewfield, and we consider all the elements of $\mathrm{Aut}\,\mathscr{P}$ which are perspectivities with incident centre and axis; the subgroup of $\mathrm{Aut}\,\mathscr{P}$ generated by all such elements will be called either the *little projective group* or the *unimodular group*. In view of Exercise 2.25, the little projective group is a subgroup of $PSL(V)$ in the case that K is a field: *in fact, they are equal*, but we do not prove this somewhat tricky result. However, it is easy to see that the little projective group is always two-transitive on points, and we sketch a proof here. If A_1, B_1 are distinct points, and A_2, B_2 are another pair of distinct points, then let W_i be the line $A_i + B_i$; suppose that $W_1 \neq W_2$. Then if we let X be the point $(A_1 + A_2) \cap (B_1 + B_2)$ (when all the points are distinct), we can use Lemma 2.19 with the same terminology to show that there is an element of the little projective group which sends W_1 to W_2 in such a way as to send A_1 to A_2 and B_1 to B_2. The various special cases are easy to handle, and in this way we can show that the little projective group is two-transitive on points.

Exercise 2.26. Let K be a field. Show that $PSL(V)$ is two-transitive on the lines of $\mathscr{P}(V) = \mathscr{P}_2(K)$.

We now state, without proof (although the reader might wish to regard the proof as a difficult exercise) the improvement referred to after the proof of Lemma 2.17. The mapping α from W_1 to W_2, where W_1 and W_2 are distinct lines of $\mathscr{P}(V)$, induced by an element of $PGL(V)$ is in fact also induced by a product of just two perspectivities of $\mathscr{P}_2(K)$ (which however might not lie in $PSL(V)$ – cf. Exercise 2.20).

In this chapter we have been discussing the projective line over a field K embedded in the projective plane. We now return for a moment to the line "in itself", studying intrinsic definitions of additional structure. Let $\mathscr{P}(V)$ be a projective line (always over a field), and suppose that P_1, P_2, P_3, P_4 are four points on the line, such that at least the first three are distinct. We define the *cross-ratio* $(P_1, P_2; P_3, P_4)$ to be the parametric coordinate of P_4 in the frame given by P_1, P_2, P_3. Later we give a more general definition of cross-ratio but meanwhile we answer the following obvious question: if we are given an arbitrary four points on a projective line how do we find their cross ratio?

Theorem 2.21. *Suppose P_1, P_2, P_3 are three distinct points in $\mathscr{P}(V)$ (where V is a two dimensional vector space over a field K), with parametric coordinates t_1, t_2, t_3 in some fixed frame. Then if $P_4 \neq P_1$ has the parametric coordinate t_4,*

$$(P_1, P_2; P_3, P_4) = \frac{(t_3 - t_1)(t_4 - t_2)}{(t_4 - t_1)(t_3 - t_2)}.$$

Proof. We must choose a basis e_1, e_2 for V such that $P_1 = \langle e_1 \rangle$, $P_2 = \langle e_2 \rangle$, $P_3 = \langle e_1 + e_2 \rangle$, and then find x and y such that $P_4 = \langle x e_1 + y e_2 \rangle$. Since $\langle x e_1 + y e_2 \rangle = \langle y^{-1} x e_1 + e_2 \rangle$ it follows that the cross-ratio $r = (P_1, P_2; P_3, P_4) = y^{-1}x$.

If no P_i coincides with the first point of the given frame then the points P_i can be written $P_i = \langle (t_i, 1) \rangle$ in terms of the given basis. We shall assume this to be the situation and leave the other cases as an exercise.

Let a and b have the property $a(t_1, 1) + b(t_2, 1) = (t_3, 1)$; this is equivalent to demanding that

$$a t_1 + b t_2 = t_3 \quad \text{and} \quad a + b = 1. \tag{1}$$

Multiply the second equation of (1) by t_2 and subtract to get $a = (t_3 - t_2)(t_1 - t_2)^{-1}$; similarly, we find $b = (t_3 - t_1)(t_2 - t_1)^{-1}$. Now, if we choose $e_1 = a(t_1, 1)$, $e_2 = b(t_2, 1)$ as our basis, we have $P_1 = \langle e_1 \rangle$, $P_2 = \langle e_2 \rangle$ and $P_3 = \langle e_1 + e_2 \rangle$. We must now determine x and y so that $P_4 = \langle x e_1 + y e_2 \rangle$. It will suffice to solve $(t_4, 1) = x e_1 + y e_2$. But this implies

$$x a t_1 + y b t_2 = t_4 \quad \text{and} \quad x a + y b = 1 \tag{2}$$

or

$$y^{-1} x a t_1 + b t_2 = y^{-1} t_4 \quad \text{and} \quad y^{-1} x a t_4 + b t_4 = y^{-1} t_4. \tag{3}$$

Subtracting the second equation of (3) from the first, we have $y^{-1} x = b(t_4 - t_2)(t_1 - t_4)^{-1} a^{-1}$. Substituting the values of a and b above, we have the result of the theorem. $\square$

Exercise 2.27. Investigate all the cases of Theorem 2.21 when the given expression is not well-defined (i.e., one of the t_i is ∞, or t_4 equals t_1).

Exercise 2.28. Suppose we are given the coordinates of the P_i in some basis as $P_i = (x_i, y_i)$, and we define D_{ij} to be the determinant of the matrix

$$\begin{bmatrix} x_i & y_i \\ x_j & y_j \end{bmatrix}.$$

Then show that $(P_1, P_2; P_3, P_4) = D_{13} D_{24} / D_{14} D_{23}$.

We can use Theorem 2.21 above to define the cross-ratio r in the "degenerate" cases that arise if only two of the first three P_i are distinct but P_4 is not equal to any of P_1, P_2, P_3: thus if $P_1 = P_2 \neq P_3$, then we define $r = 1$; if $P_1 = P_3 \neq P_2$, then we define $r = 0$; if $P_2 = P_3 \neq P_1$, then we define $r = \infty$. We avoid making a definition if as many as three of the P_i are equal.

Exercise 2.29. Let $(P_1, P_2; P_3, P_4) = r$ be a cross-ratio in a projective line $\mathscr{P}_1(K)$ over a field K. Show that the 24 possible permutations of the four points P_i give only six different cross-ratios and that the values of these are r, $1 - r$, $1/r$, $1 - (1/r)$, $1/(1 - r)$, $r/(r - 1)$. Find the subgroup of the symmetric group on four symbols which, acting on the four points P_i, fixes the cross-ratio r. (Note that for some values of r the six possible cross-ratios are not all distinct. Special consideration is needed for those values of r.)

The cross ratio has another simple and valuable property:

Theorem 2.22. *Let V and W be vector spaces of dimension two over the field K, let β be the mapping from $\mathscr{P}(V)$ onto $\mathscr{P}(W)$ induced by a non-singular semi-linear transformation β from V to W and let ϕ be the automorphism of K associated with β. Then for any choice of four points P_1, P_2, P_3, P_4 in $\mathscr{P}(V)$, no three of them equal, we have:*

$$(P_1^\beta, P_2^\beta; P_3^\beta, P_4^\beta) = (P_1, P_2; P_3, P_4)^\phi .$$

Proof. We omit the various degenerate cases, since they are all trivial, and assume in particular that the first three P_i are distinct; then their images are distinct as well. So in each projective line we have a ready-made frame; let the standard basis for the frame P_1, P_2, P_3 in $\mathscr{P}(V)$ be e_1, e_2, and define a frame f_1, f_2 of the P_i^β in $\mathscr{P}(W)$ by $f_i = e_i^\beta$, $i = 1, 2$. Then if $(P_1, P_2; P_3, P_4) = r$, we know that $P_4 = \langle re_1 + e_2 \rangle$, and so $P_4^\beta = r^\phi f_1 + f_2$. Thus $(P_1^\beta, P_2^\beta; P_3^\beta, P_4^\beta) = r^\phi$ by the definition of cross-ratio. $\square$

Corollary. *Let β be an element of $P\Gamma L(V)$, where β is induced by a semi-linear transformation whose associated automorphism is ϕ. For any choice of P_1, P_2, P_3, P_4 in $\mathscr{P}(V)$, $(P_1^\beta, P_2^\beta; P_3^\beta, P_4^\beta) = (P_1, P_2; P_3, P_4)^\phi$.*

Theorem 2.23. *Let V and W be vector spaces of dimension two over the field K, and let β be a one-to-one mapping of the points of $\mathscr{P}(V)$ onto the points of $\mathscr{P}(W)$, with the property that there is an automorphism ϕ of K for which $(P_1^\beta, P_2^\beta; P_3^\beta, P_4^\beta) = (P_1, P_2; P_3, P_4)^\phi$ for all choices of the P_i in $\mathscr{P}(V)$ (at least three of them distinct). Then β is induced by a semi-linear transformation α of V onto W.*

Proof. If P_1, P_2, P_3 is a frame for $\mathscr{P}(V)$ then they are all distinct. Thus, since β is a one-to-one mapping P_1^β, P_2^β, P_3^β are all distinct and hence, by Exercise 2.8, P_1^β, P_2^β, P_3^β, is a frame for $\mathscr{P}(W)$. Let α be a semi-linear transformation of V onto W which sends the frame P_1, P_2, P_3 onto the frame P_1^β, P_2^β, P_3^β, and whose associated automorphism is ϕ.

(Clearly such an α exists: there are many linear transformations from V onto W and the group of linear transformations of W is three-transitive on its one-dimensional subspaces (see Theorem 2.15). Hence P_1, P_2, P_3 can be mapped onto P_1^β, P_2^β, P_3^β by a linear transformation. But the mapping $(x, y) \to (x^\phi, y^\phi)$ is a semi-linear transformation with at least three fixed vectors and thus any three one dimensional subspaces of W are fixed by a semi-linear transformation with companion automorphism ϕ.)

If X is an arbitrary point of $\mathscr{P}(V)$ with parametric coordinate r in the P_i frame, then by the definition of cross ratio, X^β must have parametric coordinate r^ϕ in the P_i^β frame. But by the previous theorem, X^α also has parametric coordinate r^ϕ in the P_i^β frame, and hence $X^\beta = X^\alpha$; i.e. β is induced by α. $\square$

Corollary. *A one-to-one mapping β of the projective line $\mathscr{P}(V)$ over a field K onto itself is an element of $P\Gamma L(V)$ if and only if there is an automorphism ϕ of the underlying field K such that $(P_1^\beta, P_2^\beta; P_3^\beta, P_4^\beta) = (P_1, P_2; P_3, P_4)^\phi$ for all (appropriate) choices of the points P_i.*

The student may feel that there is nothing very surprising about the corollary to Theorem 2.23, since it was really built into the definition of cross-ratio. However it allows us to see clearly the basic role played by cross-ratio, since it shows that it is something intrinsic to the line (at least over a field), whose preservation is almost the criterion for a mapping to be an automorphism. Certainly the elements of $PGL(V)$ are exactly the mappings of $\mathscr{P}_1(K) = \mathscr{P}(V)$ which preserve cross-ratio, and this could have served as yet another definition of automorphism, in the case that K is a field. In many ways this might even have been a more satisfactory definition. The cross-ratio also plays a special role in the following (note that part of this result is valid for arbitrary skewfields).

Theorem 2.24 (The theorem of the complete quadrilateral). *Let V be a three-dimensional left vector space over the skewfield K, let W be a line in $\mathscr{P}(V)$, and A_1, A_2, A_3 three distinct points on W. Choose any point P not on W and any point Q on the line $A_3 + P$, but Q distinct from both A_3 and P. Construct the quadrilateral P, Q, S, T as follows: $S = (A_1 + P) \cap (A_2 + Q)$, $T = (A_2 + P) \cap (A_1 + Q)$, and let the line $S + T$ meet W in the point A_4. Then A_4 is independent of the choice of P and Q; when K is a field, A_4 has the property $(A_1, A_2; A_3, A_4) = -1$ (see Fig. 3).*

Proof. We can choose a frame for $\mathscr{P}(V)$ consisting of A_1, A_2, P, and Q, so $A_1 = \langle e_1 \rangle$, $A_2 = \langle e_2 \rangle$, $P = \langle e_3 \rangle$, $Q = \langle e_1 + e_2 + e_3 \rangle$. Then it is easy to see that $A_3 = \langle e_1 + e_2 \rangle$, so A_1, A_2, A_3 are a frame for W, using the basis e_1, e_2 in the standard manner. Since S is on both $A_1 + P$ and

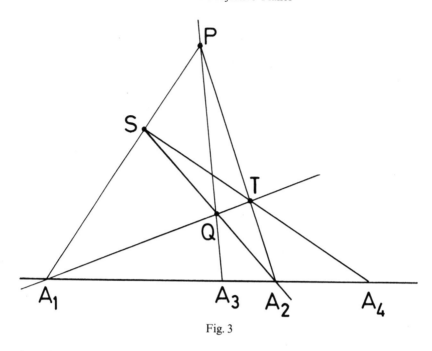

Fig. 3

$A_2 + Q$, we must have $S = \langle e_1 + e_3 \rangle$, and similarly we find $T = \langle e_2 + e_3 \rangle$. Since $\langle e_1 - e_2 \rangle$ is on both $\langle e_1 \rangle + \langle e_2 \rangle$ and on $\langle e_1 + e_3 \rangle + \langle e_2 + e_3 \rangle$, it follows that $A_4 = \langle e_1 - e_2 \rangle$. This point A_4 depends only on A_1, A_2, A_3, that is, on the frame for the line W. Finally, if K is a field, then the cross-ratio is -1, as claimed. ☐

From the theorem above, the particular cross-ratio -1 has some special geometric significance; we say that A_4 is the *harmonic conjugate of A_3 with respect to A_1 and A_2* when $(A_1, A_2; A_3, A_4) = -1$. This definition is only valid for a field, but in light of Theorem 2.24 we could even define a harmonic conjugate in the case of a skewfield, and in fact most of the properties that are interesting about harmonic conjugates are also valid in that case: but often their proofs are slightly different. So in what follows, in all references to harmonic conjugate we shall be assuming that we are working over a field.

Exercise 2.30. If A_4 is the harmonic conjugate of A_3 with respect to A_1 and A_2, then show that A_3 is the harmonic conjugate of A_4 with respect to A_1 and A_2 (so we may speak of A_3 and A_4 as being harmonic conjugates with respect to A_1 and A_2). Are there any other pairs among the A_i which are harmonic conjugates with respect to the complementary pair?

Now we note that if K has characteristic two a peculiar thing will happen. The harmonic conjugate of A_3 with respect to A_1 and A_2 will be the point A_4 in the frame A_1, A_2, A_3 whose parametric coordinate is $+1 = -1$. But this is A_3 itself, so every point is its own harmonic conjugate with respect to any other pair – and nothing else can happen. (The reader should draw another diagram for Theorem 2.24 in this case.) Conversely, if the characteristic is not two, then $-1 \neq +1$, and so A_4 is never equal to A_3.

Theorem 2.25. *Let A and B be distinct points in the projective line $\mathscr{P}(V)$ over a field K of characteristic not two. Then the points X and Y are harmonic conjugates with respect to A and B if and only if there is an element of order two in $PGL(V)$ which fixes A and B and interchanges X and Y.*

Proof. Choose a frame A, B, X for $\mathscr{P}(V)$. Then Y is the harmonic conjugate of X if and only if Y has parametric coordinate -1 in this frame. But any element of $PGL(V)$ fixing $A = \infty$ and $B = 0$ has the form $x \to \dfrac{xa + 0}{x0 + d}$ so that the unique involution fixing A and B is $\alpha : x \to -x$. Since this interchanges the points with parameters $+1$ and -1, Y is the harmonic conjugate of X if and only if α interchanges X and Y. $\square$

Exercise 2.31. Let K be a skewfield of characteristic two, V a two-dimensional left vector space over K. Show that no element of $PGL(V)$ which fixes two points can also have order two.

Hence we could have phrased Theorem 2.25 without restriction on characteristic.

6. Polarities

In this section we shall study correlations and polarities and certain distinguished subsets of points and lines associated with them. We begin by proving a result which is very similar to the fundamental theorem for collineations (Corollary to Theorem 2.8).

Theorem 2.26. *Let V be a left vector space of dimension three over a skewfield K, and let a_V and $a_{V'}$ be the annihilator mappings from $\mathscr{P}(V)$ to $\mathscr{P}(V')$ and from $\mathscr{P}(V')$ to $\mathscr{P}(V)$, respectively. If β is any correlation of $\mathscr{P}(V)$ then there is a semi-linear transformation θ from V to V' such that $\beta = \theta a_{V'}$.*

Proof. Since β is an anti-isomorphism from $\mathscr{P}(V)$ onto itself, the product βa_V is an isomorphism from $\mathscr{P}(V)$ to $\mathscr{P}(V')$. If we put $\theta = \beta a_V$ then $\theta a_{V'} = \beta a_V a_{V'} = \beta$. The corollary to Theorem 2.9 now completes the proof. $\square$

For the remainder of this section we shall simplify matters by considering vector spaces V over a field K. In this case V is uniquely determined by its dimension and the field K.

There is an alternative way of considering Theorem 2.26. Once we have chosen a basis for V then θ maps the points and lines of $\mathscr{P}(V)$ onto the points and lines, respectively, of $\mathscr{P}(V')$ in the "standard" manner. Using the homogeneous coordinates of Section IV, we may "dispense" with the mapping $a_{V'}$ in the following way. If a point of $\mathscr{P}(V')$ has given homogeneous coordinates then we identify that point with the line of $\mathscr{P}(V)$ which has the same homogeneous coordinates. We then simply say that β is the mapping which sends the points and lines of $\mathscr{P}(V)$ onto the lines and points of $\mathscr{P}(V)$, by the rule given by θ. From the discussion in Section 4 of Chapter I we can choose a canonical basis for V' in terms of the one already chosen for V. Having done this, θ has the form

$$v^{\theta} = A(v^{\phi})' \quad \text{for all } v \text{ in } V \tag{1}$$

where A is some matrix, and ϕ is an automorphism of K, and hence of $\mathscr{P}(V)$ in the given basis.

The expression (1) gives the effect of β on the points of $\mathscr{P}(V)$ and it is now natural to try to determine its action on the lines. Since the lines of $\mathscr{P}(V)$ are merely the points of $\mathscr{P}(V')$ we do this by considering β as a mapping from the points of $\mathscr{P}(V')$ to the points of $\mathscr{P}(V)$. Repeating the above argument β is induced by the mapping ψ where

$$(v')^{\psi} = v^{\phi} B \quad \text{for all } v' \text{ in } V' \tag{2}$$

and B is some matrix and ϕ is the same automorphism of K (why?). Since the correlation β preserves incidence, its action on the points of $\mathscr{P}(V)$ must determine its action on the lines. The point $\langle v \rangle$ is on the line $\langle w' \rangle$ if and only if $vw' = 0$. But v being on w' implies $(w')^{\beta}$ is on v^{β}. Thus we have

$$\text{if} \quad vw' = 0 \quad \text{then} \quad w^{\phi} BA(v^{\phi})' = 0. \tag{3}$$

Transposing the second half of (3) and putting $C = (A'B')^{\phi^{-1}}$ (recall that we denote the transpose of a matrix S by S'), we have

$$\text{if} \quad vw' = 0 \quad \text{then} \quad vCw' = 0. \tag{4}$$

From (4) it follows that any vector which annihilates v also annihilates vC. Thus $\langle v \rangle = \langle vC \rangle$ for all v and so, by Result 1.24, $C = kI$ where k is a field element and I is the identity. Thus $A'B' = k^{\phi}I$. But $(k^{\phi}I)' = k^{\phi}I$ so that $A'B' = k^{\phi}I = (A'B')' = BA$ which gives $(k^{-1})^{\phi}B = A^{-1}$. But since hI induces the identity on $\mathscr{P}(V)$ for any h in K, the matrices B and $(k^{-1})^{\phi}B$ induce the same mapping of $\mathscr{P}(V)$ and so either one may be used in (2).

Thus we have proved:

Theorem 2.27. *Let V be a vector space of dimension three over a field K. Using homogeneous coordinates and any fixed basis for V a correlation β can be represented as follows,*

$$\beta : \langle v \rangle \rightarrow \langle A(v^\phi)' \rangle$$
$$\langle w' \rangle \rightarrow \langle w^\phi A^{-1} \rangle .$$

where ϕ is some automorphism of K and A is a non-singular matrix. ☐

Now when is β a polarity? We apply β twice, using the formula of Theorem 2.27, and find that $\langle v \rangle^{\beta^2} = \langle v^{\phi^2}(A')^\phi A^{-1} \rangle$. If this last is to equal $\langle v \rangle$ for all subspaces $\langle v \rangle$ of V, then clearly the semi-linear transformation $v \rightarrow v^{\phi^2}(A')^\phi A^{-1}$ must induce the identity automorphism of $\mathscr{P}(V)$. By Result 1.24 the required conditions are $\phi^2 = 1$ and $(A')^\phi A^{-1} = kI$, where k is some element in K. Thus we have established the first claim of the following theorem.

Theorem 2.28. *Let β be a polarity of $\mathscr{P}(V)$. Then in the notation of Theorem 2.27, $\phi^2 = 1$ and $(A')^\phi = kA$ for some k in K. If $\phi = 1$ then $k = 1$, while if $\phi \neq 1$ then $kk^\phi = 1$.*

Proof. We need only prove the last sentence. But if $(A')^\phi = kA$ then, since $M'^\phi = (M^\phi)'$ for any matrix M, $([(A')^\phi]')^\phi = (A^{\phi^2})'' = A$ on the one hand while $([(A')^\phi]')^\phi$ is also equal to $((kA)')^\phi = (kA')^\phi = k^\phi(A')^\phi = k^\phi kA$. Hence $k^\phi k = 1$. If $\phi = 1$ then $k^2 = 1$ and so $k = \pm 1$. But if $k = -1$ then A is a skew symmetric matrix and so, since any three by three skew symmetric matrix is singular (see for example [1]), this case does not arise from a polarity. If $\phi = 1$ and $k = 1$ then A is symmetric and this can certainly occur. ☐

Clearly replacing A by mA, for any m in K^*, does not affect the geometrical mapping β of Theorem 2.27. In fact although replacing A by mA has the effect of changing the image of any given basis of V it does not alter the image of any frame in $\mathscr{P}(V)$. This simple obervation gives an immediate improvement of Theorem 2.28.

Theorem 2.29. *Let V be a vector space of dimension three over a field K. If β is a polarity of $\mathscr{P}(V)$ then, in any frame of $\mathscr{P}(V)$, β can be represented in the form of Theorem 2.27 where A satisfies $(A')^\phi = A$.*

Proof. Suppose A is a matrix with the property $A'^\phi = kA$ where $kk^\phi = 1$. If we let $B = mA$, then $B'^\phi = m^\phi A'^\phi = m^\phi kA = m^\phi m^{-1} kB$. So the problem is to choose m so that $m^\phi k = m$. If $k \neq -1$, let $m = 1 + k$, so $m^\phi k = (1 + k^\phi)k = k + k \cdot k^\phi = k + 1 = m$. If $k = -1$, then we can choose first an element t such that $B = tA$ has the property $B'^\phi = kB$, where $k \neq -1$: for $B'^\phi = t^\phi(-A) = -t^\phi t^{-1}B$, and if $-t^\phi t^{-1} = -1$ for *all* t in K, then $t^\phi = t$ for all t, and so ϕ is the identity. But if $\phi = 1$ then, by Theorem 2.28, $k = 1$. This proves the theorem. $\square$

We now note that the last three theorems are essentially true for any $\mathscr{P}(V)$ of arbitrary dimension; the only difference being that, if V has even dimension, the situation $\phi = 1$, $k = -1$ can occur. The proofs given here work for arbitrary dimension, although the word line must be replaced by hyperplane.

In the projective plane $\mathscr{P}(V)$ we have two types of polarity to study: (1) the *orthogonal* polarity β which can be represented by a symmetric matrix A and for which $\phi = 1$, and (2) the *unitary* or *hermitian* polarity β which can be represented by a matrix A with the property $A'^\phi = A$, where ϕ is an involutory automorphism of K. We shall show that the matrix A for either kind of polarity can, after a suitable choice of basis, be made diagonal, and we will see that other simplifications can be made. For any polarity β of any projective geometry, we define an *absolute* element to be one which is incident with its image under β. For the projective plane $\mathscr{P}(V)$, the point $\langle v \rangle$ is absolute with respect to the orthogonal polarity β, with matrix A, if and only if $v A v' = 0$; while if it is unitary, with matrix A and automorphism ϕ, then $\langle v \rangle$ is absolute if and only if $v A (v^\phi)' = 0$. Suppose we write $A = (a_j)$, $e = (x, y, z)$; then

Lemma 2.30. *If $A = (a_{ij})$ is the matrix for the polarity β then the point $\langle (x, y, z) \rangle$ is absolute if and only if:*

(1) $a_{11} x^2 + a_{22} y^2 + a_{33} z^2 + 2 a_{12} xy + 2 a_{13} xz + 2 a_{23} yz = 0$ *in the case β is orthogonal;*

(2) $a_{11} x^{1+\phi} + a_{22} y^{1+\phi} + a_{33} z^{1+\phi} + a_{12} xy^\phi + a_{21} x^\phi y + a_{13} xz^\phi$
$+ a_{31} x^\phi z + a_{23} yz^\phi + a_{32} y^\phi z = 0$, *in the case β is unitary with automorphism ϕ.*

Proof. A simple matter of computation which we leave as an exercise. $\square$

The Eqs. (1) and (2) of Lemma 2.30 remind us of quadratic and hermitian forms, which indeed they are. Conversely, given a quadratic or hermitian form with non-singular matrix (i.e. a *non-degenerate* form) it is clear how we could construct an orthogonal or unitary polarity from it, in such a way that the absolute points of the polarity would be exactly the "solutions" of the given form. We do not phrase this important fact as a theorem, since we have not given a precise definition of quadratic and hermitian forms.

Lemma 2.30 appears to attach special significance to the absolute points as opposed to the absolute lines. However the following lemmas shows that each absolute point determines a unique absolute line and vice versa.

Lemma 2.31. *Let β be a polarity of the projective plane $\mathscr{P}(V)$. Then an absolute line contains exactly one absolute point.*

Proof. Let P and Q be distinct absolute points and suppose that the line $P + Q$ is absolute. Then the lines P^β and Q^β are also absolute. Since $P + Q$ is absolute, $(P + Q)^\beta = P^\beta \cap Q^\beta$ is on $P + Q$. Now P^β contains P and $P^\beta \cap Q^\beta$, so either $P^\beta = P + (P^\beta \cap Q^\beta)$ or $P = P^\beta \cap Q^\beta$, and similarly either $Q^\beta = Q + (P^\beta \cap Q^\beta)$ or $Q = P^\beta \cap Q^\beta$. If $P^\beta \cap Q^\beta$ is not P or Q, then $P^\beta = P + (P^\beta \cap Q^\beta)$ which, since $P^\beta \cap Q^\beta$ is on $P + Q$, is the line $P + Q$. Similarly $Q^\beta = P + Q$, which implies $P^\beta = Q^\beta$; a contradiction. Thus either $P^\beta \cap Q^\beta = P$ or $P^\beta \cap Q^\beta = Q$, but not both. Suppose $P^\beta \cap Q^\beta = P$; then $Q^\beta = P + Q$ and $Q = (Q^\beta)^\beta = (P + Q)^\beta = P^\beta \cap Q^\beta = P$. This is another contradiction and proves the lemma. ☐

The same proof gives us:

Corollary 1. *Let β be a polarity of the projective plane $\mathscr{P}(V)$. Then an absolute point is on exactly one absolute line.*

Corollary 2. *If β is a polarity of a projective plane $\mathscr{P}(V)$ then β must have non-absolute elements.*

If ABC is a triangle of points in $\mathscr{P}(V)$ such that, for a given polarity β, $A^\beta = BC$, $B^\beta = CA$ and $C^\beta = AB$ then the triangle ABC is said to be *self-polar* with respect to β.

Lemma 2.32. *If β is a polarity of the projective plane $\mathscr{P}(V)$ then there exists a triangle which is self-polar with respect to β.*

Proof. Let X be any non-absolute point of β. If Y is any non-absolute point incident with X^β then X, Y, $X^\beta \cap Y^\beta$ form a self-polar triangle. Suppose every point of X^β is absolute then, by Lemma 2.31, every point not on X^β is non-absolute. It is now a simple exercise to complete the proof of Lemma 2.32. ☐

As a result of Lemma 2.32 we can now see how to choose a basis of V so that the matrix representing β in the expression of Theorem 2.27 is diagonal.

Theorem 2.33. *Let V be a vector space of dimension three over a field K, and let β be a polarity of the projective plane $\mathscr{P}(V)$. Then there is a basis of V such that β can be represented in the form of Theorem 2.27 where A is diagonal and satisfies $A'^{\,\phi} = A$.*

Proof. As a result of Theorem 2.29 we need only prove the claim that we can choose a basis to make A diagonal. Let E_1, E_2, E_3 be the vertices of a self-polar triangle of β and choose a basis of V so that, relative to this basis, $E_1 = \langle(1, 0, 0)\rangle$, $E_2 = \langle(0, 1, 0)\rangle$, $E_3 = \langle(0, 0, 1)\rangle$. Then, clearly, $E_2 + E_3 = \langle(1, 0, 0)'\rangle$, $E_3 + E_1 = \langle(0, 1, 0)'\rangle$ and $E_1 + E_2 = \langle(0, 0, 1)'\rangle$. The theorem now follows by substituting $\langle(1, 0, 0)\rangle^\beta = \langle(1, 0, 0)'\rangle$, $\langle(0, 1, 0)\rangle^\beta = \langle(0, 1, 0)'\rangle$ and $\langle(0, 0, 1)\rangle^\beta = \langle(0, 0, 1)'\rangle$ in the expression of Theorem 2.27. ☐

7. Conics

For various reasons, some of which will become evident as we go along, geometries over fields with characteristic two cause considerable extra trouble when dealing with orthogonal polarities, and so we shall often exclude them from our treatment. (But see the appendix to this chapter.) First, if $\mathscr{C}$ is the set of absolute points of an orthogonal polarity and if the characteristic $\neq 2$, then $\mathscr{C}$ is called a *conic*. Our first aim is to study these in some little detail, and whenever we refer to a conic in the following, it will be understood that the characteristic is not two.

Lemma 2.34. *Let $\mathscr{C}$ be a conic with matrix A. Then any line of $\mathscr{P}(V)$ meets $\mathscr{C}$ in at most two points.*

Proof. An arbitrary point on the line joining $\langle v \rangle$ and $\langle w \rangle$ is of the form $\langle xv + yw \rangle$ and such a point is on $\mathscr{C}$ if and only if

$$(xv + yw) A (xv + yw)' = 0. \tag{1}$$

The left hand side of (1) multiplies out to $x^2(vAv') + xy(vAw') + yx(wAv') + y^2(wAw')$. But since it is a field element (or one-by-one matrix), $wAv' = (wAv')' = vA'w'$ which, by the symmetry of A, is vAw'. Thus (1) becomes

$$x^2(vAv') + 2xy(vAw') + y^2(wAw') = 0. \tag{2}$$

Now $y = 0$ is a solution of (2) if and only if $vAv' = 0$, which is equivalent to saying that $\langle v \rangle$ is on $\mathscr{C}$. If there is a solution $y \neq 0$, then we may re-write (2) as

$$(x/y)^2 (vAv') + 2(x/y)(vAw') + wAw' = 0. \tag{3}$$

If $vAv' \neq 0$, then (3) has at most two solutions, and since $y = 0$ is not a solution, the line meets $\mathscr{C}$ at most twice. (Note that it is only the *ratio* of x to y that concerns us, since $\langle xv + yw \rangle = \langle kxv + kyw \rangle$.) If $vAv' = 0$, then besides $y = 0$, (3) has only one other solution, so long as it is really then a linear equation. But could we have $vAv' = vAw' = wAw' = 0$, so that the entire line $\langle v \rangle + \langle w \rangle$ is in $\mathscr{C}$? Consider the points $\langle v \rangle$ and $\langle w \rangle$; since they are distinct, so are their images under β, that is, $\langle Av' \rangle$ and $\langle Aw' \rangle$. But from $vAw' = 0$, we have as above that $wAv' = 0$, and so the points $\langle v \rangle$ and $\langle w \rangle$ are on $\langle Av' \rangle$ and on $\langle Aw' \rangle$. This contradicts Theorem 2.2, and proves the lemma. (Alternatively we could appeal to Lemma 2.31.) $\square$

Eq. (2) in the proof of the last lemma is often called *Joachimsthal's ratio equation* or the *J-ratio equation*.

Lemma 2.35. *A line meets $\mathscr{C}$ in exactly one point if and only if the line is absolute.*

Proof. Suppose $\langle v \rangle$ is absolute, so that $vAv' = 0$. Then certainly $\langle v \rangle$ is on the line $\langle Av' \rangle$, so the absolute line $\langle Av' \rangle$ meets $\mathscr{C}$ at least once.

But, by Lemma 2.32, any absolute line contains exactly one absolute point. Thus for any v in V such that $\langle v \rangle$ is absolute, i.e. is a point of $\mathscr{C}$, the line $\langle A v' \rangle$ is absolute and meets $\mathscr{C}$ just once.

Suppose a line $\langle v \rangle + \langle w \rangle$ meets $\mathscr{C}$ just once, in the point $\langle v \rangle$. Then $v A v' = 0$, and the J-ratio equation has only one solution; but the J-ratio equation, in the case $v A v' = 0$, becomes:

$$2xy(v A w') + y^2(w A w') = 0,$$

which simplifies to: $y = 0$ or $2(x/y)(v A w') + (w A w') = 0$. But the latter of these possibilities must not lead to a new solution, and the only way this can be avoided is that the coefficient of (x/y) is zero. So $v A w' = 0$ which means that $w A v' = 0$, and so both $\langle v \rangle$ and $\langle w \rangle$ are on the line $\langle A v' \rangle$. Hence $\langle A v' \rangle = \langle v \rangle + \langle w \rangle$, and the line is absolute. $\square$

Corollary 1. *If a conic is not empty then its points are in one-to-one correspondence with the points on a line.*

Corollary 2. *In a finite projective plane $\mathscr{P}_2(q)$ a conic is either empty or has $q + 1$ points.*

Proof. Corollary 2 will be an immediate consequence of Corollary 1 and Exercise 2.11. If P is on a conic $\mathscr{C}$ which is associated with a polarity β then P is an absolute point whose image contains no other point of $\mathscr{C}$. Since the other lines through P cannot be absolute they must contain another point of $\mathscr{C}$. Hence by Lemma 2.34 any other line through P contains exactly one other point on $\mathscr{C}$. But this means that the points of $\mathscr{C}$ are in one-to-one correspondence with the lines through P and, as P was an arbitrary point of $\mathscr{C}$, proves the first corollary. $\square$

Exercise 2.32. Show that an absolute line of an orthogonal polarity must have the form $\langle A v' \rangle$ for some absolute point $\langle v \rangle$.

*Exercise 2.33. Let β be an orthogonal polarity in a projective plane $\mathscr{P}(V)$ over a field of characteristic two. Show that if β has at least two absolute points then they form the points of a line.

Given a conic $\mathscr{C}$, the absolute lines (of the polarity) are sometimes called *tangent lines* to $\mathscr{C}$, since they have the property of containing only one point of the conic.

It is possible for a conic $\mathscr{C}$ to be empty. For example, if the polarity β has for its matrix the identity I, and if the underlying field is the field of reals, or rationals, then the Eq. (1) of Lemma 2.30 becomes $x^2 + y^2 + z^2 = 0$, and the only solution is $x = y = z = 0$. This gives no point in $\mathscr{P}(V)$, and so the conic associated with β is empty.

But suppose $\mathscr{C}$ is not empty, and let X and Z be two of its points (see Corollary 1 to Lemma 2.35). Choose a frame for $\mathscr{P}(V)$ as follows: its first point is X, its second is $Y = (X + Z)^\beta$, its third is Z, and its fourth is some point J on the conic. Using homogeneous coordinates we have:

$X = \langle(1, 0, 0)\rangle$, $Z = \langle(0, 0, 1)\rangle$, and so $X + Z = \langle(0, 1, 0)'\rangle$; hence $Y = \langle(0, 1, 0)\rangle$ is the image of $\langle(0, 1, 0)'\rangle$ under β, while finally $J = \langle(1, 1, 1)\rangle$ is a point on $\mathscr{C}$. Let A be the matrix for $\mathscr{C}$, and write $A = (a_{ij})$, not forgetting that A is symmetric. Then since X is on $\mathscr{C}$, we have

$$a_{11} = 0.$$

While the fact that Z is on $\mathscr{C}$ forces:

$$a_{33} = 0.$$

The image of $\langle(0, 1, 0)\rangle$ under β is $\langle A(0, 1, 0)'\rangle = \langle(a_{12}, a_{22}, a_{32})'\rangle = \langle(0, 1, 0)'\rangle$, and so:

$$a_{12} = a_{32} = 0.$$

Finally, $\langle(1, 1, 1)\rangle$ is on $\mathscr{C}$, and so

$$2a_{13} + a_{22} = 0.$$

Hence we can choose A to be the matrix:

$$A = \begin{bmatrix} 0 & 0 & -1 \\ 0 & 2 & 0 \\ -1 & 0 & 0 \end{bmatrix}.$$

Thus we have proved

Theorem 2.36. *Let $\mathscr{C}$ be a non-empty conic in a projective plane $\mathscr{P}(V)$. Then a basis for V can be chosen so that the matrix A for $\mathscr{C}$ has the form above, and the points of $\mathscr{C}$ are the points $\langle(x, y, z)\rangle$ satisfying $y^2 = xz$.* ☐

Corollary. *If $\mathscr{C}_1$ and $\mathscr{C}_2$ are two non-empty conics in a projective plane $\mathscr{P}(V)$, then there is an element of $PGL(V)$ which maps $\mathscr{C}_1$ onto $\mathscr{C}_2$.*

Proof. We have seen that the frame determined the special form for A above, and thus, since $PGL(V)$ is transitive on the frames of $\mathscr{P}(V)$, we can map the "right" frame for $\mathscr{C}_1$ onto that for $\mathscr{C}_2$ by an element of $PGL(V)$. This must map $\mathscr{C}_1$ onto $\mathscr{C}_2$. ☐

Two conics, or indeed any two subsets of points, $\mathscr{C}_1$ and $\mathscr{C}_2$ are said to be *equivalent* if there is a collineation α of $\mathscr{P}(V)$ with $\mathscr{C}_1^\alpha = \mathscr{C}_2$. Thus the corollary to Theorem 2.36 says that any two non-empty conics in a projective plane are equivalent.

Now we simplify some more. Using the canonical form for $\mathscr{C}$ given above, we see that $\mathscr{C}$ consists of the following points: the point $X = \langle(1, 0, 0)\rangle$ and points of the form $\langle(y^2, y, 1)\rangle$ for any y in K. We may then identify the points of the conic with the symbols of the underlying field K, plus a new symbol ∞ as follows: y is identified with $\langle(y^2, y, 1)\rangle$, and ∞ is identified with $(1, 0, 0)$. These are the *parametric coordinates* of the non-empty conic $\mathscr{C}$.

The natural next question might be: what are the elements of $PGL(V)$ (or of $\text{Aut}\,\mathscr{P}(V)$) which fix a given conic? With this in mind we set the following exercise.

Exercise 2.34. Let β be an arbitrary orthogonal polarity of $\mathscr{P}(V)$, whose set $\mathscr{C}$ of absolute points is not empty. If α is a element of $\text{Aut}\,\mathscr{P}(V)$ show that $\alpha\beta = \beta\alpha$ if and only if $\mathscr{C}^{\alpha} = \mathscr{C}$.

The subgroup of $PGL(V)$ fixing a non-empty conic $\mathscr{C}$ will be called the *orthogonal group*; since all such non-empty conics are equivalent, any two orthogonal groups are conjugate in $PGL(V)$ and so we need not refer to any particular conic in the definition. (In the light of Exercise 2.34, we could have defined the orthogonal group as the set of elements in $PGL(V)$ commuting with β; this would also serve as a definition for the orthogonal group of an empty conic.)

Theorem 2.37. *Let $\mathscr{C}$ be a non-empty conic in the projective plane $\mathscr{P}(V)$, with a frame chosen to represent $\mathscr{C}$ in parametric form. Then the orthogonal group (for $\mathscr{C}$) is triply transitive on the points of $\mathscr{C}$, is faithfully represented on $\mathscr{C}$ and, in the parametric coordinates, induces the group of permutations:*

$$y \to \frac{ay+b}{cy+d}, \quad \text{where } ad - bc \neq 0.$$

Thus the orthogonal group of a non-empty conic is isomorphic to the group $PGL(W)$, where W is a two-dimensional vector space over the same field as V.

Proof. We first show that the mappings given are actually induced by collineations of $\mathscr{P}(V)$. Consider the matrix:

$$B = \begin{bmatrix} a^2 & ac & c^2 \\ 2ab & ad+bc & 2cd \\ b^2 & bd & d^2 \end{bmatrix}.$$

After a certain amount of straightforward but tedious calculation the reader will find that

$$\det(B) = (ad - bc)^3,$$

and so B is non-singular as long as $ad - bc \neq 0$. Let β be the element of $PGL(V)$ induced by B.

The effect of β on a point y of $\mathscr{C}$, that is, on $\langle (y^2, y, 1) \rangle$, can be computed easily:

$$(y^2, y, 1)^{\beta} = (a^2 y^2 + 2aby + b^2,\, acy^2 + (ad+bc)\,y + bd,\, c^2 y^2 + 2cdy + d^2)$$
$$= ((ay+b)^2,\, (ay+b)(cy+d),\, (cy+d)^2),$$

and so:

$$\beta : \langle (y^2, y, 1) \rangle \to \langle ((ay+b)^2, (ay+b)(cy+d), (cy+d)^2) \rangle$$

and
$$= \langle (ay+b)/(cy+d)^2, (ay+b)/(cy+d), 1) \rangle \quad \text{if} \quad cy+d \neq 0$$
$$= \langle (ay+b)^2, 0, 0) \rangle = \langle (1, 0, 0) \rangle \quad \text{if} \quad cy+d = 0.$$

This proves that β has the desired effect, since the image of any point $\langle (y^2, y, 1) \rangle$ on $\mathscr{C}$ is also a point of $\mathscr{C}$, and in fact is the point whose parametric coordinate is $(ay+b)/(cy+d)$ (recall $1/0 = \infty$).

Thus we have shown that β is a collineation of $\mathscr{P}(V)$ which maps any point, other that $\langle (1, 0, 0) \rangle$, of $\mathscr{C}$ onto $\mathscr{C}$. The reader should examine the effect on $\langle (1, 0, 0) \rangle$.

Exercise 2.35. Examine the effect of the mapping β on the point ∞, i.e. the point $\langle (1, 0, 0) \rangle$.

Since the group of collineations inducing the given mappings is triply transitive on the points of $\mathscr{C}$ (being the same on $\mathscr{C}$ as $PGL(W)$ is on the line W), it only remains to show that every element in the orthogonal group induces one of the given mappings on $\mathscr{C}$ and that the group acts faithfully on $\mathscr{C}$. Suppose α is any collineation in the orthogonal group and that X, Y, Z are any three distinct points of $\mathscr{C}$. Then, by the triple transitivity, there is a collineation β which induces one of the given mappings on $\mathscr{C}$ such that $X^\alpha = X^\beta$, $Y^\alpha = Y^\beta$, $Z^\alpha = Z^\beta$. Thus $\alpha\beta^{-1}$ is an element in the orthogonal group fixing X, Y, Z. The following important exercise now completes the proof of the theorem.

Exercise 2.36. Show that the identity is the only collineation in the orthogonal group which fixes three distinct points of a non-empty conic in a projective plane $\mathscr{P}(V)$. (Hint: use Theorem 2.12.) $\square$

The theorem above justifies the attitude that the conic (at least a non-empty one) is somehow "like" the line; the subgroups of $PGL(V)$ induced on each are not only abstractly isomorphic, they even act isomorphically as permutation groups. In fact, the line and the (non-empty) conic share other interesting properties, some of which we can give here. We can define the *cross-ratio* of four points on a conic, as before: suppose P_1, P_2, P_3 are distinct, then $(P_1, P_2; P_3, P_4)$ is the parametric coordinate of P_4 in the system for which P_1, P_2, P_3 are respectively ∞, 0, and 1. The harmonic conjugate of P_3 with respect to P_1 and P_2 is that point X for which $(P_1, P_2; P_3, X) = -1$, and naturally all the theorems and results about other pairs among P_1, P_2, P_3, X being harmonic conjugates, as proved for the line, carry over immediately. Also we have, in precisely the same way as we had for the line:

Lemma 2.38. *Let P_1, P_2, P_3 be three distinct points on a conic $\mathscr{C}$. Then P_3 and X are harmonic conjugates with respect to P_1 and P_2 if and*

only if there is an involutory collineation in the orthogonal group of
$\mathscr{C}$ which fixes P_1 and P_2, and interchanges P_3 and X. ☐

We can prove something interesting about the element of order two that occurs in Lemma 2.38. Since it fixes the points P_1 and P_2, it must fix the tangent lines at those points, and hence must fix the intersection of these two tangent lines; in fact, it is a perspectivity.

Lemma 2.39. *Suppose P_1 and P_2 are two distinct points on the conic $\mathscr{C}$. Then there is a unique perspectivity α in $PGL(V)$ which fixes $\mathscr{C}$ and fixes the two points P_1 and P_2. The centre of α is the point V which is the intersection of the tangent lines at P_1 and P_2, and the axis of α is the line $P_1 + P_2$. Finally α has order two.*

Proof. We may choose a frame so that P_1 and P_2 are ∞ and 0 respectively; that is, are the points $\langle(1, 0, 0)\rangle$ and $\langle(0, 0, 1)\rangle$ respectively. Since any point A on the conic $\mathscr{C}$ is on a unique tangent line, L say, any collineation which leaves the conic invariant and fixes A must also fix L. Thus any perspectivity which fixes $\mathscr{C}$ and the points P_1 and P_2 must fix the point V.

Looking back to Section 4, or alternatively looking forward to Chapter IV, we see that a non-identity perspectivity can fix at most one point not on its axis and that, if such a point exists, it must be the centre. Thus any perspectivity α which fixes $\mathscr{C}$ and the points P_1, P_2 must either have centre V with axis $P_1 + P_2$ or have centre one of the P_i, P_1 say, and have the tangent line $V + P_2$ for its axis. If L is any line through P_1 other than $V + P_1$ then L meets the conic in a second point Q of $\mathscr{C}$. But this means that any collineation fixing P_1, L and $\mathscr{C}$ must also fix Q so that the identity is the only perspectivity which can have centre P_1 and leave $\mathscr{C}$ invariant. Hence the perspectivity α, if it exists, must have centre V and axis $P_1 + P_2$. If M is any line through V such that M meets $\mathscr{C}$ but is not a tangent line then M contains exactly two points, R_1 and R_2 say, of $\mathscr{C}$. Since α leaves $\mathscr{C}$ and M fixed, α must interchange R_1 and R_2 so that α^2 is a perspectivity fixing R_1 and R_2. Clearly α^2 must be a perspectivity with axis $P_1 + P_2$ and hence, since R_1 and R_2 are not on its axis, α^2 is the identity i.e. α has order two.

We must now show that such a perspectivity exists. To do this we merely show that the collineation in the orthogonal group which induces $y \to -y$ on $\mathscr{C}$ is the desired perspectivity α. If we construct the matrix B that induces $y \to -y$, we find:

$$B = \begin{bmatrix} 1 & 0 & 0 \\ 0 & -1 & 0 \\ 0 & 0 & 1 \end{bmatrix}.$$

A direct computation will show that B induces a collineation α that fixes every line that passes through $V = \langle (0, 1, 0) \rangle$, and every point on the line $\langle (0, 1, 0)' \rangle$. Finally $B^2 = 1$ so that α is of order 2. $\square$

Exercise 2.37. Decide what the conic analogue should be of the theorem of the complete quadrilateral (Theorem 2.24), and see if it is true.

A final analogue with the line is a theorem similar to Pappus' theorem:

Theorem 2.40 (Pascal's theorem). *Let* P_1, P_2, P_3, Q_1, Q_2, Q_3 *be six distinct points on a conic* $\mathscr{C}$, *and construct the points* $R_1 = (P_3 + Q_2) \cap (P_2 + Q_3)$, $R_2 = (P_1 + Q_3) \cap (P_3 + Q_1)$, $R_3 = (P_1 + Q_2) \cap (P_2 + Q_1)$. *Then* R_1, R_2, R_3 *are collinear.*

Exercise 2.38. Prove Theorem 2.40. (See Fig. 4.) $\square$

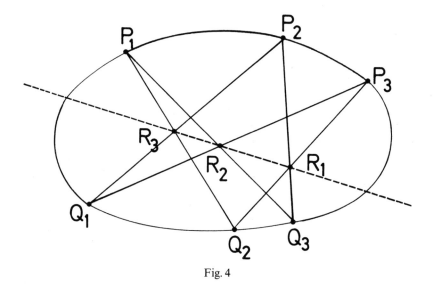

Fig. 4

In fact, had we considered conics over skewfields, instead of fields, we could have proved a theorem just like Theorem 2.6: K is commutative if and only if the three points R_i in Theorem 2.40 are collinear, where $\mathscr{C}$ is some fixed conic. This is really the better form of Pascal's theorem. Finally we note.

Theorem 2.41. *If* K *is a finite field then all conics in* $\mathscr{P}_2(K)$ *are non-empty.*

Proof. We know that by coordinatizing with three points E_1, E_2, E_3 of a self-polar triangle, the conic's A has diagonal form, say $A = \mathrm{diag}(a, b, c)$, and so $\langle (xe_1 + ye_2 + ze_3) \rangle$ is in $\mathscr{C}$ if and only if $ax^2 + by^2 + cz^2 = 0$ (see Theorem 2.33). Clearly we may assume $a = 1$ and then either both of b

and c are non-squares, or at least one of them, b say, is a square. In the latter case, let $b = d^2$, and so we may write the equation of the conic as

$$x^2 + (dy)^2 + cz^2 = 0.$$

By Result 1.4 every element of a finite field is a sum of two squares, and hence for every z, the equation $x^2 + (dy)^2 = -cz^2$ has solutions for x and y.

Suppose instead that both b and c are non-squares. If $\mu \neq 0$ is a fixed non-square then, since the squares form a subgroup of index 2 in K^*, there are elements d and e in K such that $b = d^2\mu$, $c = e^2\mu$, so our equation becomes

$$x^2 + \mu((dy)^2 + (ez)^2) = 0.$$

But the same argument tells us that for every x in K, there exist solutions y and z for $(dy)^2 + (ez)^2 = -x^2/\mu$. $\Box$

8. Unitals

Now we pass on to unitary polarities, about which we shall have somewhat less to say, since not so much is known about them, and many of the known results lie rather deeper than the results about conics. We do not make any restrictions on the characteristic of K, and the structure that we want to consider is that made up of the absolute points and the non-absolute lines, which we shall call a *unital*.

To investigate unitals, we need some lemmas about the different sort of equations that arise there. First let us suppose that K is a field with an involutory automorphism ϕ, and let F be the set of elements in K fixed by ϕ. Then the degree of K over F is necessarily two (see Result 1.6). We will write $x^{1+\phi}$ for $x \cdot x^\phi$. Let N be the subset of K of all elements n such that $n^{1+\phi} = 1$. For example if F is a subfield of the reals and d is an element of F which has no square-root in F, then K can be the subfield of the complex numbers consisting of all elements of the form $x + y\sqrt{d}$, where x and y range over F. The automorphism ϕ has the form $(x + y\sqrt{d})^\phi = x - y\sqrt{d}$, and N consists of all elements $x + y\sqrt{d}$ such that $x^2 - dy^2 = 1$. (The reader should verify all these remarks, and should also convince himself that, if $K \neq GF(4)$, N is not the whole of K^*.)

The set N is important in establishing the properties of unitals, and we now prove that N is always non-trivial.

Lemma 2.42. (a) *Let ϕ be an involutory automorphism of a field K and let $N = \{n \in K \mid n^{1+\phi} = 1\}$. Then $|N| > 1$.*

(b) *Let $F = GF(q)$ and $K = GF(q^2)$ so that K has a (unique) involutory automorphism ϕ given by $x^\phi = x^q$. Then N consists of the $q+1$ elements of K which are $(q-1)$-st powers.*

Proof. (a) Since $\phi \neq 1$ there is an element k of K with $k^\phi \neq k$. But $(k^{\phi-1})^{\phi+1} = (k^{\phi-1})^\phi k^{\phi-1} = k^{1-\phi} k^{\phi-1} = 1$.

(b) Since K is a cyclic group of order $q^2 - 1$, this part is a trivial application of the theory of cyclic groups. $\square$

Lemma 2.43. *Let K be a field with an involutory automorphism ϕ and let F be the set of elements fixed by ϕ. For any $a \neq 0$ in K the equation $x^{1+\phi} = a$ can only have solutions if a is in F, in which case if it has solutions then the set of solutions is a coset of N in the multiplicative group of K^*.*

Proof. If $x^{1+\phi} = a$, then clearly $a^\phi = x^{\phi+\phi^2} = x^{1+\phi} = a$, so a must be in F. Suppose the equation has solutions, let x be one of them and let n be any element of N. Then $(xn)^{1+\phi} = x^{1+\phi} n^{1+\phi} = a$, so xn is a solution. Conversely, if x and y are solutions, then $(y/x)^{1+\phi} = y^{1+\phi}/x^{1+\phi} = a/a = 1$, so $y/x = m$ is in N, hence $y = xm$. $\square$

Lemma 2.44. *Let a, b, c be elements of K, where a, c are in F and $a \neq 0$. The equation*

$$x^{1+\phi} a + x^\phi b + x b^\phi + c = 0 \qquad (*)$$

has solutions only if $b^{1+\phi} - ac = d$ is in F. The number of solutions of $(*)$ is zero if $w^{1+\phi} = d$ has no solutions, it is one if $d = 0$, and the set of solutions is of the form $rN + s$ for some r, $s \in F$, $r \neq 0$, if $w^{1+\phi} = d$ has solutions.

Proof. We divide $(*)$ through by a and use the "factorization" $(x + b/a)^{1+\phi} = (x + b/a)(x + b/a)^\phi = (x + b/a)(x^\phi + b^\phi/a^\phi) = x^{1+\phi} + x^\phi b/a + x b^\phi/a^\phi + b^{1+\phi}/a^{1+\phi}$, plus the fact that $a = a^\phi$ to give:

$$(x + b/a)^{1+\phi} - (b/a)^{1+\phi} + c/a = 0. \qquad (1)$$

Finally since $a^2 = a^{1+\phi}$, we can rewrite (1) as

$$(x + b/a)^{1+\phi} = (b^{1+\phi} - ac)/a^{1+\phi},$$

which proves the lemma. $\square$

Lemma 2.45. *An equation of the form $(*)$ of Lemma 2.44 is satisfied for all x in K if and only if $a = b = c = 0$, or if $K = GF(4)$.*

Proof. Certainly $(*)$ is satisfied for all x if all the coefficients are zero. Conversely, if all x satisfy $(*)$, then we must have $a = 0$ from Lemma 2.44, since N cannot include all the elements of K^* if $K \neq GF(4)$. Putting $x = 0$ in $(*)$, we find that $c = 0$ as well. Now if we let $x = 1$, we have: $b + b^\phi = 0$, or $b^\phi = -b$, and so for all $x \neq 0$, $(*)$ tells us that $x^\phi b - x b = 0$. Thus if $b \neq 0$, then $x^\phi = x$ for all x, which implies $\phi = 1$, a contradiction. We leave the case $K = GF(4)$ as a simple exercise. $\square$

Now suppose we return to our unitary situation. Let $\mathcal{U}$ be the unital associated with a unitary polarity which has matrix A and associated involutory automorphism ϕ. By Theorem 2.29 we assume $A'^{\phi} = A$. If $\langle v \rangle$ and $\langle w \rangle$ are two distinct points in the plane, then the line $\langle v \rangle + \langle w \rangle$ meets the unital $\mathcal{U}$ in those points $\langle x\,v + y\,w \rangle$ which satisfy:

$$x^{1+\phi}(v\,A\,v'^{\phi}) + x^{\phi}y(w\,A\,v'^{\phi}) + x\,y^{\phi}(v\,A\,w'^{\phi}) + y^{1+\phi}(w\,A\,w'^{\phi}) = 0. \quad \text{(JU)}$$

If we write $a = v\,A\,v'^{\phi}$, $b = w\,A\,v'^{\phi}$, $c = w\,A\,w'^{\phi}$, then we note that $b^{\phi} = (w\,A\,v'^{\phi})^{\phi} = w^{\phi}\,A^{\phi}\,v' = (w^{\phi}\,A^{\phi}\,v')'$ (since the matrix in question is one-by-one), and so $b^{\phi} = v\,A\,w'^{\phi}$. By a similar calculation, we see that a and c are fixed by ϕ. Hence they are in F, and so (JU) can be written $x^{1+\phi}a + x^{\phi}yb + x\,y^{\phi}b^{\phi} + y^{1+\phi}c = 0$. From this we can establish:

Lemma 2.46. *There is no line all of whose points lie in $\mathcal{U}$.*

Proof. If such a line exists, then in (JU) all values of x and y are solutions, and so, putting $y = 1$, we see that an equation of the form $(*)$ of Lemma 2.44 is satisfied for all x, which implies $a = b = c = 0$ by Lemma 2.45, if $K \neq GF(4)$. But then $\langle v \rangle$ is on $\langle A\,w'^{\phi} \rangle$, the image of $\langle w \rangle$ under the polarity, and $\langle w \rangle$ is on the same line, so the line $\langle v \rangle + \langle w \rangle$ is the line $\langle A\,w'^{\phi} \rangle$. But similarly we show that it is the line $\langle A\,v'^{\phi} \rangle$, which is a contradiction (compare the proof of Lemma 2.34). (The student should analyze the case $K = GF(4)$.) $\square$

We now study lines which contain points of the unital. If a line is absolute, then it meets $\mathcal{U}$ just once (Lemma 2.31), so let us assume that the line $\langle v \rangle + \langle w \rangle$ is not absolute. The point $\langle v \rangle$ is in $\mathcal{U}$ if and only if $y = 0$ is a solution of (JU), which is to say that $a = 0$; are there any other solutions in this case? Since it is not possible for the entire line in question to lie in $\mathcal{U}$, we may choose our points $\langle v \rangle$ and $\langle w \rangle$ in such a way that $\langle w \rangle$ is not a point of $\mathcal{U}$, which means that $c \neq 0$ in (JU). So (JU) can be written as $y^{1+\phi}c + y^{\phi}xb^{\phi} + yx^{\phi}b = 0$. But this is an equation of the type $(*)$ if we use the variable $z = y/x$, and has $d = b^{1+\phi}$. So by Lemma 2.44 it has solutions which are in one-to-one correspondence with the elements of N (why is $d \neq 0$?). Note that the solution $y = 0$, which we already knew about, is included in this set of solutions for y/x.

Now if we suppose that $y = 0$ is not a solution, we may divide (JU) through immediately by $y^{1+\phi}$, obtaining again an equation of type $(*)$ in the variable x/y. This may have no solutions, one solution, or its solutions may be in one-to-one correspondence with the elements of N. We now show that there cannot be exactly one solution.

Lemma 2.47. *If a line meets $\mathcal{U}$ just once then it is absolute (and is the image of the point where it meets $\mathcal{U}$).*

Proof. Supposing that a line $\langle v \rangle + \langle w \rangle$ meets $\mathcal{U}$ in one point only, we may assume it is the point $\langle w \rangle$ and hence $c = 0$ in (JU). Thus (JU) has only the solution $x = 0$. But the form of (JU) now is

$$(x/y)^{1+\phi} a + (x/y)^{\phi} b + (x/y) b^{\phi} = 0,$$

and it is an equation of the type (∗) of Lemma 2.44, satisfying the conditions of that lemma (for $a = 0$ would imply that $\langle v \rangle$ was also a point of $\mathcal{U}$, a contradiction). The element d of Lemma 2.44 is $b^{1+\phi}$, and since (∗) can only have one solution in this case, we must have $d = 0$, or in other words $b = 0$. But then $v A w'^{\phi} = 0$, and so both $\langle v \rangle$ and $\langle w \rangle$ are on the line $\langle A w'^{\phi} \rangle$. Hence $\langle A w'^{\phi} \rangle$ must be the line $\langle v \rangle + \langle w \rangle$; i.e. the given line is absolute. ⬚

Theorem 2.48. *Let $\mathcal{U}$ be a unital. Then:*
 (i) *a line meets $\mathcal{U}$ exactly once if and only if the line is absolute;*
 (ii) *a non-absolute line either meets $\mathcal{U}$ not at all, or meets it in a set of points which is in one-to-one correspondence with the elements of N. Using the notation that precedes the theorem, a non-absolute line $\langle v \rangle + \langle w \rangle$ meets the unital if and only if there is a solution to*

$$z^{1+\phi} = (w A v'^{\phi})(v A w'^{\phi}) - (v A v'^{\phi})(w A w'^{\phi}).$$

The theorem has all been proved; the last part is simply an expression of the element d of (∗) and an application of Lemma 2.44 with the values of a, b, c from (JU) substituted in the expression for d. ⬚

Exercise 2.39. Show that Theorem 2.48 could be sharpened to read: the line $\langle v \rangle + \langle w \rangle$ is absolute if and only if

$$(w A v'^{\phi})(v A w'^{\phi}) - (v A v'^{\phi})(w A w'^{\phi}) = 0.$$

Hence if we call the left hand side of this expression the *discriminant* of the line, and say that an element x of K is a *norm* if $z^{1+\phi} = x$ for some z in K, then the theorem can read:
 (i) a line is absolute if and only if its discriminant is zero
 (ii) a line fails to meet $\mathcal{U}$ if and only if its discriminant is not a norm
 (iii) a line meets $\mathcal{U}$ in a set of points in one-to-one correspondence with N if and only if its discriminant is a non-zero norm.

Corollary. *If $K = GF(q^2)$ is a finite field then every non-absolute line meets the non-empty unital $\mathcal{U}$ in $q + 1$ points.*

Proof. In a finite field $K = GF(q^2)$, $F = GF(q)$ and thus, since every element of F is a norm, all non-absolute lines meet $\mathcal{U}$ (see the exercise above, or Theorem 2.48) in $|N|$ points. But the number of elements in N is the number of solutions of $x^{q+1} = 1$, and since $q + 1$ divides $q^2 - 1$, the order of the cyclic multiplicative group of K, this equation has exactly $q + 1$ solutions. ⬚

∗Exercise 2.40. If K is finite, show that all unitals are non-empty.

Exercise 2.41. Let $K = GF(q^2)$. Then show that the number of points in a unital $\mathcal{U}$ is $q^3 + 1$, the number of lines is $q^2(q^2 - q + 1)$, and every point of $\mathcal{U}$ is on q^2 lines of $\mathcal{U}$.

Exercise 2.42. Using the terminology of Exercise 2.39 show that the discriminant of a line $\langle v \rangle + \langle w \rangle$ can only vary by a non-zero norm factor for different choices of the points $\langle v \rangle$ and $\langle w \rangle$.

Let $\mathcal{U}$ be a non-empty unital in the projective plane $\mathcal{P} = \mathcal{P}_2(K)$, and let X, Z, J be any three non-collinear points of $\mathcal{U}$. If β is the polarity defining $\mathcal{U}$, then let Y be the intersection of X^β and Z^β. Since the lines X^β and Z^β are both absolute, neither of them contains J so we may choose X, Y, Z, J as a frame for $\mathcal{P}$. Thus we may write $X = \langle (1, 0, 0) \rangle$, $Y = \langle (0, 1, 0) \rangle$, $Z = \langle (0, 0, 1) \rangle$, $J = \langle (1, 1, 1) \rangle$. Imitating the proof of Theorem 2.36 we find that β can be represented by the matrix

$$A = \begin{bmatrix} 0 & 0 & c \\ 0 & b & 0 \\ c^\phi & 0 & 0 \end{bmatrix}$$

where ϕ is the associated automorphism of β, b is an element of its fixed field F, and $b + c + c^\phi = 0$. A point $P = \langle (x, y, z) \rangle$ is on $\mathcal{U}$ if and only if

$$b y^{1 + \phi} + c^\phi x^\phi z + c x z^\phi = 0. \tag{1}$$

Now if $z = 0$, then $y = 0$, and so x can be taken to be 1; this is the point X. If $z \neq 0$, we can assume that $z = 1$, so x and y must satisfy:

$$b y^{1 + \phi} + c^\phi x^\phi + c x = 0. \tag{2}$$

Suppose x_1 and x_2 are solutions of (2) for the same value of y; then $c^\phi x_1^\phi + c x_1 = -b y^{1 + \phi} = c^\phi x_2^\phi + c x_2$, so $(c x_1 - c x_2)^\phi + (c x_1 - c x_2) = 0$.

Let M be the subset of K consisting of all elements k satisfying $k^\phi + k = 0$; M is closed under addition. We have shown that $c x_1 = k + c x_2$, for an element k in M. Conversely, if x_2 is a solution of (2) for a certain value of y, then $x_1 = c^{-1} k + x_2$ is also a solution, for every k in M.

Theorem 2.49. *Let c be any element of K such that c is not in the set M defined above. Then any non-empty unital in $\mathcal{P}_2(K)$ can be chosen (after the appropriate choice of frame) to consist of the following set of points: $\langle (1, 0, 0) \rangle$ plus all points $\langle (y^{1 + \phi} + c^{-1} k, y, 1) \rangle$ as y ranges over K and k ranges over M.*

Proof. Using the fact that $b + c + c^\phi = 0$, it is easy to see that the point $\langle (y^{1 + \phi}, y, 1) \rangle$ satisfies (2). Thus since the discussion before the theorem shows that all points of $\mathcal{U}$ have the required form, the theorem is proved. □

Corollary. *Any two non-empty unitals in $\mathscr{P}$ are equivalent.* □

The subgroup of $PGL(V)$ which fixes $\mathscr{U}$, called the *unitary group*, is not represented quite so simply as the orthogonal group. But we can deduce rather easily some of its properties. Suppose that $\mathscr{U}$ is non-empty; then, by the above corollary, the unitary group is independent of the unital $\mathscr{U}$. Let Γ be the unitary group fixing $\mathscr{U}$, where $\mathscr{U}$ is non-empty. We shall show that the subgroup of Γ fixing a point Q in $\mathscr{U}$ is transitive on the remaining points of $\mathscr{U}$, and hence, since Q was an arbitrary point of $\mathscr{U}$, the full group Γ is 2-transitive. Choose $\mathscr{U}$ to have the canonical form of Theorem 2.49 and choose Q to be the point $X = \langle (1, 0, 0) \rangle$. Consider the matrix:

$$S = \begin{bmatrix} 1 & 0 & 0 \\ s & 1 & 0 \\ a^{1+\phi} & a & 1 \end{bmatrix}$$

where a is an arbitrary element of K and $s = a^{\phi}(1 + c^{\phi}c^{-1})$. The effect of this matrix is to fix X and to send the point $(y^{1+\phi} + c^{-1}k, y, 1)$ to $(y^{1+\phi} + c^{-1}k + sy + a^{1+\phi}, y + a, 1)$. To show that this last point is in $\mathscr{U}$, it is necessary to show that $y^{1+\phi} + c^{-1}k + sy + a^{1+\phi} - (y + a)^{1+\phi}$ is in $c^{-1}M$. But this is a matter of straightforward calculation; upon simplification, the given expression becomes $c^{-1}k + a^{\phi}c^{\phi}c^{-1}y - y^{\phi}a$, and multiplying by c, we have $k + a^{\phi}c^{\phi}y - acy^{\phi}$, which, since k is in M, is clearly in M.

So far we have shown that we can map $\mathscr{U}$ onto itself by an element of Γ_X in such a way that the middle coordinate y can be mapped to any value $y + a$. In order to prove that Γ_X is transitive on the points of $\mathscr{U}$ other than X we must try to "adjust" the first coordinate. But this is easy. Consider the matrix

$$T = \begin{bmatrix} 1 & 0 & 0 \\ 0 & 1 & 0 \\ c^{-1}m & 0 & 1 \end{bmatrix}$$

where m is an arbitrary element of M. Clearly T is in Γ_X. Furthermore T sends the point $\langle (y^{1+\phi} + c^{-1}k, y, 1) \rangle$ onto the point $\langle (y^{1+\phi} + c^{-1}(k + m), y, 1) \rangle$. Thus we can fix the second coordinate and map the first coordinate onto all other possible values and so we have proved that Γ_X is transitive on the points of $\mathscr{U}$ other than X; since X is arbitrary, this demonstrates that Γ is two-transitive. We can even prove slightly more:

Theorem 2.50. *The group Γ is two-transitive on the points of $\mathscr{U}$, and the subgroup $\Gamma_{X,Y}$ fixing two points in $\mathscr{U}$ is isomorphic to the multiplicative group of K^*.*

Proof. If we fix the points $\langle(1,0,0)\rangle$ and $\langle(0,0,1)\rangle$, then we must fix the β-image lines of these two points (since they are the only lines through the points which contain just one point of $\mathcal{U}$). Hence, we must fix the intersection of the two lines, which is the point $\langle(0,1,0)\rangle$. Now the only elements of $PGL(V)$ which fix the three points given are diagonal matrices, and we may assume that the third diagonal entry is 1 in any such diagonal matrix. So let

$$R = \begin{bmatrix} a & 0 & 0 \\ 0 & d & 0 \\ 0 & 0 & 1 \end{bmatrix}$$

be such a matrix. It will send the point $(y^{1+\phi}+c^{-1}k, y, 1)$ onto the point $\langle(ay^{1+\phi}+ac^{-1}k, dy, 1)\rangle$ which will be in $\mathcal{U}$ if and only if $ay^{1+\phi}+ac^{-1}k-d^{1+\phi}y^{1+\phi}$ is in $c^{-1}M$, for all y in K and all k in M. If $y=0$, this says that ak is in M for all k in M, and hence it is easy to prove that a must be fixed by ϕ. If $k=0$, then we must have $(a-d^{1+\phi})cy^{1+\phi}$ in M for all y in K; this condition is equivalent to insisting that $(a-d^{1+\phi})c$ is in M, and hence

or
$$(a-d^{1+\phi})^\phi c^\phi + (a-d^{1+\phi})c = 0,$$
$$(a-d^{1+\phi})(c^\phi+c) = 0,$$

since $a-d^{1+\phi}$ is fixed by ϕ. But $c^\phi+c$ cannot be zero, otherwise $b=0$ and the matrix A of the polarity β would be singular, and so $a=d^{1+\phi}$. Since d is an arbitrary non-zero element of K, it is easy to see that the set of all such matrices is a group isomorphic to the multiplicative group of K^*. $\Box$

Corollary. *If* $K=GF(q^2)$, *then the order of the group* Γ *is* $(q^3+1)(q^3)(q^2-1)$.

The corollary follows from the fact that Γ is two-transitive on q^3+1 points and the stabilizer of two points has order q^2-1. $\Box$

9. Affine Planes

Let $\mathcal{P}=\mathcal{P}(V)$ be a projective geometry and W a hyperplane in $\mathcal{P}$. We define the *affine geometry* $\mathcal{A}=\mathcal{P}^W$ to be the set of objects of $\mathcal{P}$ which are not contained in W. In other words, we discard W and all its subspaces. We have already seen (Theorem 2.12) that the collineation group of $\mathcal{P}$ is transitive on frames from which it is easy to see that it is transitive on hyperplanes, so there is an obvious isomorphism between $\mathcal{P}^W$ and $\mathcal{P}^U$, for any two different hyperplanes W and U. So sometimes we will refer to $\mathcal{A}$ as $\mathcal{A}(V)$; if K is a field and V has dimension $n+1$, then we can also write $\mathcal{A}_n(K)=\mathcal{A}(V)$. In any case the g-dimension of $\mathcal{A}$ is the g-dimension of $\mathcal{P}$.

The automorphism or collineation group of $\mathscr{A}$, Aut $\mathscr{A}$, is the subgroup of Aut $\mathscr{P}$ which fixes W (where $\mathscr{A} = \mathscr{P}^W$). We shall confine our attention to the *affine line* of g-dimension one, and the *affine plane* of g-dimension two.

Let $\mathscr{P} = \mathscr{P}(V)$ have g-dimension one. Then a hyperplane is just a point, and so $\mathscr{A}(V)$ is $\mathscr{P}(V)$ with a point discarded. If we choose parametric coordinates for $\mathscr{P}(V)$ so that the discarded point is ∞, then when K is a field we may say that the affine line *consists of a set of points in one-to-one correspondence with the elements of the underlying field K.* The group Aut $\mathscr{A}(V)$ will be the subgroup of $P\Gamma L(V)$ which fixes the point ∞.

Theorem 2.51. *The group* Aut $\mathscr{A}$ *of the affine line* $\mathscr{A}$ *over the field K is (in parametric form) the group of all mappings*

$$x \to k\,x^\alpha + b$$

of the elements of K, where k and b are arbitrary elements of K, $k \neq 0$, and α is an arbitrary automorphism of K. This group has a normal subgroup Λ_1 consisting of all mappings for which $\alpha = 1$, and Λ_1 has a normal subgroup Λ_2 consisting of all elements for which $\alpha = 1$ and $k = 1$. The factor groups are:

$$\text{Aut}\,\mathscr{A}/\Lambda_1 \cong \text{Aut}\,K$$

and finally:
$$\Lambda_1/\Lambda_2 \cong (K^*, \,.\,) \quad \text{the multiplicative group of } K$$
$$\Lambda_2 \cong (K, +) \quad \text{the additive group of } K.$$

Proof. Most of the theorem is obvious, considering the formula for elements of $P\Gamma L(V)$ given in Section 4, and the subgroup of that group which fixes ∞. If (k, b, α) is the element of Aut $\mathscr{A}$ which sends x to $k x^\alpha + b$, then it is easy to see that $(k, b, \alpha)\,(m, c, \beta) = (mk^\beta,\ mb^\beta + c,\ \alpha\beta)$. The mapping $(k, b, \alpha) \to \alpha$ is thus a homomorphism whose kernel is the set of mappings $(k, b, 1)$; that is, Λ_1. Now if we map the elements $(k, b, 1)$ of Λ_1 onto k, this is also a homomorphism whose kernel is the set Λ_2 of elements $(1, b, 1)$. The images of these two homomorphisms are Aut K and $(K^*, \cdot)$, as claimed. Finally it is easy to see that Λ_2 is isomorphic to $(K, +)$. ☐

Corollary. *If K is a finite field and $\mathscr{A} = \mathscr{A}_1(K)$, then* Aut $\mathscr{A}$ *is soluble.*

Proof. For any field K, Λ_2 is abelian, so that Aut $\mathscr{A}$ is only non-soluble if either Aut K or $(K^*, \cdot)$ is non-soluble. But if K is finite both of these groups are even cyclic. ☐

Exercise 2.43. Show that the subgroup Λ_2 in Theorem 2.51 is always a normal subgroup of Aut $\mathscr{A}$.

Exercise 2.44. Show that Aut $\mathscr{A}$ in Theorem 2.51 splits over both Λ_1 and over Λ_2.

Exercise 2.45. Find all the elements of order two in Aut $\mathscr{A}$, and for each one determine its fixed elements.

When we want to study the affine plane, we shall find it necessary to spend more time on the coordinatization. This will have the advantage of giving us our first really concrete connection between the geometries studied in this chapter, and the geometries that we feel ourselves familiar with. Suppose V is a three-dimensional left vector space over the skewfield K, $\mathscr{P} = \mathscr{P}(V)$ and $\mathscr{A} = \mathscr{A}(V) = \mathscr{P}^W$. We may choose a frame for $\mathscr{P}$ such that its first two points are on W; let the frame be $E_1 = \langle e_1 \rangle$, $E_2 = \langle e_2 \rangle$, $E_3 = \langle e_3 \rangle$, $E_4 = \langle e_1 + e_2 + e_3 \rangle$. Then since E_1 and E_2 are on W, a frame for W is E_1, E_2 and $\langle e_1 + e_2 \rangle$. An arbitrary point on W has the form $\langle e_1 + m e_2 \rangle$, or is E_2 itself. A point not on W has the form $\langle x_1 e_1 + x_2 e_2 + x_3 e_3 \rangle$, where $x_3 \neq 0$; hence we may represent it as $\langle x e_1 + y e_2 + e_3 \rangle$. A line of $\mathscr{P}$ is $\langle f_1' a_1 + f_2' a_2 + f_3' a_3 \rangle$ where the f_i' form the basis of V' associated with the e_i. The line W, in this form, is $\langle f_3' \rangle$, so a line not equal to W "contains" a vector of the given form in V where not both a_1 and a_2 are zero. If $a_2 \neq 0$, then we may assume it is the identity, and so we have a line which we may represent as $\langle f_1' m + f_2' + f_3' b \rangle$; its points will be all the points $\langle x e_1 + y e_2 + e_3 \rangle$ such that $x m + y + b = 0$ together with the point $\langle e_1 - m e_2 \rangle$ on the line W. On the other hand, if $a_2 = 0$, then we may assume that $a_1 = 1$, so we have a line $\langle f_1' + f_3' b \rangle$, whose points are E_2 plus the points $\langle -b e_1 + y e_2 + e_3 \rangle$, for any y in K. Thus:

Theorem 2.52. *Let K be a skewfield and let $\mathscr{A}$ be the set of points and lines constructed as follows: the points of $\mathscr{A}$ are the ordered pairs (x, y) and the lines are the ordered pairs $[m, b]$ and the symbols $[b]$, with an incidence relation given by*

(x, y) is on $[m, b]$ if and only if $y = xm + b$

(x, y) is on $[b]$ if and only if $x = b$,

for all x, y, m, b in K.
Then $\mathscr{A}$ is isomorphic to the affine plane $\mathscr{A}(V)$ where V is a three-dimensional left vector space over K.

Proof. It is only necessary to set up correspondences as follows: (x, y) is the point $\langle x e_1 + y e_2 + e_3 \rangle$; $[m, b]$ is the line $\langle f_1'(-m) + f_2' + f_3'(-b) \rangle$; $[b]$ is the line $\langle f_1' + f_3'(-b) \rangle$. Then our remarks just before the theorem provide the rest of the proof. $\square$

If K is a familiar field, such as the reals, then this representation of $\mathscr{A}_2(K)$ gives us exactly the object that we have always thought of as being a familiar plane; in the case of the reals, this is the "blackboard plane" in which school geometry is done. Thus we can see that our familiar blackboard plane actually has some sort of natural embedding in a projective plane, and by continuing the analysis preceding Theorem 2.52, we can see what this projective plane "is". The difference between $\mathscr{A}$ and $\mathscr{P}$

is that $\mathscr{P}$ has an additional line, and a certain set of additional points, all of which are on the new line. What are these new points? A point $\langle e_1 + m e_2 \rangle$ is on every line $\langle f_1'(-m) + f_2' + f_3'(-b) \rangle$, and on no others; that is, it is on every line of the sort $[m, b]$, for all b. So it is the point where all the "parallel" lines with slope m (that is, the lines with equation $y = xm + b$) meet; similarly, E_2 is on each line $[b]$, and no others, so it is the point where all the lines of slope ∞ meet. This intuitive approach explains a very common notation: these new points are so "far away" that parallel lines meet there, and the line that contains them all is therefore equally far away. So it is often called the "line at infinity", and its points are "points at infinity"; also it has been called the "ideal line", and its points "ideal points". From an affine point of view these terminologies make sense; but from the point of view of the projective plane $\mathscr{P}$ they do not mean much, and anyway, any line whatsoever could have been chosen as the "line at infinity" W. So, from the projective point of view, the line at infinity is nothing special. Affine geometry has among its several virtues the fact that it resembles the geometries that we think we know about already, and it is often useful as an intuitive guide to theorems. But projective geometry is more homogeneous, there are no special elements, and it has many unifying features, some of which we explore here. But first note that whereas not every pair of lines meet in an affine plane, each pair of lines always meet in a projective plane; we can speak of anti-isomorphisms between projective geometries, and should not expect to be able to in affine geometries.

Now we want to investigate the group $\operatorname{Aut}\mathscr{A}$ of an affine plane $\mathscr{A}$. Clearly it is the subgroup of $\operatorname{Aut}\mathscr{P}$ fixing the line W, where $\mathscr{A} = \mathscr{P}^W$, and we have already encountered this group before, since it helped us give one of our alternate definitions of the automorphism group of a projective line. But when we met the group earlier, we were only concerned with its action on the line W, so we were not interested in those elements fixing every element of W. In other words we were able to ignore the kernel of the representation of $(\operatorname{Aut}\mathscr{P})_W \cong \operatorname{Aut}\mathscr{A}$ on W. But from the point of view of $\mathscr{A}$ we must preserve this kernel. If we call it Σ, then we know that $\operatorname{Aut}\mathscr{A}/\Sigma$ is isomorphic to $P\Gamma L(W)$, and that $\operatorname{Aut}\mathscr{A}$ splits over Σ. Furthermore we know that Σ consists of the perspectivities of $\mathscr{P}$ with axis W. In order to study the structure of Σ we really need some elementary results which will be proved in Chapter IV. But we can do without these, simply using two facts which we noted, but did not prove, in an earlier section: namely that if an element fixes all the points on a line (or all the lines on a point) then it is a perspectivity, and that if it fixes any elements other than its centre or axis or the lines and points incident with these, then it is the identity.

For each a, b in K, let $\phi_{a,b}$ be the mapping of $\mathscr{P}$ given below:

$$\phi_{a,b} \colon \langle x_1 e_1 + x_2 e_2 + x_3 e_3 \rangle \to \langle (x_1 + x_3 a) e_1 + (x_2 + x_3 b) e_2 + x_3 e_3 \rangle.$$

This mapping is induced by a non-singular linear transformation (note that if we were to write ax_3 instead of $x_3 a$ then the mapping would not be linear, in general). Any point on W, that is either E_2 or $\langle e_1 + m e_2 \rangle$, is clearly fixed by $\phi_{a,b}$, and no point off W (that is, with $x_3 \neq 0$) is fixed provided a, b are not both zero. So $\phi_{a,b}$ is a perspectivity with axis W whose centre lies on W.

Lemma 2.53. *The set of all the $\phi_{a,b}$ is a group which is regular on the points of $\mathscr{A}$.*

Proof. It is immediate that $\phi_{a,b}\phi_{c,d} = \phi_{a+c, b+d}$, and that $\phi_{a,b}^{-1} = \phi_{-a, -b}$. So the set of all the $\phi_{a,b}$ is certainly a subgroup of $\mathrm{Aut}\,\mathscr{A}$. To show transitivity, we note that the point $E_3 = \langle e_3 \rangle$ is sent to $\langle a e_1 + b e_2 + e_3 \rangle$ by $\phi_{a,b}$, and so E_3 can be sent to any point of $\mathscr{A}$. Since $\phi_{0,0}$ (i.e. the identity) is the only element in the group fixing E_3 it follows that the group is regular.   □

Lemma 2.54. *Let α be a perspectivity of $\mathscr{P}$ whose axis is W and whose centre lies on W, and let $\mathscr{A} = \mathscr{P}^W$. Then $\alpha = \phi_{a,b}$ for some a and b.*

Proof. If $\alpha = 1$ then $\alpha = \phi_{0,0}$. Suppose $\alpha \neq 1$ and that the effect of α on E_3 is to send it to $E_3^{\phi_{a,b}}$ which, of course, is $\langle a e_1 + b e_2 + e_3 \rangle$, and let X be the point on W which is collinear with E_3 and $E_3^{\phi_{a,b}}$. Then if U is the line $X + E_3$, $U^\alpha = X^\alpha + E_3^\alpha = X^\alpha + E_3^{\phi_{a,b}}$. But since X is on the axis of α, $X^\alpha = X$ so that $U^\alpha = U$. As we have just noted, the only lines fixed by α are its axis and the lines through its centre. Thus the centre of α is X. By the same reasoning, the centre of $\phi_{a,b}$ is X, and so $\phi_{a,b}\alpha^{-1}$ is a perspectivity with axis W which fixes all lines passing through X and also fixes the point E_3. Since X is the centre of $\phi_{a,b}\alpha^{-1}$ and E_3 is a fixed point which is not on W we must have $\phi_{a,b}\alpha^{-1} = 1$, or $\phi_{a,b} = \alpha$.   □

If we let Φ be the subgroup of Σ consisting of all the $\phi_{a,b}$, then we have:

Lemma 2.55. *Φ is normal in $\mathrm{Aut}\,\mathscr{A}$.*

Proof. As always we regard $\mathrm{Aut}\,\mathscr{A}$ as the subgroup of $\mathrm{Aut}\,\mathscr{P}$ fixing the line W where $\mathscr{A} = \mathscr{P}^W$.

Since Φ is the set of all perspectivities in $\mathrm{Aut}\,\mathscr{A}$ whose centres lie on W we merely have to show that conjugating $\phi_{a,b}$ by an element of $\mathrm{Aut}\,\mathscr{A}$ gives another perspectivity whose centre is on W. Let A be the centre of $\phi_{a,b}$ and let β be any element of $\mathrm{Aut}\,\mathscr{A}$; then since A is on W and $W^\beta = W$, A^β is on W. If L is any line through A^β then, using the incidence preserving property of β, $L^{\beta^{-1}}$ passes through A. The perspectivity $\phi_{a,b}$ fixes every line through its centre A and so $L^{\beta^{-1}\phi_{a,b}} = L^{\beta^{-1}}$ which immediately implies $L^{\beta^{-1}\phi_{a,b}\beta} = L^{\beta^{-1}\beta} = L$. Thus $\beta^{-1}\phi_{a,b}\beta$ fixes every line through A^β. But, clearly, $\beta^{-1}\phi_{a,b}\beta$ fixes every point of W so that $\beta^{-1}\phi_{a,b}\beta$ is in Φ and Φ is normal in $\mathrm{Aut}\,\mathscr{A}$.   □

Corollary. Φ *is normal in* Σ. $\square$

The determination of the factor group Σ/Φ is slightly more complicated than one might expect. Since Φ is transitive on the points of $\mathscr{A}$, it follows that Σ is also transitive, and so, since the identity is the only element of Σ fixing two points of $\mathscr{A}$, Σ acts as a Frobenius permutation group on the points of $\mathscr{A}$. Thus, by Result 1.18 $\Sigma = \Phi \cdot \Sigma_P$, where P is any point of $\mathscr{A}$. Also since Φ is regular we have $\Phi \cap \Sigma_P = 1$ so that $\Sigma/\Phi \cong \Sigma_P$. Now Σ_P is just the set of perspectivities in Σ with centre P, and, by the transitivity of Σ, we may suppose $P = E_3$.

Lemma 2.56. *If* $P = E_3$, *then* Σ_P *is the group of all mappings defined as follows: for each element* a *in* K, $a \neq 0$, *define a mapping* λ_a *by*

$$\lambda_a : \langle x_1 e_1 + x_2 e_2 + x_3 e_3 \rangle \rightarrow \langle x_1 e_1 + x_2 e_2 + x_3 a e_3 \rangle .$$

Proof. It is easy enough to see that λ_a is a perspectivity with axis W and centre E_3, and we leave that as an exercise. If there were any more perspectivities with this property, then the same argument used in Lemma 2.54 would show that we had a contradiction, since the group of mappings $\{\lambda_a\}$ is transitive on the points of a line passing through E_3, excepting the point E_3 itself and the intersection of the line with $E_1 + E_2$. $\square$

Since $\lambda_a \lambda_b = \lambda_{ab}$ the mapping which sends k onto λ_k is an isomorphism from the multiplicative group of K to Σ_P. Equally the group Φ is isomorphic to the direct product of the additive group of K with itself.

The preceeding lemmas and discussion are summarized by:

Theorem 2.57. *Let* V *be a three dimensional vector space over a skewfield* K *and let* $\mathscr{A} = \mathscr{P}^W$ *where* $\mathscr{P} = \mathscr{P}(V)$ *and* W *is a line of* $\mathscr{P}$. *Then* $\mathrm{Aut}\,\mathscr{A}$ *has a normal subgroup* Σ, *and* Σ *has a normal subgroup* Φ *with the following properties:*

$\mathrm{Aut}\,\mathscr{A}/\Sigma \cong P\Gamma L(W)$, *where* W *is a two-dimensional subspace of* V,
$\Sigma/\Phi \cong \Sigma_P \cong$ *the multiplicative group of* K,
$\Phi \cong$ *the direct product of the additive group of* K *with itself.* $\square$

Exercise 2.46. Determine the subgroup of $PGL(V)$ which fixes W; i.e., find the "linear" part of $\mathrm{Aut}\,\mathscr{A}$.

We conclude our investigation into affine planes by seeing what happens to conics. So from now on we assume that the characteristic of K is not two, and that K is a field. If $\mathscr{C}$ is a (non-empty) conic of $\mathscr{P} = \mathscr{P}_2(K)$ and if $\mathscr{A} = \mathscr{P}^W$, then $\mathscr{C}$ meets the line W in at most two points. Thus in $\mathscr{A}$ we may speak of three sorts of *affine conics*: the *ellipses* which do not meet W, the *parabolas* which meet it once, and the *hyperbolas* which meet it twice. Generally we shall use the same symbol $\mathscr{C}$ for the affine as for the projective conic. We already know that all non-empty conics are equivalent in $\mathscr{P}$, in the sense that by proper choice of basis any conic can be made to have the projective form $y^2 = xz$, or alternatively, that given any two non-empty conics, there is an element of Aut $\mathscr{P}$ mapping one to the other. If we define *equivalence* between affine conics in the same way as for the projective case, i.e. $\mathscr{C}_1$ is equivalent to $\mathscr{C}_2$ if there is a collineation α in Aut $\mathscr{A}$ with $\mathscr{C}_1^\alpha = \mathscr{C}_2$, then there must be at least three classes of affine conics, (since if $\mathscr{C}$ is a conic of $\mathscr{P}$ and α is in Aut $\mathscr{A}$ then the number of points of $\mathscr{C} \cap W$ is equal to the number of points of $\mathscr{C}^\alpha \cap W$). But the group Aut $\mathscr{A}$ is 3-transitive on W, and so if we have two inequivalent conics of the same sort (that is, both with the same number of points on W) we may assume that the two conics meet W in the same set of points. In our construction of parametric coordinates for a projective conic we saw that if a conic meets a line twice, we may choose the two points to be $\langle (1,0,0) \rangle$ and $\langle (0,0\ 1) \rangle$, and find that the conic can always have the given form $y^2 = xz$. In other words any two conics through the given two points are equivalent. Similarly, if a conic meets a line once we may take the line as $\langle (0,0,1)' \rangle$ and then again we find that we can put the conic in the required form. This proves:

Lemma 2.58. *All affine hyperbolas are equivalent, and all affine parabolas are equivalent.* ☐

The situation for the ellipse is not so straightforward. If we choose a self-polar triangle as part of our frame then we have a simple form for a conic which involves picking out a definite line not meeting it. However we do not know if this form depends on the chosen frame. To determine this we first need to know if Aut $\mathscr{A}$ is transitive on the conics not meeting W. Before discussing this we set an exercise.

Exercise 2.47. Prove that Aut $\mathscr{A}$ is transitive on the ellipses of $\mathscr{A}$ if and only if the orthogonal group of any conic $\mathscr{C}$ in $\mathscr{P}$ is transitive on the lines not meeting $\mathscr{C}$.

In fact the situation for ellipses is somewhat different to that for the other two families of affine conics and the number of equivalence classes depends on the underlying field K in a rather complicated way that we are not prepared to go into. In some fields (e.g. the rationals) it is possible to construct two ellipses that are not equivalent in Aut $\mathscr{A}$, while in others this cannot happen. In particular the finite fields are of the latter sort. But first we give an exercise to illustrate the difference the ground field makes.

Exercise 2.48. Consider the equation $x^2 + 2y^2 = 1$ and the equation $x^2 - 2y^2 = 1$. Decide what sort of affine conics they represent over (*a*) the field of reals, (*b*) the field of rationals, *(*c*) a finite field $GF(q)$.

To study the question of equivalence of ellipses in finite planes, we coordinatize the plane $\mathscr{P}$ in a special way. Suppose W is a given line and $\mathscr{C}$ is a given conic, not meeting W: and let ϱ be the polarity associated with $\mathscr{C}$. We shall choose a frame E_1, E_2, E_3, E_4 for $\mathscr{P}$ such that E_1, E_2, E_3 form a self-polar triangle of $\mathscr{C}$ (with $W = E_1 + E_2$). This gives, for orthogonal polarities, a stronger form of Theorem 2.33.

Since ϱ is the polarity associated with $\mathscr{C}$, ϱ interchanges the points of $\mathscr{C}$ with its tangent lines. Thus, since W contains no point of $\mathscr{C}$, there is no tangent line through W^ϱ and we choose $E_3 = W^\varrho$, so that every line through E_3 meets $\mathscr{C}$ either twice or not at all. Now we choose a line through E_3 which meets $\mathscr{C}$, hence meets it twice, at X and X_1 say. The tangent lines at X and X_1 are X^ϱ and X_1^ϱ, and since the line $X + X_1$ passes through E_3, its image under ϱ is on W. Similarly, since X and X_1 are on the line $X + X_1$, their image lines pass through $(X + X_1)^\varrho$. Hence X^ϱ and X_1^ϱ meet on W, and their point of intersection is the image of $X + X_1$.

If we put $E_1 = (X + X_1)^\varrho$, $E_2 = (X + X_1) \cap W$, then $E_2^\varrho = E_1 + E_3$, $E_3^\varrho = E_1 + E_2$, and $E_1^\varrho = E_2 + E_3$. So the triangle formed by E_1, E_2, E_3 is self-polar. Finally we choose E_4 on the line $E_1 + X$, and this can be done so that $X = \langle e_2 + e_3 \rangle$, in terms of the usual basis that goes with the frame. This choice of our frame will give us a simple equation for $\mathscr{C}$.

Writing as usual $\langle (x, y, z) \rangle$ for $\langle xe_1 + ye_2 + ze_3 \rangle$ etc. we now see what our choice of frame has done for us. Since $\langle (0, 0, 1) \rangle^\varrho = \langle (0, 0, 1)' \rangle$, we have

$$a_{13} = a_{23} = 0 .$$

Since $\langle (0, 1, 1) \rangle$ is on the conic, we have

$$a_{22} = - a_{33} .$$

Since the point $X = \langle (0, 1, 1) \rangle$ is sent by ϱ to $E_1 + X = \langle (0, 1, -1)' \rangle$ we have

$$a_{12} = 0 .$$

So finally A has the form

$$A = \begin{bmatrix} a & 0 & 0 \\ 0 & +b & 0 \\ 0 & 0 & -b \end{bmatrix} .$$

The equation of the conic is now $ax^2 + by^2 - bz^2 = 0$; clearly we can assume $b = 1$, and so we have

Lemma 2.59. *If K is a field and $\mathscr{C}$ is an ellipse in $\mathscr{A} = \mathscr{P}^W$, where $\mathscr{P} = \mathscr{P}_2(K)$, then there is a frame for $\mathscr{P}$ in which E_1 and E_2 are on W such that $\mathscr{C}$ has the equation*

$$a x^2 + y^2 = 1 .$$

(The equation is in affine form.) In addition a must have the property that $- a$ is not a square.

Proof. A point $\langle (x, y, 1) \rangle$ not on W is on $\mathscr{C}$ if and only if $ax^2 + y^2 = 1$, as claimed. Since no point on W is on $\mathscr{C}$, there must be no x and y in K such that $\langle (x, y, 0) \rangle$ is on $\mathscr{C}$. But this implies $ax^2 + y^2 = 0$ has only the trivial solution, or $a = -(y/x)^2$ has no solution, so $-a$ is not a square, as claimed. $\square$

Now consider the situations in which $K = GF(q)$ and $-a$ is not a square. If -1 is a square in K (i.e. $q \equiv 1 \pmod 4$) then a must be a non-square, while if -1 is a non-square in K (i.e. $q \equiv 3 \pmod 4$), then a must be a square. We examine the two cases separately. Suppose a is a square, $a = r^2$; then (see Exercise 2.12) we change the frame of P by letting $E_1^* = \langle r^{-1} e_1 \rangle = E_1$, $E_2^* = \langle e_2 \rangle = E_2$, $E_3^* = \langle e_3 \rangle = E_3$, $E_4^* = \langle r^{-1} e_1 + e_2 + e_3 \rangle$. The point whose coordinates were (x, y, z) now has coordinates (xr, y, z); if it satisfied $ax^2 + y^2 - z^2 = 0$ before, now it satisfies $x^2 + y^2 - z^2 = 0$.

Suppose a is a non-square; then in terms of a fixed non-square λ, we can write $a = \lambda r^2$ and if we change the frame again, as above, we find that our conic has (projective) equation $\lambda x^2 + y^2 - z^2 = 0$. So:

Theorem 2.60. *If $K = GF(q)$, then in $\mathscr{A}_2(q)$ every ellipse is equivalent to every other.* $\square$

Corollary. *In $\mathscr{P}_2(q)$ the orthogonal group of the conic $\mathscr{C}$ has three orbits of lines: the tangent lines, the lines meeting $\mathscr{C}$ twice, and the lines not meeting $\mathscr{C}$.*

Proof. Certainly the orthogonal group of $\mathscr{C}$ is transitive on tangent lines, since it is transitive on the points of the conic. Exercise 2.47 implies that the lines not meeting $\mathscr{C}$ form another orbit. Finally, since the group is doubly transitive on the points of $\mathscr{C}$, it is transitive on the set of lines meeting $\mathscr{C}$ twice. $\square$

Exercise 2.49. Find the point orbits of the orthogonal group of $\mathscr{C}$, where $\mathscr{C}$ is a conic in $\mathscr{P}_2(K)$ for a finite field K.

Exercise 2.50. If K is the field of complex numbers, then show that there are no ellipses in $\mathscr{A}_2(K)$, while if K is the field of real numbers, then all ellipses in $\mathscr{A}_2(K)$ are equivalent.

10. Singer Groups

In this section we want to investigate a particular subgroup of the automorphism group of certain projective and affine planes (and in fact the reader should be able to see how this approach generalizes immediately to arbitrary dimensions). First, let K be a skewfield with the property that there exists another skewfield F containing K such that F is a three-dimensional left vector space over K (any skewfield is a vector space in a natural manner over any sub-skewfield). Since F is a three-dimensional left vector space over K we may construct the projective plane $\mathscr{P} = \mathscr{P}(F)$ in the way described in Section 2. The points of $\mathscr{P}$ are then the elements of F^*, modulo K^*, that is, the element x of F^* represents the same point of $\mathscr{P}$ as the element kx, where k is any element of K^*.

For each element f in F^*, define the mapping R_f of F onto F by $R_f : x \rightarrow xf$.

Lemma 2.61. *The mappings R_f, for $f \neq 0$, are non-singular linear transformations of the left vector space F over K.*

Proof. The non-singularity follows from the obvious fact that R_f must be one-to-one and onto, by its definition. To show that it is linear, we must verify $(x + y) R_f = x R_f + y R_f$, and $(kx) R_f = k(x R_f)$ for all x and y in F and all k in K. But these are trivial, when written out. []

Now clearly the mappings R_f form a group isomorphic to F^*. So it is a subgroup of $GL(F)$, and induces a subgroup Σ of $PGL(F)$.

Theorem 2.62. *The group Σ is transitive on the points of $\mathscr{P}$.*

Proof. Certainly for any two elements x and y of F^*, there is an element $f = x^{-1} y$ such that $x R_f = y$, and so the group of R_f is transitive on the elements of F^*, i.e. the non-zero vectors. So the group it induces on $\mathscr{P}$ must be transitive. []

What is the kernel of the representation of the group of all R_f as Σ? We can use the theory of Chapter I, or simply compute directly: a mapping R_f induces the identity in $\mathscr{P}$ if and only if for every x in F^*, the point $\langle x \rangle$ is fixed by R_f. That is, xf must lie in $\langle x \rangle$ for all x; this means that $xf = kx$, for some k in K^* so xfx^{-1} is in K^*, for all x in F^*. Finally, this is equivalent to demanding that f shall be in every subfield $x^{-1} K x$. So the kernel of the representation is the set of non-zero elements in the intersection of K with all its conjugates in F: in the special case of a field, we have:

Lemma 2.62. *If F is a field (so K is a field with a cubic extension field), then the group Σ is isomorphic to F^*/K^*.* []

In a projective geometry, a *Singer group* is a subgroup of the automorphism group which is regular on the points and hyperplanes. We have:

Theorem 2.64. *If K is a field which possesses a cubic extension field, then the projective plane $\mathscr{P}_2(K)$ possesses a Singer group.*

Proof. The group Σ is certainly transitive on points, and in the case that F is a field, its kernel is K^* itself. But the stabilizer of the point $\langle 1 \rangle$ is exactly the set of collineations induced by those R_f for which $\langle 1 R_f \rangle = \langle f \rangle = \langle 1 \rangle$. But $\langle f \rangle = \langle 1 \rangle$ if and only if f is in K, which implies that R_f induces the identity mapping in Σ. So Σ is regular on points.

To finish the theorem, we note that Σ is abelian, and hence the following three lemmas provide the proof.

Lemma 2.65. *If Γ is a non-trivial collineation group of a projective plane $\mathscr{P}$, and if Γ is abelian and is semi-regular on points, then Γ has at most one fixed line and is semi-regular on the non-fixed lines.*

Proof. Let m be any line of $\mathscr{P}$.

If $m^\alpha = m$, where $\alpha \neq 1$, then the stabilizer of m is non-trivial; since Γ is abelian, the stabilizer is normal and so fixes all the images of m. But two distinct images of m intersect in a point which is fixed by the stabilizer, and this contradicts the semi-regularity of Γ on points. So m must be fixed by Γ. Clearly the intersection of two lines fixed by Γ would be a point fixed by Γ, and this is impossible. So there are no other lines fixed, hence no other lines are fixed by a non-identity element of Γ.

Lemma 2.66. *As in Lemma 2.65, if Γ is also transitive on points (i.e. is regular) than Γ is semi-regular on lines.*

Proof. If Γ fixed a line, then the points on that line would have all their images under Γ on the line, so Γ could not be transitive on points

Lemma 2.67. *As in Lemma 2.65, if Γ is also transitive on points, then Γ is transitive (hence regular) on lines.*

Proof. Let P be a point and m a line of $\mathscr{P}$, and let $\Delta = \{\delta \in \Gamma \mid P^\delta$ on $m\}$. Suppose α is in Γ, $\alpha \neq 1$, so that $m^\alpha \neq m$; then the point where m meets m^α is a point P^β. So β and $\beta\alpha^{-1}$ are in Δ, hence $\beta = \delta_1$, $\beta\alpha^{-1} = \delta_2$, and so $\alpha = \delta_2^{-1}\delta_1$, i.e. we have shown that every non-identity element of Γ can be represented as $\delta_2^{-1}\delta_1$, for some pair of elements δ_1, δ_2 in Δ. Now let l be an arbitrary line of $\mathscr{P}$, and choose two points X, Y on l, $X \neq Y$. Then, by the transitivity of Γ, $X = P^\lambda$, $Y = P^\mu$, for some λ, μ in Γ where $\lambda \neq \mu$, and so $\lambda\mu^{-1} \neq 1$. Thus there are elements δ_1', δ_2' in Δ such that $\lambda\mu^{-1} = \delta_2'^{-1}\delta_1'$. Hence $\delta_1' = \lambda\mu^{-1}\delta_2'$, and so $P^{\lambda\mu^{-1}\delta_2'} = P^{\delta_1'}$ is on m, and hence P^λ is on $m^{\delta_2^{-1}\mu}$; P^{δ_2} is on m, so P is on $m^{\delta_2^{-1}}$ and P^μ is on $m^{\delta_2^{-1}\mu}$. So the line $m^{\delta_2^{-1}\mu}$ must be l, and thus Γ is transitive on lines.

Corollary. *If $K = GF(q)$ then the projective plane $\mathscr{P}_2(q)$ always possesses a cyclic Singer group.*

Proof. Since $GF(q^3) = F$ always exists and is a cubic extension of K, the Singer group exists. But the factor group F^*/K^* is cyclic, since F^* is cyclic.

*****Exercise 2.51.** Let K be (a) $GF(2)$, and (b) $GF(3)$. Choose irreducible cubics over K so as to construct the cubic extension field (a) $GF(8)$

and (b) $GF(27)$. Find a representation of Σ, the Singer group, in each case and in particular find the subset Δ in Σ, where Δ is the set used in the proof of Lemma 2.67.

Now in fact we can perform a similar "trick" with some affine planes. Instead of wanting a cubic extension, we will want a quadratic extension, and we will be able to construct a group Γ which fixes the origin $(0, 0)$ of the affine plane $\mathscr{A}_2(K)$ and is transitive (and regular, if the extension is a field) on the remaining points. The reader should be able to fill in the details, giving in particular:

Exercise 2.52. Let K be a field and F a quadratic extension field of K. Then the affine plane $\mathscr{A}_2(K)$ possesses a collineation group Γ which fixes the point $(0, 0)$ and is regular on the remaining points; furthermore, $\Gamma \cong F^*/K^*$, and Γ has two orbits of lines in $\mathscr{A}_2(K)$, one consisting of the lines through $(0, 0)$, the other of the remaining lines.

11. Appendix (Characteristic Two)

As fields with characteristic two are so different to other fields it is not surprising that the planes which they coordinatize also behave rather specially. At some places in this chapter we avoided fields of even characteristic, but the difference in the behaviour of polarities is so important that we now give a solution to Exercise 2.33.

We first note that for a field of characteristic two the J-ratio equation simplifies to

$$x^2(v A v') + y^2(w A w') = 0 . \tag{J_2}$$

Lemma 2.68. *If $\mathscr{P}$ is a projective plane over a field of characteristic two, and if X and Y are distinct absolute points, then every point on the line $X + Y$ is absolute.*

Proof. Let $X = \langle v \rangle$, $Y = \langle w \rangle$, from which we know that $v A v' = w A w' = 0$. But then the J-ratio equation (see (J_2) above) tells us that every point $\langle x v + y w \rangle$ is absolute as well. $\square$

Corollary. *If $\mathscr{P}$ is a projective plane over a field of characteristic two, then the set of absolute points of an orthogonal polarity is one of (a) the empty set, (b) a single point, (c) the points of a line.*

Proof. If there were more absolute points than all the points on a line, then the absolute points would include all the points on at least two distinct lines, and hence all the points of the plane. But the quadratic form (Lemma 2.30) for an orthogonal polarity, in the case of characteristic two, becomes $a_{11} x^2 + a_{22} y^2 + a_{33} z^2 = 0$. If this is to be satisfied for all x, y, z, then it is easy to see that $a_{11} = a_{22} = a_{33} = 0$. But now a trivial computation of the determinant of the matrix $A = (a_{ij})$ (which is symmetric) shows that A is singular, which is a contradiction. $\square$

Now we give some examples:

Example 1. *Let F be a field of characteristic two and K its subfield of squares. Suppose that the dimension of F over K is greater than two (e.g., let $F = Z_2(u, v)$, the field of all rational forms in two variables over the field of two elements); then if b is a fixed non-square in F, there must be a non-square a in F such that $a = bx^2 + y^2$ has no solution for x and y in F. Let*

$$A = \begin{bmatrix} a & 0 & 0 \\ 0 & b & 0 \\ 0 & 0 & 1 \end{bmatrix};$$

then the polarity β defined by A has no absolute points.

Example 2. *Let K be a field of characteristic two in which not all the elements are squares, and let $a = b$ be a non-square. Using the same matrix A as in the example immediately above, the polarity β will have exactly one absolute point, namely $\langle (1, 1, 0) \rangle$.*

But in contrast to the two examples above, we have:

Theorem 2.69. *If K is a field of characteristic two in which every element is a square, then the absolute points of an orthogonal polarity are always exactly the points of a line.*

Proof. Using the quadratic form again, we have $a_{11}x^2 + a_{22}y^2 + a_{33}z^2 = 0$. But for each a_{ii} there is an element b_i in K such that $b_i^2 = a_{ii}$, and so we may write the quadratic form as $(b_1 x + b_2 y + b_3 z)^2 = 0$, which is satisfied exactly when $b_1 x + b_2 y + b_3 z = 0$, and this is the equation of a line (namely, $\langle (b_1, b_2, b_3)' \rangle$). $\square$

Corollary. *In the projective plane $\mathscr{P}_2(2^t)$, the absolute points of an orthogonal polarity are exactly the points of some line.*

Proof. By Result 1.4. $\square$

References

Any results on matrices which are quoted in this chapter will be found in [3]. The reader who is interested in further reading on projective geometry should find [1], [2] or [4] both helpful and interesting.

1. Artin, E.: Geometric algebra. New York: Interscience 1957.
2. Baer, R.: Linear algebra and projective geometry. New York: Academic Press 1952.
3. Birkhoff, G., MacLane, S.: A survey of modern algebra. London-New York: Macmillan 1965.
4. Gruenberg, K., Weir, J.: Linear geometry. New York-London: Van Nostrand 1967.

III. Elementary Properties of Projective and Affine Planes

1. Introduction

For the rest of the book, we shall be concerned only with projective planes and affine planes; but not just the planes $\mathscr{P}(V)$ and $\mathscr{A}(V)$ constructed in Chapter II, for some three dimensional vector space V over a skewfield. Instead, we shall make an abstract, incidence-type definition of projective plane, and later of affine plane. This definition will include all the planes of Chapter II, but will include many other examples as well. If K is a skewfield and V is a three-dimensional left or right vector space over K, then the plane $\mathscr{P}(V)$ will be called the *desarguesian projective plane* over K, or simply a *desarguesian* projective plane (the reason for this will become clear later; see the summary of Chapter VI.). If K is a field, we often say that $\mathscr{P}(V)$ is a *pappian* projective plane. In Chapter II we treated left vector spaces as the most "natural" (since then mappings appeared naturally on the right, the vectors most commonly used were row vectors, and so on), but clearly right vector spaces are just as good; in the case of fields (i.e. pappian planes) the two concepts lead to isomorphic planes anyway. But for the rest of the book, whenever we refer to "the projective plane over the skewfield (or field) K", or to the "desarguesian projective plane $\mathscr{P}_2(K)$", etc., *we shall always mean the one constructed over the right vector space over K.* The real effect of this will be that, once a basis is chosen, points will be (projectively) $\langle (x, y, z)' \rangle$ and lines will be $\langle (a, b, c) \rangle$ with incidence given by $ax + by + cz = 0$. Affinely we still have points (x, y) but lines will have equations either $x = k$ or $y = mx + k$ (or $mx + y = k$ or $mx + y + k = 0$). In other words, slopes appear on the left. Recall that we write $\mathscr{P}_2(q)$ for the plane $\mathscr{P}_2(GF(q))$, and $\mathscr{A}_2(q)$ for $\mathscr{A}_2(GF(q))$.

In this chapter we begin the study of these more "arbitrary" or general projective and affine planes. While no use is made of any results in Chapter II we frequently refer back to that chapter for motivation and examples. However a reader who has not read the earlier chapter will not be at any serious disadvantage.

The basic combinatorial properties of projective and affine planes are established, including the celebrated Bruck-Ryser Theorem on the non-existence of projective planes with certain orders. The concept of isomorphism between planes is introduced and there is a lengthy discussion on the relation between projective and affine planes. Finally incidence matrices are introduced and some of their elementary properties are established.

2. Basic Concepts

From now on a *projective plane* $\mathcal{P}$ is a set of points and lines, called the elements of $\mathcal{P}$, together with an incidence relation between the points and lines such that
  (i) Any two distinct points are incident with a unique line.
  (ii) Any two distinct lines are incident with a unique point.
  (iii) There exist four points such that no three are incident with one line.

Any set of points which satisfies conditions (i) and (ii) is called a *closed configuration*, and any closed configuration which does not satisfy condition (iii) is called a *degenerate plane*.

Exercise 3.1. List all degenerate planes.

The following are typical expressions which will be used to express the fact that a point P is incident with a line l: PIl, lIP, P is on l, l contains P or l passes through P. Often it will be convenient to identify a line l with the set of points incident with it. In this situation we may also write $P \in l$.

If l is a line containing two distinct points A and B, we say that l joins A to B and write $l = AB$. Similarly if P is a point on two distinct lines l, m we write $P = lm$ or $P = l \cap m$ and call P the intersection of l with m.

Exercise 3.2. If $\mathcal{B}$ is any set of points and lines with an incidence relation such that any two distinct points are on a unique common line, show that two lines intersect in at most one point.

Any set of points incident with a common line are said to be *collinear*, and any set of lines passing through a common point are *concurrent*. An ordered set of three distinct non-collinear points A, B, C, together with the lines BC, CA, AB, is called a *triangle*. Any ordered set of four points, no three of which are collinear is called a *quadrangle*. An ordered set of four lines such that no three are concurrent is called a *quadrilateral*. (But often the ordering will be irrelevant.)

In the definition of a projective plane, axiom (iii) suggests that the points assume a special role. This is not in fact the case, as is shown by the following lemma.

Lemma 3.1. *Any projective plane contains a quadrilateral.*

Proof. From axiom (iii) every projective plane contains a quadrangle. Let A, B, C, D be any four distinct points which form a quadrangle. Consider the lines AB, BC, CD, DA and assume that three of them are concurrent. Without any loss of generality we make take these three lines as AB, AD and CD. Let T be the point common to these lines.

Since any two lines have a unique common point and since A is on both AB and AD we must have $A = T$. Similarly, by considering the lines AD, CD, we have $T = D$. This contradicts the choice of A, B, C, D as distinct. Thus AB, BC, CD, DA form a quadrilateral. $\square$

If $\mathscr{P}$ is any projective plane, let $\mathscr{P}^*$ be a set of points and lines together with an incidence relation such that the points (lines) of $\mathscr{P}^*$ are the lines (points) of $\mathscr{P}$ and two elements are incident in $\mathscr{P}^*$ if and only if they are incident in $\mathscr{P}$. As an immediate consequence of Lemma 3.1, $\mathscr{P}^*$ is a projective plane which we call the *dual plane* of $\mathscr{P}$. Clearly $(\mathscr{P}^*)^* = \mathscr{P}$ so that every plane is a dual plane. This observation establishes one of the most important results in the study of projective planes:

Theorem 3.2 (The Principle of Duality). *Let A be any theorem about projective planes. If A^* is the statement obtained by interchanging the words "points" and "lines", then A^* is a theorem about dual planes. Hence A^* is also a theorem about projective planes.* $\square$

Clearly, from the definition, a projective plane must contain at least four points and, by Lemma 3.1, four lines. In fact, as we shall soon prove, each projective plane must have at least seven points and lines. Before proving this, however, we give an example of a projective plane $\mathscr{P}$ which has exactly seven points and lines. If we label the points of $\mathscr{P}$ as $P_1, \dots, P_7$, then the lines of $\mathscr{P}$ are:

$$l_1 = \{P_1, P_2, P_4\}, \qquad l_2 = \{P_2, P_3, P_5\}, \qquad l_3 = \{P_3, P_4, P_6\},$$
$$l_4 = \{P_4, P_5, P_7\}, \qquad l_5 = \{P_5, P_6, P_1\}, \qquad l_6 = \{P_6, P_7, P_2\},$$
$$l_7 = \{P_7, P_1, P_3\}.$$

Incidence in $\mathscr{P}$ is given by $P_i I l_j$ if and only if $P_i \in l_j$. Straightforward verification of the axioms shows that $\mathscr{P}$ is a projective plane.

Exercise 3.3. Show that $\mathscr{P}_2(K)$, for any field K, is a projective plane according to our new definition.

Before we can prove our claim that we have exhibited the smallest possible projective plane, we prove a simple lemma.

Lemma 3.3. *Let l and m be any two distinct lines of a projective plane $\mathscr{P}$. Then there is a point X of $\mathscr{P}$ such that X is not on either l or m.*

Proof. Suppose the lemma is false.

Let A, B, C, D be any quadrangle of $\mathscr{P}$. By assumption each of them is incident with either l or m and thus, since no three of them are collinear, two of them $(A, B$ say), belong to l and the others belong to m. Let $X = AC \cap BD$.

Suppose X is on l. Since l passes through A and B, $l = AB$. Thus X is on AB. Since A, B, C are not collinear, $AB \neq AC$ and thus, since X is on AC, X must be the unique intersection of the lines AB and AC. But A

is also incident with both these lines. Thus $A = X$. However, this is impossible since X is on the line BD and A, B, D are not collinear. Hence X is not on l.

Similarly X is not on m and the lemma is proved. $\square$

Theorem 3.4. *Let $\mathcal{P}$ be a projective plane. If l and m are any two lines of $\mathcal{P}$ there is a one-to-one correspondence between the points of l and m. Furthermore, if l is any line and P is any point of $\mathcal{P}$ then there is a one-to-one correspondence between the points of l and the lines through P.*

Proof. Let l and m be any two lines of $\mathcal{P}$. By Lemma 3.2 there is a point O not on either l or m. For any point X of l define a point X^{α} of m by $X^{\alpha} = OX \cap m$. Since, given a point X, OX is a unique line and any two lines intersect in a unique point, X^{α} is uniquely determined by X. For any point Y on m, $Y = Z^{\alpha}$ where $Z = OY \cap l$. Thus α is a one-to-one correspondence between the points of l and the points of m.

Let P be any point and l any line of $\mathcal{P}$. There are two possibilities.

Case (a) P is not on l

For any X on l, define $X^{\alpha} = PX$. Clearly α is a one-to-one correspondence between the points of l and the lines through P.

Case (b) P is on l

Since $\mathcal{P}$ contains a quadrilateral there is a line m such that P is not on m. Case (a) gives a one-to-one correspondence between the lines through P and the points of m. By the first part of the theorem there is a one-to-one correspondence between the points of m and the points of l. The composition of these two correspondences is the desired one. $\square$

As an immediate corollary to Theorem 3.4 we see that if one line of a projective plane $\mathcal{P}$ contains only a finite number of points, then the number of points on any line is finite. A projective plane $\mathcal{P}$ with this property is called *finite*.

Theorem 3.5. *Let $\mathcal{P}$ be a finite projective plane. There exists a positive integer $n \geq 2$, such that*

   (i) *each line contains exactly $n + 1$ points,*
   (ii) *each point is on exactly $n + 1$ lines,*
   (iii) *$\mathcal{P}$ contains $n^2 + n + 1$ points and $n^2 + n + 1$ lines.*

Proof. (i) and (ii) are immediate corollaries of Theorem 3.4.

Let Q be any point of $\mathcal{P}$. By (ii) there are $n + 1$ lines through Q. Since any two points of $\mathcal{P}$ are joined by a unique line, every point of $\mathcal{P}$ except Q is on exactly one of these $n + 1$ lines. By (i) each of these lines contains n points distinct from Q. Thus the total number of points is $(n + 1)n + 1 = n^2 + n + 1$.

The dual argument proves the similar statement about lines.

In order to show $n \geq 2$ it is sufficient to observe that if A, B, C, D is any quadrangle of $\mathscr{P}$ then the point $AB \cap CD$ is on AB and is distinct from A and B. □

If $\mathscr{P}$ is any finite projective plane then the integer n of Theorem 3.5 is called the *order* of $\mathscr{P}$. If no such integer exists then $\mathscr{P}$ is said to have *infinite order* or, alternatively, to be an *infinite plane*.

Since $n \geq 2$, the number of points in a projective plane must be at least $2^2 + 2 + 1 = 7$. Thus we have proved our claim that the plane we exhibited had the smallest possible order. The following exercise shows that given any prime power there is a finite projective plane with that order.

Exercise 3.4. Show that the order of $\mathscr{P}_2(p^a)$ is p^a.

A very interesting problem immediately suggests itself: given a composite integer n does there exist a projective plane of order n? So far all the known examples of finite projective planes have prime power order. However the only result showing the non-existence of any finite projective plane with a given order is the following remarkable theorem.

Theorem 3.6 (The Bruck-Ryser Theorem). *If* $n \equiv 1$ *or* 2 (mod 4) *there cannot be a projective plane of order* n *unless* n *can be expressed as a sum of two integral squares*[1].

Proof. The proof of this theorem is not difficult but uses some elementary results and techniques from number theory which are superfluous to the rest of this book. For completeness a proof is given in the appendix at the end of this chapter. But failure to read the proof will not hinder the reader in understanding the rest of this chapter. □

The smallest composite integer not covered by the Bruck-Ryser Theorem is 10, and it is still unknown whether or not a plane of order 10 exists.

So far we appear to have exhibited two planes of order 2. One was $\mathscr{P}_2(2)$ and for the other we wrote down the seven points and gave the lines as subsets of the points. However, it is not clear whether the two planes are essentially "different" or whether we have merely found two ways of representing the "same" plane. But, of course, before we can discuss such problems we must, as in most other branches of mathematics, introduce the concept of isomorphism. Two projective planes $\mathscr{P}$ and $\mathscr{P}'$ are said to be *isomorphic* if there exists a one-to-one mapping θ from the points of $\mathscr{P}$ onto the points of $\mathscr{P}'$ and the lines of $\mathscr{P}$ onto the lines of $\mathscr{P}'$ such that, for any point A and line l of $\mathscr{P}$, A is on l if and only if A^θ is on l^θ. If $\mathscr{P}$ is isomorphic to $\mathscr{P}'$ we write $\mathscr{P} \cong \mathscr{P}'$. The mapping θ is called an *isomorphism*.

---

[1] It is well known (see for instance [1]) that an integer n can be expressed as a sum of two integral squares if and only if the square free part of n has no prime divisor $\equiv 3$ (mod 4). This gives an alternative formulation of the theorem.

Exercise 3.5. If $\mathscr{P}$ is any projective plane of order 2 show that $\mathscr{P} \cong \mathscr{P}_2(2)$.

Note that since an isomorphism is a one-to-one correspondence between the elements of two planes, any pair of isomorphic projective planes must have the same order.

3. Subplanes

A *subplane* $\mathscr{P}_0$ of $\mathscr{P}$ is a subset of the elements of $\mathscr{P}$ which form a projective plane having the same incidence relation as $\mathscr{P}$.

With this definition $\mathscr{P}$ is a subplane of itself. A subplane $\mathscr{P}_0$ of $\mathscr{P}$ is called *proper* if $\mathscr{P}_0 \neq \mathscr{P}$.

Note that a proper subplane of $\mathscr{P}_2(K)$ is an incomplete sub-geometry. Furthermore $\mathscr{P}_2(p^a)$ has a proper subplane of order p^b for all integers b such that $b\,|\,a$ (see Result 1.5).

If $\mathscr{P}$ is any finite projective plane then the possible orders for subplanes of $\mathscr{P}$ are restricted by the following theorem.

Theorem 3.7 (Bruck). *Let $\mathscr{P}$ be a finite projective plane of order n with a proper subplane $\mathscr{P}_0$ of order m. Then either $n = m^2$ or $n \geq m^2 + m$.*

Proof. Let l be any line of $\mathscr{P}_0$. Then l contains $m + 1$ points of $\mathscr{P}_0$ and, hence, $n + 1 - (m + 1) = n - m$ points of $\mathscr{P} \backslash \mathscr{P}_0$. Since any two lines of $\mathscr{P}_0$ intersect in a point of $\mathscr{P}_0$ and since $\mathscr{P}_0$ has $m^2 + m + 1$ lines, there are $(m^2 + m + 1)(n - m)$ points of $\mathscr{P} \backslash \mathscr{P}_0$ which are incident with a line of $\mathscr{P}_0$. Thus there are at least $m^2 + m + 1 + (m^2 + m + 1)(n - m)$ points in $\mathscr{P}$. This gives:

$$n^2 + n + 1 \geq (m^2 + m + 1)(n - m) + m^2 + m + 1 \,,$$

$$n^2 + n + 1 \geq m^2 n - m^3 + mn + n + 1 \,,$$

$$0 \geq m^2 n - m^3 + mn - n^2 \,,$$

$$0 \geq (m^2 - n)(n - m) \,.$$

But $n - m > 0$. Thus $n \geq m^2$.

Note that if $n = m^2$, then $n^2 + n + 1 = m^2 + m + 1 + (m^2 + m + 1)(n - m)$; i.e. every point of $\mathscr{P}$ is on a line of $\mathscr{P}_0$.

Suppose that $n \neq m^2$. Then there is a point A of $\mathscr{P}$ such that A is not incident with any line of $\mathscr{P}_0$. Hence every line through A contains at most one point of $\mathscr{P}_0$. Thus, since every point of $\mathscr{P}_0$ is on a line through A, the total number of lines through A is at least as great as the total number of points in $\mathscr{P}_0$. This gives $n + 1 \geq m^2 + m + 1$ or $n \geq m^2 + m$. □

There is no instance known of a finite projective plane of order n having a subplane of order m such that $n = m^2 + m$. So it is possible that the conclusion of Theorem 3.7 may not be the best possible.

A subset $\mathscr{C}$ of the points and lines of a projective plane $\mathscr{P}$ is called a *Baer subset* (or *dense subset*) if every element of $\mathscr{P}$ is incident with an element of $\mathscr{C}$.

If $\mathscr{C}$ is a subplane then $\mathscr{C}$ is called a *Baer subplane*. From the proof of Theorem 3.5 we have;

Theorem 3.8. *Let $\mathscr{P}$ be a finite projective plane of order n with a subplane $\mathscr{P}_0$ of order m. Then $\mathscr{P}_0$ is a Baer subplane if and only if $n = m^2$.* $\square$

Exercise 3.6. List all Baer subsets which are also closed configurations. (We call such subsets *closed Baer subsets*.)

4. Affine Planes

If $\mathscr{P}$ is a projective plane and l is any line of $\mathscr{P}$ then let $\mathscr{P}^l$ be the set of points and lines of $\mathscr{P}$ obtained by deleting the line l and all the points incident with it. $\mathscr{P}^l$ has a natural incidence, namely the incidence of $\mathscr{P}$. Clearly any two points of $\mathscr{P}^l$ are incident with a unique line of $\mathscr{P}^l$. However it is no longer true that any two lines of $\mathscr{P}^l$ intersect in a point of $\mathscr{P}^l$. This is readily seen as follows; let X be any point of l and let m_1, m_2 be any pair of distinct lines through X such that neither is l. Then m_1, m_2 are lines of $\mathscr{P}^l$, and m_1 and m_2 do not intersect in $\mathscr{P}^l$. But clearly if A is any point of $\mathscr{P}^l$ not on m_1, and if m is the line joining A to X in $\mathscr{P}$, then m is the only line of $\mathscr{P}^l$ through A with the property that it does not intersect m_1.

We shall now define an affine plane to be a set of points and lines with an incidence relation having the properties just established for $\mathscr{P}^l$, and will then show that any affine plane is of the form $\mathscr{P}^l$ for some projective plane $\mathscr{P}$ and some line l of $\mathscr{P}$.

An *affine plane* $\mathscr{A}$ is a set of points and lines together with an incidence relation between the points and lines such that

(i) any two distinct points lie on a unique line,

(ii) given any line l and any point P not on l there is a unique line m such that P is on m, and l and m have no common point,

(iii) there are three non-collinear points.

The preceding discussion shows that for any projective plane $\mathscr{P}$ and any line l of $\mathscr{P}$, $\mathscr{P}^l$ is an affine plane. (The existence of three non-collinear points follows immediately from the existence of a quadrangle and Lemma 3.3.)

In any affine plane $\mathscr{A}$ there is a natural way to define a "parallelism" between the lines. We will call two lines l, m *parallel*, and write $l \parallel m$, if $l = m$ or l and m do not intersect.

Lemma 3.9. *In any affine plane parallelism is an equivalence relationship. Furthermore each point is on exactly one line from each parallel class.*

Proof. Clearly the reflexive and symmetric laws hold, so we only have to show transitivity. Let l, m, h be any three lines such that $l \parallel m$ and $m \parallel h$. If $l = m, m = h$ or $h = l$ then it follows trivially that $l \parallel h$. Thus we may assume $l \neq m \neq h \neq l$.

Suppose $l \nparallel h$. Then there is a point A on both l and h. But now we have two distinct lines through A and not intersecting m. Since this contradicts axiom (ii) of the definition of an affine plane we have shown $l \parallel h$.

The second claim of the lemma follows immediately from axiom (ii) in the definition of an affine plane. []

Isomorphism between affine planes is defined in a similar way to isomorphism between projective planes: thus two affine planes $\mathscr{A}$ and $\mathscr{A}'$ are isomorphic if there exists a one-to-one mapping θ from the points of $\mathscr{A}$ onto the points of $\mathscr{A}'$ and from the lines of $\mathscr{A}$ onto the lines of $\mathscr{A}'$ such that a point A is on a line l if and only if A^θ is on l^θ.

Exercise 3.7. If θ is an isomorphism from an affine plane $\mathscr{A}$ onto an affine plane $\mathscr{A}'$ show that, for any two lines l, m of $\mathscr{A}$, $l \parallel m$ if and only if $l^\theta \parallel m^\theta$.

Exercise 3.8. Let $\mathscr{P} = \mathscr{P}_2(K)$ for any field K and let l be the line $z = 0$. Show $\mathscr{P}^l \cong \mathscr{A}_2(K)$.

We have already seen that given a projective plane $\mathscr{P}$ and a line l then $\mathscr{P}^l$ is an affine plane. It is worth pointing out that if l and m are distinct lines of $\mathscr{P}$ then $\mathscr{P}^l$ need not be isomorphic to $\mathscr{P}^m$ (see Lemma 3.11). However, we have

Theorem 3.10. *Let $\mathscr{A}$ be any affine plane. Then there is, up to isomorphism, a unique projective plane $\mathscr{P}$ such that $\mathscr{A} = \mathscr{P}^l$ for some line l of $\mathscr{P}$.*

Proof. We first construct a projective plane $\mathscr{P}$ with a line l_∞ such that $\mathscr{A} = \mathscr{P}^{l_\infty}$.

The points of $\mathscr{P}$ will be the points of $\mathscr{A}$, which we denote by upper case Latin letters, and the parallel classes of $\mathscr{A}$, which we denote by asterisked upper case Latin letters. The lines of $\mathscr{P}$ are the lines of $\mathscr{A}$, which we denote by lower case Latin letters, plus a new line l_∞.

Incidence in $\mathscr{P}$ is defined as follows:

(i) P is on l in $\mathscr{P}$ if and only if P is on l in $\mathscr{A}$;

(ii) P^* is on l in $\mathscr{P}$ if and only if $l \in P^*$ in $\mathscr{A}$ (recall that P^* is a parallel class of $\mathscr{A}$);

(iii) P^* is on l_∞ for all "parallel class" points P^*, and there is no point of $\mathscr{A}$ on l_∞.

Clearly, by construction, $\mathscr{A} = \mathscr{P}^{l_\infty}$ but we must still verify that $\mathscr{P}$ is a projective plane.

Any two points A, B of $\mathscr{P}$ are joined by the line AB of $\mathscr{A}$ which is also a line of $\mathscr{P}$. Any two points A, B^* are joined in $\mathscr{P}$ by the unique line of $\mathscr{A}$ which belongs to the parallel class B^* and contains A. (Such a line exists by Lemma 3.9.) Any two points A^*, B^* are joined by l_∞. Thus any two points of $\mathscr{P}$ are on a unique line of $\mathscr{P}$. (Why unique?)

Clearly the line l_∞ of $\mathscr{P}$ intersects every other line of $\mathscr{P}$ in a unique point. Let l and m be any two lines of $\mathscr{P}$. If l and m intersect in $\mathscr{A}$ then they certainly intersect in $\mathscr{P}$. (Furthermore this intersection is unique by Exercise 3.2.) Suppose l does not intersect m in $\mathscr{A}$. Then $l \parallel m$ in $\mathscr{A}$ so that l and m both belong to the same parallel class, A^* say, of $\mathscr{A}$. Thus l and m intersect in the point A^* of $\mathscr{P}$. Hence any two lines of $\mathscr{P}$ intersect in a unique point of $\mathscr{P}$.

Since $\mathscr{A}$ is an affine plane, $\mathscr{A}$ contains three non-collinear points X, Y, Z. Let $XY \cap l_\infty = A^*$ and $YZ \cap l_\infty = B^*$ in $\mathscr{P}$. Since XY is not parallel to YZ in $\mathscr{A}$, $A^* \ne B^*$ and X, Z, A^*, B^* form a quadrangle in $\mathscr{P}$. This proves that $\mathscr{P}$ is a projective plane.

Suppose now that $\mathscr{A} = \mathscr{P}^l = \mathscr{P}'^{l'}$. Then the identity mapping is an isomorphism from $\mathscr{P}^l$ to $\mathscr{P}'^{l'}$. We conclude the proof of Theorem 3.10 by proving the following lemma.

Lemma 3.11. *Two affine planes $\mathscr{P}^l, \mathscr{P}'^{l'}$ are isomorphic if and only if there is an isomorphism α from the projective plane $\mathscr{P}$ onto the projective plane $\mathscr{P}'$ with $l^\alpha = l'$.*

Proof. Suppose θ is an isomorphism from $\mathscr{P}^l$ onto $\mathscr{P}'^{l'}$. Define α by $X^\alpha = X^\theta$ for all X not on l, and for any Y on l pick any line m through Y different from l and define $Y^\alpha = m^\theta \cap l'$. The action of α on the lines of $\mathscr{P}$ is given by $h^\alpha = h^\theta$ for all $h \ne l$ and $l^\alpha = l'$.

In order to show that α is an isomorphism it is only necessary to show that, for Y on l, Y^α is independent of the choice of m. But this is clearly so, since θ preserves parallel classes, by Exercise 3.7.

Now suppose that α is an isomorphism between two projective planes $\mathscr{P}$, $\mathscr{P}'$ such that $l^\alpha = l'$. Clearly α induces a one-to-one mapping from the points (lines) of $\mathscr{P}^l$ onto the points (lines) of $\mathscr{P}'^{l'}$. Since α is an isomorphism from $\mathscr{P}$ onto $\mathscr{P}'$, this induced mapping must preserve incidence. Thus we have $\mathscr{P}^l \cong \mathscr{P}'^{l'}$. $\square$

If $\mathscr{A} = \mathscr{P}^l$ then we say that $\mathscr{A}$ is *the affine plane associated with $\mathscr{P}$ and l* or, sometimes, that $\mathscr{A}$ is *an affine plane associated with $\mathscr{P}$*. Once again we emphasize that non-isomorphic affine planes may be associated with the same projective plane. We also say that $\mathscr{P}$ is associated with $\mathscr{A}$ and, as the last theorem showed, this association is unique, up to isomorphism. This fact enables us to define the *order* of an affine plane as the order of the projective plane associated with it.

Theorem 3.12. *Let $\mathscr{A}$ be a finite affine plane of order n. Then*

(i) *$\mathscr{A}$ has n^2 points and $n^2 + n$ lines,*

(ii) *every line of $\mathscr{A}$ contains n points,*

(iii) *each point of $\mathscr{A}$ is on $n + 1$ lines,*

(iv) *each parallel class of $\mathscr{A}$ has n lines, and there are $n + 1$ parallel classes.*

Proof. We leave the proof as an exercise. $\square$

If $\mathscr{A} = \mathscr{P}^l$ then we often refer to the points and lines of $\mathscr{A}$ as its *affine elements*. We then call l the *special line* of $\mathscr{A}$ and the points of l are its *special points*.

If $\mathscr{A} = \mathscr{P}^l$ for some projective plane $\mathscr{P}$ and if $\mathscr{Q}$ is a subplane of $\mathscr{P}$ then there are three possibilities for the intersection of l with $\mathscr{Q}$; (a) l a line of $\mathscr{Q}$, (b) l contains exactly one point of $\mathscr{Q}$ or (c) l contains no point of $\mathscr{Q}$. If case (a) occurs then $\mathscr{Q}^l$ is an affine plane whose affine elements are elements of $\mathscr{A}$ and whose special points are special points of $\mathscr{A}$. In this case we say that $\mathscr{Q}^l$ is an *affine subplane*, or *subplane*, of $\mathscr{A}$. In case (c) the elements of $\mathscr{Q}$ are all special elements of $\mathscr{A}$ and we say that $\mathscr{Q}$ is a *projective subplane* of $\mathscr{A}$. Case (b) will be of no interest to us in this book and we do not give it a name.

5. Incidence Matrices

Let $\mathscr{B}$ be a finite plane of order n (projective or affine). Let $P_1, P_2, ..., P_v$ be a labelling of the points and $l_1, l_2, ..., l_b$ be a labelling of the lines. (By Theorems 3.5 and 3.12, $v = b = n^2 + n + 1$ if $\mathscr{B}$ is projective and $v = n^2$, $b = n^2 + n$, if $\mathscr{B}$ is affine.) An *incidence matrix* A of $\mathscr{B}$ is a $v \times b$ matrix of zeros and ones such that $a_{ij} = 1$ if and only if P_i is on l_j.

Example. With the labelling of Section 2 the incidence matrix of the seven-point plane is

$$\begin{pmatrix} 1 & 0 & 0 & 0 & 1 & 0 & 1 \\ 1 & 1 & 0 & 0 & 0 & 1 & 0 \\ 0 & 1 & 1 & 0 & 0 & 0 & 1 \\ 1 & 0 & 1 & 1 & 0 & 0 & 0 \\ 0 & 1 & 0 & 1 & 1 & 0 & 0 \\ 0 & 0 & 1 & 0 & 1 & 1 & 0 \\ 0 & 0 & 0 & 1 & 0 & 1 & 1 \end{pmatrix}.$$

(Note that the incidence matrix is definitely not determined uniquely by the plane. It depends on the labelling of the elements.)

Incidence matrices have some interesting elementary properties.

Theorem 3.13. *Let A be an incidence matrix of a finite projective plane of order n. Then $AA' = nI_v + J_v$ where I_v is the v by v identity matrix and J_v is the v by v matrix with a 1 in every position.*

Proof. Let $AA' = (b_{ij})$.

Consider first the diagonal terms b_{ii}. This entry is the scalar product of the ith row of A with itself. Thus it is the sum of the non-zero entries of the ith row of A. But the number of non-zero entries in the ith row of A is equal to the number of lines through P_i. Thus, by Theorem 3.5, $b_{ii} = n + 1$ for $i = 1, 2, ..., n^2 + n + 1$.

Similarly b_{ij} $(i \neq j)$ is the scalar product of the ith row of A with the jth row of A. This is equal to the number of values of k such that $a_{ik} = a_{jk} = 1$. But $a_{ik} = 1$ if and only if P_i is on l_k and $a_{jk} = 1$ if and only if P_j is on l_k. Thus since P_i and P_j determine a unique line, $b_{ij} = 1$ for $i \neq j$.

This proves the theorem. $\square$

We now compute $\det(AA')$.

$$\det(AA') = \begin{vmatrix} n+1 & 1 & \ldots\ldots\ldots\ldots & 1 \\ 1 & n+1 & & \vdots \\ \vdots & & \ddots & \vdots \\ 1 & 1 & & n+1 \end{vmatrix} = \det(nI + J).$$

Subtracting the first row from each of the others gives

$$\begin{vmatrix} n+1 & 1 & \ldots\ldots\ldots\ldots & 1 \\ -n & n & 0\ldots\ldots\ldots & 0 \\ -n & 0 & n & \\ \vdots & \vdots & & \ddots & \\ & & & & 0 \\ -n & 0\ldots\ldots\ldots & 0 & n \end{vmatrix}$$

Thus, by adding each column to the first, we get:

$$\det(AA') = \begin{vmatrix} (n+1)^2 & 1 & \ldots\ldots\ldots\ldots & 1 \\ 0 & n & 0\ldots\ldots\ldots & 0 \\ \vdots & 0 & \ddots & \\ \vdots & \vdots & & \ddots & \\ & & & & 0 \\ 0 & 0\ldots\ldots\ldots & 0 & n \end{vmatrix} = (n+1)^2 n^{n^2+n}.$$

Thus we have shown

Theorem 3.14. *If A is an incidence matrix of a finite projective plane of order n then A is non-singular, over the rational field.*

Exercise 3.9. If $\mathscr{B}$ is a set of $n^2 + n + 1$ points and an unknown number of lines such that any two distinct points are on a unique line and each line contains $n + 1$ points, show that each point is on $n + 1$ lines.

Exercise 3.10. Show that a set of points satisfying the conditions of Exercise 3.9 with $n \geq 2$ is a projective plane. (Hint: assume $\mathscr{B}$ has two non-intersecting lines and show that any point on one of these lines must lie on at least $n + 2$ lines.)

Exercise 3.11. If $n \geq 2$ and $\mathcal{B}$ is a set of $n^2 + n + 1$ points and $n^2 + n + 1$ lines such that any line is incident with $n + 1$ points and any two distinct lines have a unique common point, show that $\mathcal{B}$ is a projective plane.

Exercise 3.12. Let A be an incidence matrix for a set of points and lines satisfying the conditions of Exercise 3.9. Show that $AA' = nI + J$ and $AJ = JA = (n + 1)J$. Hence show that there exist x, y such that $A^{-1} = xA' + yJ$. Now use this to show $AA' = A'A$ (i.e. that A is *normal*) and deduce that A is an incidence matrix of a projective plane for $n \geq 2$. (Thus we have an alternative proof of Exercise 3.10.)

*Exercise 3.13. Find the eigenvalues of $nI_v + J_v$ where $v = n^2 + n + 1$.

Exercise 3.14. If $n \geq 2$ and $\mathcal{B}$ is a set of n^2 points and an unknown number of lines such that any two points are on a unique line and each line contains exactly n points, show that $\mathcal{B}$ is an affine plane.

Exercise 3.15. Let A be the incidence matrix of an affine plane of order n. Show that $AA' = nI_{n^2} + J_{n^2}$.

Exercise 3.16. If $\mathcal{P} = \mathcal{P}_2(K)$ and if $\mathcal{P}^*$ is the dual of $\mathcal{P}$, show that $\mathcal{P} \cong \mathcal{P}^*$.

*Exercise 3.17. If $\mathcal{P} = \mathcal{P}_2(K)$, where K is the field of complex numbers, find nine points and twelve lines of $\mathcal{P}$ which form an affine plane of order 3.

6. Appendix (Proof of the Bruck-Ryser Theorem)

We first state two number theoretic results. For proofs the reader should consult [1].

Every positive integer is the sum of four integral squares.   (1)

If an integer is a sum of two rational squares then it is the sum of two integral squares.   (2)

We shall also use the following elementary identity.

$$(a^2 + b^2 + c^2 + d^2)(x^2 + y^2 + z^2 + w^2)$$
$$= (ax - by - cz - dw)^2 + (bx + ay - dz + cw)^2 \qquad (3)$$
$$+ (cx + dy + az - bw)^2 + (dx - cy + bz + aw)^2 .$$

Let $\mathcal{P}$ be a finite plane of order n and let $v = n^2 + n + 1$. Let $P_1, P_2, \ldots, P_v$ be a labelling of the points of $\mathcal{P}$ and $l_1, l_2, \ldots, l_v$ be a labelling of the lines.

Let $x_1, x_2, \ldots, x_v$ be v indeterminates and for $i = 1, 2, \ldots, v$ write $L_i = \Sigma x_j$ where, for a given i, the sum is taken over all those j for which P_j is on l_i.

Then $\displaystyle\sum_{i=1}^{v} L_i^2 = 2 \sum_{i \neq j} x_i x_j + (n+1) \sum_{i=1}^{v} x_i^2$.

$$\sum_{i=1}^{v} L_i^2 = n \sum_{i=1}^{v} x_i^2 + \left(\sum_{i=1}^{v} x_i \right)^2. \tag{4}$$

Note that (4) is an identity satisfied by whatever values we wish to assign the x_i. Let T denote $\displaystyle\sum_{i=1}^{v} x_i$.

Now suppose $n \equiv 1$ or 2 (mod 4) so that $v \equiv 3$ (mod 4). Let x_{v+1} be one more indeterminate. Then, clearly,

$$\sum_{i=1}^{v} L_i^2 + n x_{v+1}^2 = n \sum_{i=1}^{v+1} x_i^2 + T^2. \tag{5}$$

By (1) there exist integers a, b, c, d such that $n = a^2 + b^2 + c^2 + d^2$. Let A_n be the linear transformation of the four dimensional vector space over the reals given by the matrix

$$A_n = \begin{pmatrix} a & b & c & d \\ -b & a & d & -c \\ -c & -d & a & b \\ -d & c & -b & a \end{pmatrix}.$$

Then $\det A_n = n^2 \neq 0$ since $n \neq 0$.

If $t = x^2 + y^2 + z^2 + w^2$ then we shall say that (x, y, z, w) *represents* t. Eq. (3) implies that if (x, y, z, w) represents t then $(x, y, z, w) A_n$ represents tn. Thus we can write

$$n(x_1^2 + x_2^2 + x_3^2 + x_4^2) = y_1^2 + y_2^2 + y_3^2 + y_4^2 \quad \text{where}$$
$$(y_1, y_2, y_3, y_4) = (x_1, x_2, x_3, x_4) A_n. \tag{6}$$

Note that if the x_i are independent indeterminates then, since A_n is invertible, so are the y_i. Each y_i is a linear combination of the x_i with integer coefficients. However, since A_n^{-1} may have rational entries, each x_i is a linear combination of the y_i with rational coefficients.

Since (4) is an identity involving linear expressions in the x_i we can replace x_1, x_2, x_3, x_4 by the appropriate linear combinations of the y_i and still have an identity. If we do this then, using (6) we get:

$$\sum_{i=1}^{v} L_i^2 + n x_{v+1}^2 = y_1^2 + y_2^2 + y_3^2 + y_4^2 + n \sum_{5}^{v+1} x_i^2 + T^2, \tag{7}$$

where the L_i and T are linear expressions in $y_1, y_2, y_3, y_4, x_5, \ldots, x_{v+1}$.

If we repeat this process with four x_i's at a time then, since $v + 1 \equiv 0$ (mod 4) we will eventually arrive at

$$\sum_{i=1}^{v} L_i^2 + n x_{v+1}^2 = \sum_{i=1}^{v+1} y_i^2 + T^2 \tag{8}$$

where now the L_i, x_{v+1} and T are linear expressions in the indeterminates $y_1, y_2, \ldots, y_{v+1}$.

Suppose we put $L_1 = \pm y_1$, where we use the "$+$" sign unless the coefficient of y_1 in L_1 is 1, in which case we use the "$-$" sign. Then $L_1 = \pm y_1$ can be solved to give y_1 as a linear expression in the indeterminates $y_2, \ldots, y_{v+1}$. This expression can then be substituted everywhere in (8) for y_1, to give

$$\sum_{i=2}^{v} L_i^2 + n x_{v+1}^2 = \sum_{i=2}^{v+1} y_i^2 + T^2 . \tag{9}$$

Since (8) was an identity for the indeterminates $y_1, \ldots, y_{v+1}$, (9) is an identity in $y_2, \ldots, y_{v+1}$.

Repeating this process we eventually arrive at

$$n x_{v+1}^2 = y_{v+1}^2 + T^2 \tag{10}$$

where, now, x_{v+1} and T are linear expressions in y_{v+1}; i.e. $x_{v+1} = a y_{v+1}$ and $T = b y_{v+1}$ for rationals a, b. This gives

$$n a^2 = 1 + b^2 \tag{11}$$

or

$$n = \frac{1}{a^2} + \left(\frac{b}{a} \right)^2 \tag{12}$$

where a, b are rationals.

But now, by (2), this implies that $n = u^2 + v^2$ where u, v are integers.

Thus we have shown that if there is a projective plane of order $n \equiv 1$ or 2 (mod 4) then n is the sum of two integral squares.

This is the Bruck-Ryser Theorem. □

References

As this chapter contains only the basic results and is completely self-contained there is no real need to suggest further reading. However any student who is unfamiliar with the elementary number theoretic results used in proving the Bruck-Ryser theorem will find them in [1].

1. Le Veque, W. J.: Topics in number theory. Reading/Mass.: Addison-Wesley 1956.

IV. Collineations of Projective and Affine Planes

1. Introduction

In this chapter we introduce the concepts of collineation and correlation and then proceed with a detailed study of a special family of collineations called quasiperspectivities. These collineations play a crucial role in the study of projective planes, and a number of the results proved here will be vital to the later chapters. Most of the chapter is self-contained but occasionally we need to use some very elementary permutation group theory; in particular repeated use is made of Result 1.13.

The chapter ends with a discussion of the relation between the existence of certain perspectivities and the existence of Desargues configuration.

2. Basic Concepts and Definitions

A *collineation* (or *automorphism*) of a projective, or affine, plane is an isomorphism of the plane onto itself. Under the usual definition of products of mappings it is clear that the set of all collineations of a given plane $\mathscr{B}$ form a group. This group, which we denote by $\operatorname{Aut}\mathscr{B}$, is called the *full collineation group of* $\mathscr{B}$. When we refer to a collineation group of $\mathscr{B}$ we mean a subgroup of $\operatorname{Aut}\mathscr{B}$. We use the letter 1 to denote the identity collineation.

Exercise 4.1. If α is a collineation of an affine plane $\mathscr{A} = \mathscr{P}^l$ show that there is a unique collineation β in $\operatorname{Aut}\mathscr{P}$ such that $l^\beta = l$ and β induces α on $\mathscr{A}$. (We shall say that α *extends* to β.)

Lemma 3.11 implies:

Lemma 4.1. *If l, k are any two lines of a projective plane $\mathscr{P}$ then $\mathscr{P}^l \cong \mathscr{P}^k$ if and only if there is a collineation α in $\operatorname{Aut}\mathscr{P}$ such that $l^\alpha = k$.*

A *correlation* θ of a projective plane $\mathscr{P}$ is a one-to-one mapping of the points of $\mathscr{P}$ onto the lines of $\mathscr{P}$ and the lines of $\mathscr{P}$ onto the points of $\mathscr{P}$ such that a point A is on a line l if and only if l^θ is on A^θ. For any correlation ϕ and any lines l, k of $\mathscr{P}$ the line $l^\phi k^\phi$ is the image of the point lk. This simple observation shows that an affine plane $\mathscr{A}$ cannot admit a correlation since if l and k were chosen parallel, the line joining l^ϕ to k^ϕ would not have a pre-image in $\mathscr{A}$.

Clearly the product of any two correlations of a projective plane $\mathscr{P}$ is a collineation of $\mathscr{P}$. A correlation θ such that θ^2 is the identity collineation is called a *polarity*. Polarities play an important role in the study of projective planes and will be discussed in Chapter XII. For most of this chapter we shall only be interested in collineations.

Lemma 4.2. *Let $\mathscr{P}$ be a projective plane with a collineation α. If $\mathscr{F}(\alpha)$ denotes the set of all points and lines fixed by α, then $\mathscr{F}(\alpha)$ is a closed configuration.*

Proof. Let A, B be any two distinct points of $\mathscr{F}(\alpha)$. Then, since α preserves incidence, $(AB)^\alpha = A^\alpha B^\alpha$. But, since $A, B \in \mathscr{F}(\alpha)$, $A^\alpha = A$ and $B^\alpha = B$. Thus $(AB)^\alpha = A^\alpha B^\alpha = AB$ i.e. $AB \in \mathscr{F}(\alpha)$. Dually for any two distinct lines l, m of $\mathscr{F}(\alpha)$, $lm \in \mathscr{F}(\alpha)$.

Hence $\mathscr{F}(\alpha)$ is a closed configuration. ☐

Corollary. *If α fixes a quadrangle then $\mathscr{F}(\alpha)$ is a subplane of $\mathscr{P}$.*

If $\mathscr{C}$ is any subset of the points and lines of $\mathscr{P}$ which form a closed configuration then we shall merely say that $\mathscr{C}$ is *closed*. If α is any collineation of $\mathscr{P}$ such that $\mathscr{F}(\alpha)$ is a subplane of $\mathscr{P}$ we shall call α a *planar collineation* or, alternatively, say that α is *planar*. In the special case where $\mathscr{F}(\alpha)$ is a Baer subplane, α is sometimes called a *Baer collineation*.

It seems reasonable to expect that a knowledge of $\mathscr{F}(\alpha)$ might give some information about the action of α on $\mathscr{P}$; particularly if $\mathscr{F}(\alpha)$ is in some sense "large" relative to $\mathscr{P}$. This, in fact, is often true, especially if $\mathscr{F}(\alpha)$ happens to be a Baer subset. Indeed, collineations α for which $\mathscr{F}(\alpha)$ is a Baer subset play a crucial role in the study of projective planes, and it is basically these collineations, and the groups which they generate, which we study in this chapter.

3. Quasiperspectivities

A *quasiperspectivity*, or *quasicentral collineation*, of a projective plane $\mathscr{P}$ is a collineation α such that $\mathscr{F}(\alpha)$ is a Baer subset. Note that since $\mathscr{P}$ is a Baer subset of itself, 1 is a quasiperspectivity.

While it may appear that a quasiperspectivity is a rather special type of collineation, the following simple theorem shows that they frequently occur.

Theorem 4.3. *Let $\mathscr{P}$ be a projective plane. Any collineation α of order 2 is a quasiperspectivity.*

Proof. We must show that every element of $\mathscr{P}$ is incident with an element of $\mathscr{F}(\alpha)$.

Let A be any point of $\mathscr{P}$ such that $A^\alpha \neq A$. Then, since $(AA^\alpha)^\alpha = A^\alpha A^{\alpha^2}$ $= A^\alpha A$, the line AA^α is fixed by α. Thus any non-fixed point of α is on a line of $\mathscr{F}(\alpha)$ and, dually, any non-fixed line contains a fixed point.

Let B be any fixed point of α. If α fixes a second point, C say, then α also fixes the line BC. Thus either B is on a line of $\mathscr{F}(\alpha)$ or B is the only point fixed by α. But let l be any non-fixed line not containing B. (Since any two fixed lines intersect in a fixed point, α can fix at most one line not containing B. Thus we may certainly assume that l exists.) Then, by the earlier argument, l contains a fixed point not equal to B, namely $l \cap l^\alpha$. Hence α must fix at least two points. Dually, α fixes at least two lines and the theorem is proved. □

We will call a collineation of order 2 an *involution*.

In order to study quasiperspectivities it is clearly necessary to determine all closed Baer subsets. Thus we prove:

Theorem 4.4. *A closed Baer subset of a projective plane $\mathscr{P}$ is either a Baer subplane or consists of a line l and all the points on it together with a point V and all the lines through it.*

Proof. If $\mathscr{B}$ is a closed Baer subset and is a subplane, then it is a Baer subplane, by the definition. Suppose $\mathscr{B}$ is a closed Baer subset which is not a subplane. Then, clearly, $\mathscr{B}$ cannot contain a quadrangle.

Case (a). *$\mathscr{B}$ contains a triangle ABC.*

Since $\mathscr{B}$ does not contain a quadrangle, every other point of $\mathscr{B}$ must lie on the same side, BC say, of the triangle ABC. Let D be any point of the line BC other than B or C, and let m be any line through D such that m is distinct from both BC and AD. (Both D and m exist since, by Theorem 3.5, each element of $\mathscr{P}$ is incident with at least three elements.) By the definition of a Baer subset, m must contain a point of $\mathscr{B}$. Since m does not pass through A, this point must lie on BC. Hence $D \in \mathscr{B}$, and so, since D was any point on BC, $\mathscr{B}$ consists of all the points of BC together with A and all the lines through A plus the line BC.

Case (b). *$\mathscr{B}$ contains no triangle.*

Clearly all the points of $\mathscr{B}$ are collinear, on a line l say, and the lines of $\mathscr{B}$ are concurrent at a point A. Since every point of $\mathscr{P}$ is on a line of $\mathscr{B}$, $\mathscr{B}$ contains more than one line. Hence, since A is the intersection of at least two lines of the closed configuration $\mathscr{B}$, $A \in \mathscr{B}$ and so is a point of l. Let D be any point of l other than A, and let $m \neq l$ be any line of $\mathscr{P}$ through D. Since $\mathscr{B}$ is a Baer subset, m must contain a point of B. Hence, since all the points of $\mathscr{B}$ lie on l, this point must be D. But D was any point of l which means that every point of l is in $\mathscr{B}$. Dually every line through A belongs to $\mathscr{B}$. □

We now wish to show that proper Baer subsets, and so in particular Baer subplanes, are maximal closed subsets of a projective plane, i.e. that if $\mathscr{B}$ is a closed Baer subset of a projective plane $\mathscr{P}$ and $\mathscr{C}$ is any closed configuration of $\mathscr{P}$ with $\mathscr{B} \subseteq \mathscr{C}$, then either $\mathscr{B} = \mathscr{C}$ or $\mathscr{C} = \mathscr{P}$. This will enable us to prove some results about quasiperspectivities. To do this we first prove a simple lemma.

Lemma 4.5. *Let $\mathscr{P}$ be a projective plane. If $\mathscr{P}_0$ is a subplane containing all the points of one line l of $\mathscr{P}$, then $\mathscr{P} = \mathscr{P}_0$.*

Proof. Let A be any point of $\mathscr{P}$ and let B, C be any two distinct points of $\mathscr{P}_0$ not on l such that $AB \neq AC$. Let $AB \cap l = X$ and $AC \cap l = Y$. Since $B \in \mathscr{P}_0$ and $X \in \mathscr{P}_0$, we have $BX \in \mathscr{P}_0$. Similarly CY is also a line of $\mathscr{P}_0$. Thus, since A is the intersection of two lines of $\mathscr{P}_0$, $A \in \mathscr{P}_0$, and so every point of $\mathscr{P}$ is a point of $\mathscr{P}_0$. Since every line of $\mathscr{P}$ is the join of two points of $\mathscr{P}_0$, every line of $\mathscr{P}$ is a line of $\mathscr{P}_0$ and the lemma is proved. □

Note that if $\mathscr{P}$ is finite then the lemma follows immediately from the fact that $\mathscr{P}$ and $\mathscr{P}_0$ have the same order.

Theorem 4.6. *A proper closed Baer subset $\mathscr{B}$ of a projective plane $\mathscr{P}$ is a maximal closed configuration of $\mathscr{P}$.*

Proof. Let $\mathscr{C}$ be the intersection of the closed configurations containing $\mathscr{B}$ and a single element not in $\mathscr{B}$. Without loss of generality we may assume this element is a point X. Clearly, for all possible choices of $\mathscr{B}$, $\mathscr{C}$ contains a quadrangle, so that $\mathscr{C}$ is a subplane of $\mathscr{P}$. If $\mathscr{B}$ is not a subplane then, by Theorem 4.4, $\mathscr{B}$ contains all the points of some line l of $\mathscr{P}$. Since $\mathscr{B} \subset \mathscr{C}$, $\mathscr{C} = \mathscr{P}$ by Lemma 4.5, and so either $\mathscr{B}$ is a subplane or $\mathscr{B}$ is maximal.

Suppose $\mathscr{B}$ is a Baer subplane. Then, by definition, every line of $\mathscr{P}$ contains a point of $\mathscr{B}$. So, in particular, every line through X contains a point of $\mathscr{B}$. Thus every line through X contains at least two points of $\mathscr{C}$, i.e. is a line of $\mathscr{C}$. But now $\mathscr{C}$ is subplane with the property that it contains every line through a given point X of $\mathscr{P}$. By the dual of Lemma 4.5, $\mathscr{C} = \mathscr{P}$ and the theorem is proved. □

The following exercise shows that the converse of Theorem 4.6 is false.

Exercise 4.2. If $\mathscr{P}$ is a finite projective plane of order p^3 with a subplane $\mathscr{P}_0$ of order p, show that $\mathscr{P}_0$ is a maximal subplane of $\mathscr{P}$ (see Theorem 3.7).

This is perhaps a convenient place to point out a very important consequence of the principle of duality. When we proved Lemma 4.5 we were, in effect, proving two lemmas; namely Lemma 4.5 and its dual; compare Theorem 3.2. (In fact it was the dual of the lemma which we used in the proof of Theorem 4.6.) This will be true for many of the results which we prove and we shall not, in general, state the dual result explicitly.

We now use the properties which we have established for Baer subsets to prove the following important properties of quasiperspectivities.

Theorem 4.7. *Let α be a quasiperspectivity of a projective plane $\mathscr{P}$. Then α is completely determined by $\mathscr{F}(\alpha)$ and the image of one element not in $\mathscr{F}(\alpha)$.*

Proof. Suppose α, β are two quasiperspectivities of $\mathscr{P}$ such that $\mathscr{F}(\alpha)$ $= \mathscr{F}(\beta)$ and $X^\alpha = X^\beta$ for some point $X \notin \mathscr{F}(\alpha)$. The collineation $\alpha\beta^{-1}$ fixes all of $\mathscr{F}(\alpha)$ and at least one extra element, namely X. By Lemma 4.2 $\mathscr{F}(\alpha\beta^{-1})$ is a closed configuration. But, clearly, $\mathscr{F}(\alpha\beta^{-1}) \supset \mathscr{F}(\alpha)$ which is a closed Baer subset. Thus, by Theorem 4.6, $\mathscr{F}(\alpha\beta^{-1}) = \mathscr{P}$ i.e. $\alpha\beta^{-1} = 1$ and $\alpha = \beta$. $\quad\square$

Lemma 4.8. *If α is a quasiperspectivity of a projective plane $\mathscr{P}$ then $\langle \alpha \rangle$ acts semi-regularly on the elements of $\mathscr{P}$ not in $\mathscr{F}(\alpha)$.*

Proof. We must show that if some power of α, α^t say, fixes an element not in $\mathscr{F}(\alpha)$, then $\alpha^t = 1$. But this follows immediately since $\mathscr{F}(\alpha) \subset \mathscr{F}(\alpha^t)$ and $\mathscr{F}(\alpha)$ is maximal. $\quad\square$

For any collineation group Γ let $\mathscr{F}(\Gamma) = \bigcap_{\gamma \in \Gamma} \mathscr{F}(\gamma)$.

We leave as a very easy exercise the following slight generalization of Lemma 4.8.

Exercise 4.3. If Γ is any collineation group of a projective plane $\mathscr{P}$ such that $\mathscr{F}(\Gamma)$ is a Baer subset show that Γ acts semi-regularly on the elements of $\mathscr{P}$ not in $\mathscr{F}(\Gamma)$.

The conditions of Theorem 4.7 and Lemma 4.8 may be relaxed slightly. Again we leave this as an exercise.

Exercise 4.4. If Γ is any collineation group of a projective plane $\mathscr{P}$ such that $\mathscr{F}(\Gamma)$ is a maximal closed configuration, show that Γ acts semi-regularly on the elements of $\mathscr{P}$ not in $\mathscr{F}(\Gamma)$.

Exercise 4.5. Let $\mathscr{P}$ be a projective plane. If Γ is the set of all collineations fixing a given subset (elementwise, or merely as a set) show that Γ is a subgroup of $\operatorname{Aut}\mathscr{P}$.

4. Perspectivities

Theorem 4.9. *Let $\mathscr{P}$ be a projective plane. If $\alpha \neq 1$ is a collineation fixing a line l pointwise then there is a point V fixed linewise by α. Furthermore α fixes no other point or line.*

Proof. Since every line of $\mathscr{P}$ intersects l, each line of $\mathscr{P}$ contains a point of $\mathscr{F}(\alpha)$.

Let B be any point not on l. If $B^\alpha = B$ then the line joining B to any point, A say, of l is fixed by α. So, in this case, B is incident with a line of $\mathscr{F}(\alpha)$. If $B^\alpha \neq B$ then let $P = B B^\alpha \cap l$. Now $(PB)^\alpha = P^\alpha B^\alpha = PB^\alpha$ (since α fixes all points of l). Thus $(PB)^\alpha = PB$, so that B is on a line of $\mathscr{F}(\alpha)$.

We have now shown that $\mathscr{F}(\alpha)$ is a Baer subset. Since, by Lemma 4.2, $\mathscr{F}(\alpha)$ is closed, the theorem is proved by Theorem 4.4 and Lemma 4.5. $\quad\square$

It is important to observe that of course the dual of Theorem 4.9 is also true. If a collineation α fixes a line l pointwise and a point V linewise then α is called a (V, l)-*perspectivity* or a (V, l)-*central collineation*. The point V is called the *centre* of α and l is its *axis*. Two possibilities arise; if V is on l we call α an *elation*, and if α is not on l, α is a *homology*. For convenience the identity is regarded as both a homology and an elation.

Clearly any perspectivity is also a quasiperspectivity. In fact any non-planar quasiperspectivity is a perspectivity.

By Lemma 4.8, if α is any perspectivity, $\langle \alpha \rangle$ induces a semi-regular permutation group on the non fixed points of any line (other than the axis) through the centre. Thus we have:

Lemma 4.10. *Let $\mathscr{P}$ be a finite projective plane of order n with a perspectivity α of order k. Then either*

(i) $k \mid n$ *and α is an elation or*
(ii) $k \mid n - 1$ *and α is a homology.* $\Box$

From Theorem 4.3 we know that an involution is either planar or a perspectivity.

Exercise 4.6. Let α be an involution of a finite projective plane whose order n is not a square. Show that α is an elation (homology) if and only if n is even (odd).

Since the identity is a (V, l)-perspectivity for any V and l, Exercise 4.5 and Theorem 4.9 show that the set of all (V, l)-perspectivities is a subgroup of the full collineation group of the plane. If Γ is any collineation group of a projective plane $\mathscr{P}$, let $\Gamma_{(V, l)}$ be the set of all (V, l)-perspectivities in Γ. Then, clearly, $\Gamma_{(V, l)}$ is a subgroup of Γ (the proof of Exercise 4.5 will prove this also). Lemma 4.10 can now be strengthened in an obvious way.

Exercise 4.7. Let Γ be any collineation group of a finite projective plane of order n. Suppose for a given V and l, $\Gamma_{(V, l)}$ has order k. Show that either (i) $k \mid n$ in which case V is on l or (ii) $k \mid n - 1$ in which case V is not on l.

If A is a fixed point of a collineation α, then for any collineation β we have $A^{\beta(\beta^{-1}\alpha\beta)} = A^{\alpha\beta} = A^{\beta}$, i.e. A^{β} is a fixed point of $\beta^{-1}\alpha\beta$. If we write $\gamma = \beta^{-1}\alpha\beta$ then there is an obvious mapping between the elements of $\mathscr{F}(\alpha)$ and $\mathscr{F}(\gamma)$. This fact is utilized in the following lemma.

Lemma 4.11. *Let $\mathscr{P}$ be a projective plane. If α is a (V, l)-perspectivity of $\mathscr{P}$ and β is any collineation then $\beta^{-1}\alpha\beta$ is a (V^{β}, l^{β})-perspectivity.*

Proof. Let X be any point of l^{β}. Then $X^{\beta^{-1}}$ is on l. But this implies that $X^{\beta^{-1}\alpha} = X^{\beta^{-1}}$. Hence $X^{\beta^{-1}\alpha\beta} = X$ so that $\beta^{-1}\alpha\beta$ fixes l^{β} pointwise. Similarly $\beta^{-1}\alpha\beta$ fixes V^{β} linewise and the lemma is proved. $\Box$

Corollary 1. *If α is a (V, l)-perspectivity $\neq 1$ and π is any collineation which commutes with α, then $V^{\pi} = V$ and $l^{\pi} = l$.*

Corollary 2. *If Γ is a collineation group of a projective plane $\mathcal{P}$ then, for any $\gamma \in \Gamma$ and any point V and line l, $\Gamma_{(V,l)}^{\gamma} = \gamma^{-1}\Gamma_{(V,l)}\gamma = \Gamma_{(V^{\gamma},l^{\gamma})}$.*

Exercise 4.8. If Γ is a collineation group of a projective plane $\mathcal{P}$ show that, if $\Gamma_{(V,l)} \neq 1$, then for any $\gamma \in \Gamma$, $\Gamma_{(V,l)}^{\gamma} = \Gamma_{(V,l)}$ if and only if $V^{\gamma} = V$ and $l^{\gamma} = l$.

It is clear that the set of all perspectivities with a given axis form a group. This group has many interesting subgroups; before discussing these subgroups we introduce a notation which we shall adopt for the rest of the book.

Let Γ be any collineation group of a projective plane $\mathcal{P}$. Then let:

$\Gamma_{(V,l)} = $ group of all (V, l)-perspectivities in Γ.

$$\Gamma_{(k,l)} = \bigcup_{A \,|\, k} \Gamma_{(A,l)}\,,$$

$$\Gamma_{(P,Q)} = \bigcup_{l \,|\, Q} \Gamma_{(P,l)}\,,$$

$$\Gamma_{(l)} = \bigcup_{X \in \mathcal{P}} \Gamma_{(X,l)}\,,$$

$$\Gamma_{(A)} = \bigcup_{l \in \mathcal{P}} \Gamma_{(A,l)}\,.$$

(The reader should carefully note the difference between Γ_l and $\Gamma_{(l)}$; if necessary refer back to the definition of Γ_l in Chapter I.)

We shall show that each of the sets defined above is in fact a subgroup of Γ. Before doing this however we draw attention to an elementary property of a perspectivity, which is really a re-wording of part of Theorem 4.9.

Lemma 4.12. *Let $\mathcal{P}$ be a projective plane. If $\alpha \neq 1$ is a perspectivity with axis l and if $k^{\alpha} = k$ for any line $k \neq l$, then k passes through the centre of α.* $\square$

Exercise 4.9. If α is a (A, l)-perspectivity and β is a (B, l)-perspectivity both non-trivial and with $A \neq B$, show that $\alpha\beta$ is a non-trivial (C, l)-perspectivity and $C \neq A$, $C \neq B$.

Theorem 4.13. *Let Γ be a collineation group of a projective plane $\mathcal{P}$. Then $\Gamma_{(k,l)}, \Gamma_{(P,Q)}, \Gamma_{(l)}, \Gamma_{(P)}$ are all subgroups of Γ, for all points P, Q and all lines l, k. Further $\Gamma_{(l,l)} \trianglelefteq \Gamma_{(l)}$ and $\Gamma_{(P,P)} \trianglelefteq \Gamma_{(P)}$ for all choices of P and l. (Note that the possibility that some, or even all, of these subgroups are trivial is not excluded.)*

Proof. By duality it is clearly sufficient to prove that $\Gamma_{(k,l)}$ and $\Gamma_{(l)}$ are subgroups for all lines l, k and that $\Gamma_{(l,l)} \trianglelefteq \Gamma_{(l)}$.

Since $\Gamma_{(l)}$ is merely the set of perspectivities in Γ with axis l, we have already observed that $\Gamma_{(l)}$ is a group. We now show $\Gamma_{(k,l)}$ is a group for all choices of k and l. We distinguish between the two cases $k = l$, $k \neq l$.

Clearly $\Gamma_{(k,l)} \leqq \Gamma_{(l)}$ in both cases so our proof will show that $\Gamma_{(k,l)}$ is a subgroup of $\Gamma_{(l)}$.

Case (a). $k \neq l$.

Let α, β be any two elements of $\Gamma_{(k,l)}$. Clearly $k^\alpha = k^\beta$ so that $k^{\alpha\beta^{-1}} = k$. Thus $\alpha\beta^{-1}$ is an element of $\Gamma_{(l)}$ whose centre, by Lemma 4.12, lies on k. In other words $\alpha\beta^{-1} \in \Gamma_{(k,l)}$ and the theorem is proved.

Case (b). $k = l$.

We already know that $\Gamma_{(P,l)}$ is a group for any P and l. Thus we must show that for $\alpha \in \Gamma_{(A,l)}$ and $\beta \in \Gamma_{(B,l)}$ with $A \neq B$, $A \, I \, l$, $B \, I \, l$, the collineation $\alpha\beta^{-1} \in \Gamma_{(l,l)}$. Since $\alpha\beta^{-1}$ fixes l pointwise it must have a centre and we need to show that this centre is on l. In order to do this it is sufficient to show $X^{\alpha\beta^{-1}} \neq X$ for any X not on l. If $X^{\alpha\beta^{-1}} = X$, then $X^\alpha = X^\beta$. But, clearly, since A is the centre of α, A, X, X^α are collinear. Similarly B, X, X^β are also collinear. But $X^\alpha = X^\beta$ implies that A, B are both on the line $X X^\alpha$. Since $l = AB$ and X is not on l, this is impossible. Thus $\alpha\beta^{-1}$ is an element of $\Gamma_{(l)}$ with no fixed point not incident with l. In other words $\alpha\beta^{-1} \in \Gamma_{(l,l)}$ so that $\Gamma_{(l,l)}$ is a subgroup of $\Gamma_{(l)}$.

It only remains to show $\Gamma_{(l,l)} \trianglelefteq \Gamma_{(l)}$. But this is virtually an immediate consequence of Corollary 2 to Lemma 4.11. For any $\gamma \in \Gamma_{(l)}$, $\gamma^{-1}\Gamma_{(X,l)}\gamma = \Gamma_{(X^\gamma, l^\gamma)}$ for all X on l. But for any X on l and any $\gamma \in \Gamma_{(l)}$, $X^\gamma = X$ and $l^\gamma = l$. Thus $\gamma^{-1}\Gamma_{(X,l)}\gamma = \Gamma_{(X,l)}$ for all $X \, I \, l$ and so, since $\Gamma_{(l,l)} = \bigcup_{X I l} \Gamma_{(X,l)}$, we have $\gamma^{-1}\Gamma_{(l,l)}\gamma = \Gamma_{(l,l)}$ for all $\gamma \in \Gamma_{(l)}$, i.e. $\Gamma_{(l,l)} \trianglelefteq \Gamma_{(l)}$. $\square$

Exercise 4.10. Let $\Pi = \text{Aut}\,\mathscr{P}_2(K)$ and let l be the line $z = 0$ in $\mathscr{P}_2(K)$. Determine the groups $\Pi_{(l,l)}$ and $\Pi_{(l)}$ for $K = GF(2)$, $GF(3)$, $GF(4)$.

Exercise 4.11. Using the terminology of Exercise 4.10 determine $\Pi_{(l,l)}$ and $\Pi_{(l)}$ for an arbitrary field K. Show that $\Pi_{(l,l)}$ is abelian and is elementary abelian if K is finite; show that $\Pi_{(l)}$ is soluble.

Remarkably little can be said about the possible structure for a group $\Gamma_{(P,l)}$ for a single choice of P and l. In fact, it is known that given any group Γ, there is an infinite projective plane such that Γ is contained in a subgroup of the (P, l)-perspectivities. However, if it is known that the plane admits non-trivial (P, l)-perspectivities for two distinct choices of P on l, then we have the following:

Theorem 4.14. *Let $\mathscr{P}$ be a projective plane and let Γ be any collineation group of $\mathscr{P}$. If $\Gamma_{(P,l)}$ is non-trivial for two distinct choices of P on l then $\Gamma_{(l,l)}$ is abelian and all its non-identity elements have the same order (either infinite or a prime).*

Proof. We first show that any two elations with common axis, but distinct centres, commute. Let $\alpha \in \Gamma_{(A,l)}$ and $\beta \in \Gamma_{(B,l)}$ with $A \neq B$, $\alpha \neq 1$,

$\beta \neq 1$. By Lemma 4.11, $\beta^{-1}\alpha\beta$ is an (A^{β}, l^{β})-elation and so, since β has axis l, $\beta^{-1}\alpha\beta$ is a (A, l)-elation. Similarly $\alpha^{-1}\beta^{-1}\alpha \in \Gamma_{(B,l)}$ (since $\beta^{-1} \in \Gamma_{(B,l)}$).

Consider $\alpha^{-1}\beta^{-1}\alpha\beta$. Writing $\alpha^{-1}\beta^{-1}\alpha\beta = \alpha^{-1}(\beta^{-1}\alpha\beta)$ we see that $\alpha^{-1}\beta^{-1}\alpha\beta \in \Gamma_{(A,l)}$. But similarly, writing $\alpha^{-1}\beta^{-1}\alpha\beta = (\alpha^{-1}\beta^{-1}\alpha)\beta$ we have $\alpha^{-1}\beta^{-1}\alpha\beta \in \Gamma_{(B,l)}$. Thus $\alpha^{-1}\beta^{-1}\alpha\beta \in \Gamma_{(A,l)} \cap \Gamma_{(B,l)}$. But, since any non-identity elation has a unique centre, $\Gamma_{(A,l)} \cap \Gamma_{(B,l)} = 1$. Thus $\alpha\beta = \beta\alpha$.

In order to show that $\Gamma_{(l,l)}$ is abelian it is now sufficient to show that two elations with the same centre commute. Let α_1, α_2 be any non-trivial (A, l)-elations and let $\beta \neq 1$ be a (B, l) elation for $B \neq A$, $\alpha_1, \alpha_2, \beta \in \Gamma$. Then by the above, $\alpha_1\beta = \beta\alpha_1$ and $\alpha_2\beta = \beta\alpha_2$. By Exercise 4.9 the centre of $\alpha_1\beta$ is different from A. But this implies that $\alpha_1\beta$ commutes with α_2. Thus $\alpha_2(\alpha_1\beta) = (\alpha_1\beta)\alpha_2 = \alpha_1(\beta\alpha_2) = \alpha_1\alpha_2\beta$ since $\beta\alpha_2 = \alpha_2\beta$. Cancelling β, this gives $\alpha_1\alpha_2 = \alpha_2\alpha_1$, and hence $\Gamma_{(l,l)}$ is abelian.

Suppose $\Gamma_{(l,l)}$ contains an elation of finite order. Then $\Gamma_{(l,l)}$ contains an element of prime order. Let γ be a (C, l)-elation in Γ of prime order p and let $\delta \in \Gamma_{(D,l)}$ for any D on l, $D \neq C$. Since $\gamma\delta = \delta\gamma$, $(\gamma\delta)^p = \gamma^p\delta^p = \delta^p \in \Gamma_{(D,l)}$. But by Exercise 4.9 $\gamma\delta \in \Gamma_{(E,l)}$ for some E on l, $E \neq D$ (or C). Thus $(\gamma\delta)^p \in \Gamma_{(E,l)} \cap \Gamma_{(D,l)}$, i.e. $(\gamma\delta)^p = 1 = \delta^p$. This shows that every elation in Γ with axis l and centre distinct from C has order p. A similar argument with D in the role of C completes the proof. $\square$

Corollary. *Let $\mathscr{P}$ be a finite projective plane of order n and let Γ be a collineation group of $\mathscr{P}$. If $|\Gamma_{(A,l)}| > 1$ for at least two choices of A on l then $\Gamma_{(l,l)}$ is an elementary abelian p-group where p is a prime divisor of n.*

Proof. Since $\mathscr{P}$ is finite, $|\Gamma|$ is finite. The fact that $p|n$ is a direct consequence of Lemma 4.10. $\square$

Note that since Theorem 4.14 and its corollary are true for any collineation group of $\mathscr{P}$ then, in particular, they are true for Aut $\mathscr{P}$. Thus if, for *any* collineation group Γ, $|\Gamma_{(P,l)}| > 1$ for two distinct choices of P on l then all elations with axis l in Aut $\mathscr{P}$ have the same order.

We now prove a simple lemma which gives a condition under which the requirements of Theorem 4.14 are satisfied.

Lemma 4.15. *Let $\mathscr{P}$ be a projective plane and let Γ be a collineation group of $\mathscr{P}$. If $|\Gamma_{(P,l)}| > 1$ and $|\Gamma_{(A,m)}| > 1$ with A, P on l and A on m, then $|\Gamma_{(A,l)}| > 1$.*

Proof. The case $l = m$ is trivial and so we may assume $l \neq m$.

Choose $\alpha \neq 1$, $\alpha \in \Gamma_{(P,l)}$ and $\beta \neq 1$, $\beta \in \Gamma_{(A,m)}$ and let $\gamma = \alpha^{-1}\beta^{-1}\alpha\beta$.

Clearly, since $A^{\alpha} = A^{\beta} = A$ and $l^{\alpha} = l^{\beta} = l$, we have $A^{\gamma} = A$ and $l^{\gamma} = l$. If X is any point of l then $X^{\alpha^{-1}} = X$, and thus $X^{\gamma} = X^{\beta^{-1}\alpha\beta} = (X^{\beta^{-1}\alpha})^{\beta}$. However, since $X^{\beta^{-1}}$ is on the axis l of α, $X^{\beta^{-1}\alpha} = X^{\beta^{-1}}$ so that $X^{\beta^{-1}\alpha\beta} = X$. Thus γ is a perspectivity with axis l.

Similarly γ fixes A linewise. In order to prove the lemma it is now only necessary to show that $\gamma \neq 1$. But $P^{\beta} \neq P$ which means, by Corollary 1 to

Lemma 4.11, that $\alpha\beta \neq \beta\alpha$ i.e. $\gamma = \alpha^{-1}\beta^{-1}\alpha\beta \neq 1$. Thus Lemma 4.15 is proved. $\square$

Since, for any collineation group Γ of a projective plane $\mathscr{P}$ and any line l of $\mathscr{P}$, every element of $\Gamma_{(l,l)}$ is an elation, the non identity elements of $\Gamma_{(l,l)}$ act as fixed point free permutations on the points of $\mathscr{P}^l$. Thus $\Gamma_{(l,l)}$ induces a semi-regular permutation group on the points of $\mathscr{P}^l$. Combining this with Theorem 3.12 gives:

Theorem 4.16. *Let $\mathscr{P}$ be a finite projective plane of order n and let Γ be a collineation group of $\mathscr{P}$. Then, for any line l of $\mathscr{P}$, $|\Gamma_{(l,l)}| \mid n^2$.*

Not only does $\Gamma_{(l,l)}$ induce a permutation group on the points of $\mathscr{P}^l$, it also permutes the lines of $\mathscr{P}^l$ which pass through any point A of l. This gives:

Theorem 4.17. *Let $\mathscr{P}$ be a projective plane and Γ a collineation group of $\mathscr{P}$.*

(i) *If l is any line of $\mathscr{P}$ then, for any line $m \neq l$, $(\Gamma_{(l,l)})_m = \Gamma_{(A,l)}$ where $A = lm$.*

(ii) *If $\mathscr{P}$ is finite of order n and $|\Gamma_{(l,l)}| > n$ then $|\Gamma_{(B,l)}| > 1$ for all B on l.*

(iii) *If $\mathscr{P}$ is finite of order n then $|m\Gamma_{(l,l)}| \mid n$ for any line m.*

Proof. Before proving the theorem we recall the notation being used. For any permutation group Σ on a set S, Σ_a, where $a \in S$, is the subgroup of Σ fixing a. Thus $(\Gamma_{(l,l)})_m$ is the subgroup of the elations with axis l fixing the line m. But, by Lemma 4.12, this is precisely $\Gamma_{(A,l)}$ where $A = lm$.

In order to prove the second part of the theorem, let B be any point of l. Then $\Gamma_{(l,l)}$ induces a permutation group on the n lines through B other than l. Applying Result 1.13 to this group, we have

$$|\Gamma_{(l,l)}| = |m\Gamma_{(l,l)}| \cdot |(\Gamma_{(l,l)})_m|$$

for any line m through B. By hypothesis $|\Gamma_{(l,l)}| > n$ and, clearly, $|m\Gamma_{(l,l)}| \leq n$. Thus $(\Gamma_{(l,l)})_m > 1$ i.e. $|\Gamma_{(B,l)}| > 1$.

Since $(\Gamma_{(l,l)})_m = \Gamma_{(A,l)}$ for all m through A, $m \neq l$, $|m\Gamma_{(l,l)}|$ is the same for all m through A, $m \neq l$. Thus, as there are n such lines, $|m\Gamma_{(l,l)}| \mid n$. $\square$

5. (V, l)-Transitivity

If $\mathscr{P}$ is a finite projective plane of order n with a collineation group Γ, then, by Exercise 4.7, $|\Gamma_{(V,l)}| \leq n$ if V is on l and $|\Gamma_{(V,l)}| \leq n-1$ if V is not on l. If either of these bounds is attained, then $\Gamma_{(V,l)}$ must be transitive on the non-fixed points of any line through V. A projective plane $\mathscr{P}$ is said to be (V, l)-*transitive* if, for any distinct points A, B with $VA = VB$, $A \neq V \neq B$ and $A \not\mathrel{I} l$, $B \not\mathrel{I} l$ there is a (V, l)-perspectivity α in $\text{Aut}\,\mathscr{P}$ with $A^\alpha = B$.

The above observations about the bounds on the orders of perspectivity groups in finite projective planes give:

Lemma 4.18. *Let $\mathscr{P}$ be a finite projective plane of order n and let $\Pi = \mathrm{Aut}\,\mathscr{P}$. Then*

(i) *$\mathscr{P}$ is (V, l)-transitive for V on l if and only if $|\Pi_{(V,\,l)}| = n$.*

(ii) *$\mathscr{P}$ is (V, l)-transitive for V not on l if and only if $|\Pi_{(V,\,l)}| = n - 1$.* ☐

If a projective plane $\mathscr{P}$ is (X, l)-transitive for all points X on a line m then $\mathscr{P}$ is said to be (m, l)-*transitive; (A, B)-transitivity* is defined dually.

If l is any line of $\mathscr{P}$ such that $\mathscr{P}$ is (l, l)-transitive then l is called a *translation line* of $\mathscr{P}$ and $\mathscr{P}$ is said to be a *translation plane with respect to the line l.* Dually a point A is called a *translation point* if $\mathscr{P}$ is (A, A)-transitive and $\mathscr{P}$ is said to be a *dual translation plane with respect to A.*

If $\mathscr{P}$ is a translation plane with respect to a line l then we shall often refer to $\mathscr{P}^l$ as a *translation plane*; i.e. we shall regard a translation plane as being affine. However, as was seen in Chapter III, $\mathscr{P}$ together with l uniquely determine $\mathscr{P}^l$ and vice-versa. Thus the two statements are clearly equivalent and we shall feel quite free to take whichever of the affine or projective viewpoints seems most convenient. This should not cause the reader any confusion as we shall take great care to explain which viewpoint we are utilizing at any given time. If $\mathscr{A} = \mathscr{P}^l$ is a translation plane and $\Gamma = \mathrm{Aut}\,\mathscr{A}$ then we call the group $\Gamma_{(l,\,l)}$ the *translation group* of $\mathscr{A}$.

We now prove a theorem which shows that one does not need to be given all the elations with axis l before one knows that l is a translation line.

Theorem 4.19. *If a projective plane $\mathscr{P}$ is (A, l)-transitive and (B, l)-transitive for distinct points A, B on l then l is a translation line of $\mathscr{P}$.*

Proof. Let C be any point of l and choose $X \ne Y$ such that $CX = CY$, $X, Y \, \text{I} \, l, X \ne C \ne Y$. We must show there is a (C, l)-elation mapping X onto Y. Let $Z = AX \cap BY$. Since $\mathscr{P}$ is (A, l)-transitive there is an (A, l)-elation $\alpha \in \mathrm{Aut}\,\mathscr{P}$ such that $X^\alpha = Z$. Similarly $\mathrm{Aut}\,\mathscr{P}$ contains a (B, l)-elation β with $Z^\beta = Y$. Clearly $X^{\alpha\beta} = Y$ and $\alpha\beta$ has axis l. Let the centre of $\alpha\beta$ be V. Then, since $X^{\alpha\beta} = Y$, $VX = VY$ i.e. $V = XY \cap l$. But $C = XY \cap l$, so $C = V$ and the theorem is proved. ☐

Exercise 4.12. If a projective plane $\mathscr{P}$ is (A, l)-transitive and (B, l)-transitive for $A \ne B$ show that $\mathscr{P}$ is (AB, l)-transitive. (Theorem 4.19 is a special case of this where A, B are both on l.)

Exercise 4.13. If l is a translation line of a projective plane $\mathscr{P}$ and $\alpha \in \mathrm{Aut}\,\mathscr{P}$ show that l^α is also a translation line of $\mathscr{P}$.

Exercise 4.14. For any field K show that every line of $\mathscr{P}_2(K)$ is a translation line.

Exercise 4.15. For any field K show that $\mathscr{P}_2(K)$ is (m, l)-transitive for all lines m, l and (A, B)-transitive for all points A, B.

The following theorem is almost a direct consequence of Exercise 4.13.

Theorem 4.20. (a) *If l and m are translation lines of a projective plane $\mathscr{P}$ then every line through lm is also a translation line.*

(b) *If $\mathscr{P}$ has three non-concurrent translation lines then every line of $\mathscr{P}$ is a translation line.*

Proof. Let $\Pi = \operatorname{Aut}\mathscr{P}$.

To prove (a) we merely show that $\Pi_{(l,l)}$ is transitive on the lines through lm distinct from l, and then use Exercise 4.13.

Let $Q = lm$ and let h, k be any pair of distinct lines through Q such that neither is l. Pick any point A on l, $A \neq Q$ and choose H on h, and K on k such that A, H, K are collinear. Since l is a translation line, $\mathscr{P}$ is certainly (A, l)-transitive. Thus there is an elation $\alpha \in \Pi_{(A,l)}$ such that $H^\alpha = K$. Since Q is on l, $Q^\alpha = Q$. Thus $h^\alpha = (QH)^\alpha = Q^\alpha H^\alpha = QK = k$, and hence $\Pi_{(l,l)}$ is transitive on the lines through Q and different from l. Exercise 4.13 now gives that every line through Q is a translation line.

(b) The proof of (b) is very similar. It consists of showing that Π is transitive on the lines of $\mathscr{P}$ and again applying Exercise 4.13.

We leave the proof as an important exercise. □

A projective plane with the property that every line is a translation line is called a *Moufang* plane.

6. Collineation Groups Containing Perspectivities

We have already seen in Exercise 4.6 that involutory collineations of finite projective planes of non-square order are perspectivities. Since any finite group of even order contains involutions, involutory perspectivities have been studied in considerable detail. We give here two simple lemmas.

Lemma 4.21. *Let $\mathscr{P}$ be a projective plane. If α is an involutory (A, l)-homology and β is an involutory (B, l)-homology, with $A \neq B$, then $\alpha\beta$ is an $(AB \cap l, l)$-elation.*

Proof. Clearly $\alpha\beta$ is a perspectivity with axis l. We first show that $\alpha\beta$ is an elation by showing that $P^{\alpha\beta} \neq P$ for all points P not on l. Suppose $P^{\alpha\beta} = P$ and let $P^\alpha = Q$. Then $P = P^{\alpha\beta} = Q^\beta$. Thus, since α and β are involutory, $P^\beta = Q$ and $Q^\alpha = P$ so that $Q^{\alpha\beta} = P^\beta = Q$. But since $\alpha\beta \neq 1$, $\alpha\beta$ can fix at most one point of $\mathscr{P}^l$. Therefore $P = Q$, and $P^\alpha = Q = P$, so $P = A$; similarly however $P = B$ and this is a contradiction.

Finally we must determine the centre of $\alpha\beta$. But clearly α and β both fix the line AB, thus $\alpha\beta$ fixes AB and so, by Lemma 4.12, $\alpha\beta$ is an $(AB \cap l, l)$-elation. □

Lemma 4.22. *Let $\mathscr{P}$ be a projective plane. If α is an involutory (A, a)-homology of $\mathscr{P}$ and β is an involutory (B, b)-homology of $\mathscr{P}$ such that B is on a and A is on b, then $\alpha\beta$ is an involutory (ab, AB)-homology of $\mathscr{P}$.*

Proof. Let $\gamma = (\alpha\beta)^2$.

For any point X on a, $X^\alpha = X$. Thus $X^{\alpha\beta} = X^\beta$ which is again on a, so that $X^{\alpha\beta\alpha} = X^{\beta\alpha} = X^\beta$. This gives $X^{\alpha\beta\alpha\beta} = X^{\beta\beta} = X$ for all X on a. Hence γ fixes a pointwise. Similarly γ fixes b pointwise and so, since $a \neq b$, $\gamma = 1$, and $\alpha\beta$ is an involution.

Since the points A, B, ab are non-collinear and are all fixed by $\alpha\beta$, $\alpha\beta$ cannot be an elation. Thus, by Theorem 4.3, either $\alpha\beta$ is a homology or $\alpha\beta$ is planar. However, for any point P on a, $P^{\alpha\beta} = P^\beta \neq P$ unless $P = B$ or ab. Thus, since $\alpha\beta$ fixes only two points of a, $\mathscr{F}(\alpha\beta)$ cannot be a subplane so that $\alpha\beta$ is a homology.

Since $\alpha\beta$ does not fix a quadrangle, every fixed point of $\alpha\beta$ must lie on either a, b or AB. But, as we have just seen, $\alpha\beta$ fixes exactly two points on a and exactly two points on b. Thus AB must be its axis, i.e. $\alpha\beta$ is a (ab, AB)-homology. $\square$

*Exercise 4.16. Let Σ be a finite group with two distinct elements α, β each of order 2. Show that α, β are conjugate in $\langle \alpha, \beta \rangle$ if and only if $\alpha\beta$ has odd order. If $\alpha\beta$ has even order show that $\langle \alpha, \beta \rangle$ contains an element γ of order 2 such that γ commutes with α and β. (This exercise is really an elementary result from group theory. A proof is in [3].)

Exercise 4.17. Let $\mathscr{P}$ be a finite projective plane. If α is an involutory perspectivity of $\mathscr{P}$ and β is a Baer involution show that $\alpha\beta$ has even order. (Use Exercise 4.16.)

If α, β are as in Exercise 4.17 then, by Exercise 4.16, there is an involutory collineation γ which commutes with α and β. But, as the following lemma shows, this means that γ must leave the subplane $\mathscr{F}(\beta)$ invariant and, hence, induces a collineation on $\mathscr{F}(\beta)$.

Lemma 4.23. *If α, β are two commuting permutations on a set S and if $T = \{t \in S \mid t^\alpha = t\}$ then β leaves T invariant.*

Proof. Let t be any element of T. Then, since $t^\alpha = t$, $t^{\alpha\beta} = t^\beta$, and so, since $\alpha\beta = \beta\alpha$, $t^{\beta\alpha} = t^{\alpha\beta} = t^\beta$. But this means that, for any $t \in T$, $t^\beta \in T$, i.e. that β leaves T invariant. $\square$

Thus, if α and β are the collineations of Exercise 4.18, the involution γ which commutes with α and β must induce a collineation of $\mathscr{F}(\beta)$. Our next problem then is to determine the action of γ on $\mathscr{F}(\beta)$.

Lemma 4.24. *Let $\mathscr{P}$ be a projective plane with a subplane $\mathscr{P}_0$. If $\alpha \neq 1$ is a (V, l)-perspectivity of $\mathscr{P}$ leaving $\mathscr{P}_0$ invariant, then $V \in \mathscr{P}_0$, $l \in \mathscr{P}_0$ and α induces a perspectivity on $\mathscr{P}_0$.*

Proof. Let X be any point of $\mathscr{P}_0$ then, since $\mathscr{P}_0$ is left invariant by α, $X^\alpha \in \mathscr{P}_0$. Thus XX^α is a line of $\mathscr{P}_0$. Let Y be any other point of $\mathscr{P}_0$ such that Y is not on XX^α, then a similar argument shows that YY^α is also a line of $\mathscr{P}_0$. By Lemma 4.12, V is on XX^α and YY^α, i.e. $V = XX^\alpha \cap YY^\alpha$. Hence, since XX^α and YY^α are lines of $\mathscr{P}_0$, $V \in \mathscr{P}_0$. Similarly $l \in \mathscr{P}_0$.

Since α fixes l pointwise in $\mathscr{P}$, it clearly fixes l pointwise in $\mathscr{P}_0$. The same reasoning shows that α fixes V linewise in $\mathscr{P}_0$ and proves the lemma. □

We conclude this introductory discussion of perspectivities by proving two important theorems which show how the existence of a comparatively small number of perspectivities with a given axis l implies that l is a translation line.

The first, due to André, is a much stronger form of Lemma 4.21. It has many important consequences and in particular shows that the existence of any two homologies with distinct centres but common axis l implies the existence of a non-trivial elation with axis l.

Theorem 4.25 (André). *Let $\mathscr{P}$ be a finite projective plane and let Γ be any collineation group of $\mathscr{P}$. If, for any line l, $\mathscr{H}$ is the set of points $V \in \mathscr{P}^l$ such that $\Gamma_{(V,\,l)} \neq 1$ then $\mathscr{H}$ is a point orbit under $\Gamma_{(l,l)}$.*

Proof. Let $A_1, A_2, \ldots, A_k$ be the points of $\mathscr{H}$. Note that this means that $A_1, A_2, \ldots, A_k$ are the points of $\mathscr{P}^l$ which are the centres of non-trivial homologies in Γ with axis l.

Put $h_i = |\Gamma_{(A_i,\,l)}|$ for $i = 1, 2, \ldots, k$, so that the number of non-trivial homologies in Γ with centre A_i and axis l is $h_i - 1$. Clearly the group of all perspectivities with axis l is the union of the group of elations with the set of non-identity homologies with axis l. Thus

$$|\Gamma_{(l)}| = |\Gamma_{(l,l)}| + \sum_{i=1}^{k} (h_i - 1). \tag{1}$$

Since $1 \in \Gamma_{(l,l)}$, $|\Gamma_{(l,l)}| \geq 1$. Thus, since $h_i \geq 2$ for all i, we have

$$|\Gamma_{(l)}| \geq 1 + k > k. \tag{2}$$

By Lemma 4.11, $\Gamma_{(l)}$ permutes the elements of $\mathscr{H}$ amongst themselves and, since the only element of $\Gamma_{(l)}$ fixing two points of $\mathscr{P}^l$ is the identity, the permutation group induced on $\mathscr{H}$ is faithful. If we let r be the number of orbits of $\Gamma_{(l)}$ on $\mathscr{H}$ then, using the fact that the identity fixes k elements, the homologies fix one, and the other collineations are fixed point free, we apply Result 1.14 to get

$$r|\Gamma_{(l)}| = k + \sum_{i=1}^{k} (h_i - 1). \tag{3}$$

Subtracting (1) from (3) gives

$$(r - 1)|\Gamma_{(l)}| = k - |\Gamma_{(l,l)}|. \tag{4}$$

Clearly $r \geq 1$ and $|\Gamma_{(l,l)}| \geq 1$, so that (2) and (4) imply $r = 1$ and $|\Gamma_{(l,l)}| = k$. Since $\Gamma_{(l,l)}$ induces a semiregular permutation group on the points of $\mathscr{P}^l$, each point orbit of $\Gamma_{(l,l)}$ in $\mathscr{P}^l$ has length k. But as $\mathscr{H}$ is an orbit of

$\Gamma_{(l)}$, $\Gamma_{(l,l)}$ must leave $\mathcal{H}$ invariant and so, since $|\mathcal{H}| = k = |\Gamma_{(l,l)}|$, $\Gamma_{(l,l)}$ is transitive on the points of $\mathcal{H}$. $\square$

The reader should prove as exercises the following very important corollaries of Theorem 4.25.

Corollary 1. *Let $\mathcal{P}$ be a finite projective plane and let α, β be two nontrivial homologies with distinct centres A, B and the same axis l. Then $\langle \alpha, \beta \rangle$ contains an elation mapping A onto B.*

Corollary 2. *Let $\mathcal{P}$ be a finite projective plane such that, for some line l, each point of $\mathcal{P}^l$ is the centre of a non-trivial homology with axis l. Then l is a translation line of $\mathcal{P}$.*

Theorem 4.26. *Let $\mathcal{P}$ be a finite projective plane of order n and let Γ be a collineation group of $\mathcal{P}$. Suppose there is a line l and a point Q on l such that $|\Gamma_{(A,l)}| = h > 1$ for all A in l, $A \neq Q$. Then $|\Gamma_{(Q,l)}| = n$, i.e. $\mathcal{P}$ is (Q, l)-transitive.*

Proof. Since each point of l, other than Q, is the centre of $h - 1$ non-trivial elations there are at least $n(h - 1)$ non-trivial elations in $\Gamma_{(l,l)}$. Thus $|\Gamma_{(l,l)}| \geq n(h - 1) + 1 > n$. So, by Theorem 4.17, $|\Gamma_{Q,l}| = k > 1$. Since Q is the centre of $k - 1$ non-trivial elations in $\Gamma_{(l,l)}$, we have

$$|\Gamma_{(l,l)}| = n(h - 1) + (k - 1) + 1 = n(h - 1) + k . \tag{1}$$

Let A be any point of l other than Q. If m is any line through A other than l then, by Result 1.13,

$$|\Gamma_{(l,l)}| = |(\Gamma_{(l,l)})_m| \cdot |m\Gamma_{(l,l)}| .$$

But, by Theorem 4.17, $(\Gamma_{(l,l)})_m = \Gamma_{(A,l)}$ and $|m\Gamma_{(l,l)}| \,|\, n$. Thus $|\Gamma_{(l,l)}| = hs$ where $s \,|\, n$. Using (1) we have $n(h - 1) + k = hs \,|\, hn$, i.e. $[nh - (n - k)] \,|\, hn$. But this implies that either $nh - (n - k) = nh$ or $nh - (n - k) \leq \frac{1}{2} nh$.
If $nh - (n - k) \leq \frac{1}{2} nh$, then $\frac{1}{2} nh \leq n - k$. But, since $h \geq 2$, this would imply $n \leq \frac{1}{2} nh \leq n - k$, i.e. $k \leq 0$. Since this contradicts $k > 1$, we must have $nh - (n - k) = nh$ i.e. $n = k$. Thus $|\Gamma_{(Q,l)}| = n$. $\square$

Once again this theorem has important corollaries whose proofs we leave as exercises.

Corollary 1 (Gleason). *Let $\mathcal{P}$ be a finite projective plane and let Γ be a collineation group of $\mathcal{P}$. If, for some line l, $|\Gamma_{(X,l)}| = h > 1$ for all X on l then l is a translation line.*

Corollary 2. *Let $\mathcal{P}$ be a finite projective plane and let Γ be a collineation group of $\mathcal{P}$. If, for some line l, $|\Gamma_{(l,l)}| > 1$ and Γ_l is transitive on the points of l, then l is a translation line.* (Again we remind the reader not to confuse Γ_l, (the subgroup of Γ fixing l), with $\Gamma_{(l)}$, (the subgroup of perspectivities with axis l).)

*Exercise 4.18. Let $\mathcal{P}$ be a finite projective plane and let Γ be a collineation group of $\mathcal{P}$. If, for some line l, $|\Gamma_{(l,l)}| > 1$ and Γ is transitive on the points of l, show that l is a translation line. (Hint: show that Γ_l must be transitive on the points of l and use Corollary 2.)

*Exercise 4.19. Let $\mathcal{P}$ be a finite projective plane and let Γ be a collineation group transitive on the points of $\mathcal{P}$. If Γ contains a non-trivial elation show that $\mathcal{P}$ is a Moufang plane.

Exercise 4.20. Let $\mathcal{P}$ be a finite projective plane. If $\alpha \neq 1$ is an (A, a)-elation and $\beta \neq 1$ is a (B, b)-homology with B on a but A not on b, show that $\langle \alpha, \beta \rangle$ contains a non-identity (B, a)-elation.

Exercise 4.21. Let $\mathcal{P}$ be a projective plane. If $\alpha \neq 1$ is an (A, a)-elation and $\beta \neq 1$ is a (B, b)-elation with $A \, \mathbb{I} \, b$, $B \, \mathbb{I} \, a$ show that if X is any point of $\mathcal{P}$ not on AB such that $X^{\alpha\beta} = X$, then $X = ab$.

Exercise 4.22. Let α be a (V, l)-perspectivity of a projective plane $\mathcal{P}$ and let $\mathcal{P}_0$ be a subplane of $\mathcal{P}$ such that $V \in \mathcal{P}_0$ and $l \in \mathcal{P}_0$. If, for some point A of $\mathcal{P}_0$, $A^\alpha \in \mathcal{P}_0$ and $A^\alpha \neq A$ show that $\mathcal{P}_0^\alpha = \mathcal{P}_0$, i.e. that α leaves $\mathcal{P}_0$ invariant.

Exercise 4.23. Let $\mathcal{A} = \mathcal{P}^l$ be an affine plane and let $\mathcal{A}_0 = \mathcal{P}_0^l$ for some subplane $\mathcal{P}_0$ containing l of $\mathcal{P}$. If α is any perspectivity with axis l show that either $\mathcal{A}_0^\alpha = \mathcal{A}_0$ or there is at most one point of $\mathcal{A}$ common to $\mathcal{A}_0$ and $\mathcal{A}_0^\alpha$. (What can be said about α if there is (i) one common point or (ii) no common point?)

7. Desargues Configuration

We now consider the configurational consequences of the existence of perspectivities in a projective plane $\mathcal{P}$. If α is a (V, l)-perspectivity of $\mathcal{P}$ then, by Theorem 4.7, the action of α is completely determined by the image of a single non-fixed element.

Suppose we are given P_1^α for some non-fixed point P_1. Then the image of P_2, for any non-fixed point P_2 not on $P_1 V$, may be constructed as follows: Let $P_1 P_2 \cap l = X$. Then, since $P_2 = VP_2 \cap XP_1$, $P_2^\alpha = (VP_2)^\alpha \cap (XP_1)^\alpha$ $= VP_2 \cap XP_1^\alpha$. (Note that $(VP_2)^\alpha = VP_2$ since α fixes V linewise.) For this, and all the following constructions, the reader is strongly urged to draw a rough diagram so that he can see exactly what is going on.

Now let P_3 be any point which is not on either of the lines VP_1 or VP_2. Then we may construct P_3^α by the above construction, but we now have a choice as we may use either P_1 or P_2 as the other point in the construction. Since P_3^α is unique, either choice must give the same point so that certain incidences are forced upon $\mathcal{P}$. Let $P_2 P_3 \cap l = Y$ and $P_3 P_1 \cap l = Z$. Then, from the above, $P_3^\alpha = VP_3 \cap ZP_1^\alpha = VP_3 \cap YP_2^\alpha$.

Any subset of the points and lines of $\mathcal{P}$ is called a *configuration*[2], and the configuration formed by the ten points $V, X, Y, Z, P_1, P_1^\alpha, P_2, P_2^\alpha, P_3, P_3^\alpha$

---

[2] See Chapter XI for an abstract definition.

and the ten lines l, $VP_1 P_1^\alpha$, $VP_2 P_2^\alpha$, $VP_3 P_3^\alpha$, $XP_1 P_2$, $XP_1^\alpha P_2^\alpha$, $YP_2 P_3$, $YP_2^\alpha P_3^\alpha$, $ZP_3 P_1$, $ZP_3^\alpha P_1^\alpha$ is called the *Desargues configuration*. Clearly any plane which admits a perspectivity has many Desargues configurations. However before discussing any consequences of this configuration we give a more formal definition.

Let Δ_i ($i = 1, 2$) be a triangle with vertices A_i, B_i, C_i and opposite edges a_i, b_i, c_i. If there is a point V such that $VA_1 = VA_2$, $VB_1 = VB_2$ and $VC_1 = VC_2$ then the triangles Δ_1, Δ_2 are said to be *in perspective from V*. Being in perspective from a line l is defined dually.

The earlier discussion on perspectivities may be summarized by

Theorem 4.27. *Let $\mathscr{P}$ be a projective plane. If α is a (V, l)-perspectivity and if Δ is any triangle having no side or vertex fixed by α, then the triangles Δ, Δ^α are in perspective from both V and l.* ☐

Note. Here we are using the "obvious" notation that if Δ is the triangle determined by A, B, C then Δ^α is the triangle $A^\alpha B^\alpha C^\alpha$.

We are now able to give a formal definition of the Desargues configuration. Let Δ_i ($i = 1, 2$) be any two triangles with vertices A_i, B_i, C_i and opposite edges a_i, b_i, c_i such that they are in perspective from a point V and a line l. Then the configuration formed by the ten points V, A_1, A_2, B_1, B_2, C_1, C_2, $la_1 a_2$, $lb_1 b_2$, $lc_1 c_2$ and the ten lines l, a_1, a_2, b_1, b_2, c_1, c_2, $VA_1 A_2$, $VB_1 B_2$, $VC_1 C_2$ is called the *Desargues configuration*. (See Fig. 5.) In the special situation where V is on l the configuration is often referred to as the *minor* or *little* Desargues configuration. (Draw a diagram!)

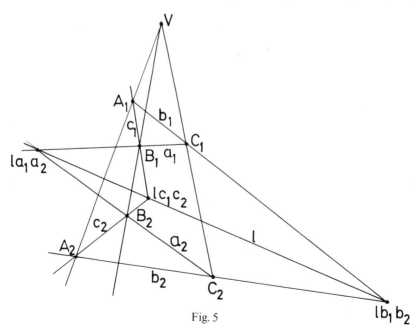

Fig. 5

This definition does not depend on the existence of any perspectivities of the plane. In Chapter XI we exhibit infinitely many infinite planes which admit no non-identity perspectivities. (Indeed we even exhibit some which admit no non-identity collineations at all.)

It is not known whether finite projective planes must admit perspectivities (or even collineations). However the following theorem, due to Ostrom, shows that every finite plane must contain a number of Desargues configurations.

Theorem 4.28. *Any finite projective plane $\mathscr{P}$ contains a pair of triangles which are in perspective from a point and a line.*

Proof. Let the order of $\mathscr{P}$ be n. We shall assume $n \geq 5$ since we shall show (see Chapter V) that, up to isomorphism, $\mathscr{P}_2(m)$ is the only plane of order m if $m \leq 4$ and the theorem is easily seen to be true for all planes $\mathscr{P}_2(K)$, where K is any field (see Exercise 4.24).

Let l be any line of $\mathscr{P}$ and let V be any point of $\mathscr{P}^l$. Let l_1, l_2, l_3 be any three distinct lines through V and let A, B be any distinct points of l such that neither is on any of the three lines l_1, l_2, l_3. (Those choices are all possible since $n > 4$.)

We now construct a mapping α from the points of $l_1 \setminus \{V, ll_1\}$ into the points of $l \setminus \{A, B, ll_1, ll_3\}$ as follows:

Pick any point $X \in l_1 \setminus \{V, ll_1\}$. Join A to X and let $X_2 = AX \cap l_2$. Now join B to X_2 and let $X_3 = BX_2 \cap l_3$. Finally join X to X_3 and define α by $X^\alpha = XX_3 \cap l$. (Once again the reader is urged to draw a diagram.)

Clearly, for any given X, X^α is unique. Also it should be clear that for any $X \in l_1 \setminus \{V, ll_1\}$ the point $X^\alpha \in l \setminus \{A, B, ll_1, ll_3\}$. The number of choices for X is $n - 1$ while the number of possible values for X^α is at most $n - 3$. Thus there exist two distinct points Y, Z on l_1 such that $Y^\alpha = Z^\alpha$. For any two points P, Q in $l_1 \setminus \{V, ll_1\}$, the triangles PP_2P_3, QQ_2Q_3 are in perspective from V. Also $PP_2 \cap QQ_2 = A$ I l and $P_2P_3 \cap Q_2Q_3 = B$ I l. Thus the triangles are in perspective from l if and only if $PP_3 \cap QQ_3$ is on l. But $PP_3 \cap l = P^\alpha$ and $QQ_3 \cap l = Q^\alpha$ so that the triangles PP_2P_3, QQ_2Q_3 are in perspective from l if and only if $P^\alpha = Q^\alpha$. Thus the triangles YY_2Y_3 and ZZ_2Z_3 are in perspective from both V and l. $\square$

It is worth noting that Theorem 4.28 implies the existence of many Desargues configurations in a given finite projective plane. There is at least one such configuration for each choice of $V, l, l_1, l_2, l_3, A, B$.

We have already seen that the existence of a single perspectivity of a projective plane $\mathscr{P}$ implies the existence of many Desargues configurations in $\mathscr{P}$. Since it is not known whether a finite plane must admit a collineation, it is not known whether the existence of a Desargues configuration in a finite projective plane guarantees the existence of a perspectivity. However we now show that the existence of sufficiently many Desargues configurations is equivalent to the existence of perspectivities.

A projective plane $\mathscr{P}$ is said to be (V, l)-*desarguesian* if, for each pair of non-degenerate triangles Δ_i $(i = 1, 2)$ with vertices A_i, B_i, C_i and opposite sides a_i, b_i, c_i such that (a) Δ_1 and Δ_2 are in perspective from V and (b) $a_1 a_2, b_1 b_2$ are on l, it follows that $c_1 c_2$ is on l.

Exercise 4.24. Show that, for any field K, $\mathscr{P}_2(K)$ is (V, l)-desarguesian for all choices of V and l.

We now prove the very important fact, originally proved by Baer, that the configurational property of being (V, l)-desarguesian is equivalent to being (V, l)-transitive, which is a collineation group property.

Theorem 4.29 (Baer). *A projective plane $\mathscr{P}$ is (V, l)-transitive if and only if $\mathscr{P}$ is (V, l)-desarguesian.*

Proof. Suppose $\mathscr{P}$ is (V, l)-transitive. Then Theorem 4.27 implies that $\mathscr{P}$ is (V, l)-desarguesian.

Suppose $\mathscr{P}$ is (V, l)-desarguesian. If A, A_1 are any two distinct points such that $A \neq V \neq A_1$, $VA = VA_1$ and neither A or A_1 is on l, then we must construct a (V, l)-perspectivity δ with $A^\delta = A_1$.

We define a mapping $\alpha = \alpha_{AA_1}$ on the points of $(\mathscr{P} \backslash A A_1) \cup \{V\}$ by:

(i) if X is on l, $X^\alpha = X$;

(ii) $V^\alpha = V$;

(iii) if B is not on l or AA_1, let $Y = l \cap AB$, $B_1 = YA_1 \cap VB$ and define $B_1 = B^\alpha$.

(Note that all we are doing here is defining α in such a way that if the required collineation δ exists then δ and α act identically on the points of $(\mathscr{P} \backslash A A_1) \cup \{V\}$.)

For any point B of $\mathscr{P} \backslash A A_1$, $B \neq V$ and not incident with l we can define a mapping $\beta = \alpha_{BB_1}$, where $B_1 = B^\alpha$, in a similar way. We now show that, for any point C not on any of the lines l, AA_1, BB_1, $C^\alpha = C^\beta$.

If C is on AB then clearly $C^\alpha = C^\beta$. (This is because $YB_1 = YA_1$ so that $C^\alpha = YA_1 \cap VB = C^\beta$. Note that this uses the fact that $B_1 = B^\alpha$.)

Suppose C is not on AB. Let $C_1 = C^\alpha$, $P = AB \cap l$, $Q = BC \cap l$ and $R = CA \cap l$. The triangles ABC, $A_1 B_1 C_1$ are in perspective from V. Furthermore, by the definition of α, $AB \cap A_1 B_1 = P$ and $CA \cap C_1 A_1 = R$. Thus since P and R both lie on l and $\mathscr{P}$ is (V, l)-desarguesian, $BC \cap B_1 C_1$ must also lie on l. But $BC \cap l = Q$. Hence Q, B_1, C_1 are collinear and $C_1 = C^\beta$.

We can now define a mapping δ on the points of $\mathscr{P}$ by means of A, A_1 and any other pair B, B^α where B is not on either VA or l. In order to extend δ to a collineation of $\mathscr{P}$ we must define its action on the lines.

Given any line m distinct from l and AA_1, pick any point T on m, but not on l or AA_1, and define m^δ to be the line joining lm to T^δ. If we define $l^\delta = l$ and $(AA_1)^\delta = (AA_1)$ then all we need to do to show that δ is a collineation is to show that m^δ is independent of the choice of T. To show this we merely need to show that S^δ is on m^δ for all S on m, i.e. if S is any other point of m we must show that lm, S^δ, T^δ are collinear. But the two triangles AST, $A_1S^\delta T^\delta$ are in perspective from V and both $AS\cap A_1S^\delta$ and $AT\cap AT^\delta$ are on l. Thus, since $\mathscr{P}$ is (V,l)-desarguesian, $S^\delta T^\delta\cap ST$ is on l. Hence δ is a collineation and the theorem is proved. $\square$

A projective plane $\mathscr{P}$ is called *desarguesian* if it is (V,l)-desarguesian for all choices of V and l. As a result of Theorem 4.29, a Moufang plane is (V,l)-desarguesian for all choices of V and l with V on l.

As a final remark, concerned with research in this area, we mention that there is a classification called the "Lenz-Barlotti Classification" which lists all possibilities for the configuration formed by the total set of points V and lines l for which a projective plane may be (V,l)-desarguesian. The list is exhibited in [1]. A great amount of research has been carried out on this list in an attempt to find examples of planes with a given configuration or, alternatively, to show that such a plane cannot exist. For a recent survey article on the work in this direction see [2].

Exercise 4.25. If l is a translation line of a projective plane $\mathscr{P}$ show that $\Gamma_{(l,l)}=\Gamma_{(A,l)}\cdot\Gamma_{(B,l)}$ for any A,B on l, $A\ne B$.

Exercise 4.26. If $\mathscr{P}$ is a Moufang plane and if Γ is the subgroup of $\mathrm{Aut}\mathscr{P}$ generated by all the elations of $\mathscr{P}$ show that Γ is transitive on the triangles of $\mathscr{P}$.

*Exercise 4.27. Let $\mathscr{P}$ be a finite projective plane of order n and let $\Pi=\mathrm{Aut}\mathscr{P}$. If there is a line l such that $|\Pi_{(l,l)}|>n$ show
(a) that $|\Pi_{(A,l)}|>1$ for all A on l,
(b) that n is a prime power.
(Part (a) is straightforward but (b) is difficult.)

*Exercise 4.28. Let $\mathscr{P}$ be a projective plane such that $\mathrm{Aut}\mathscr{P}$ fixes no point or line of $\mathscr{P}$. If $\mathscr{P}$ is (V,l)-transitive for some point V and line l with V on l, show that $\mathscr{P}$ is a Moufang plane. (Hint: refer back to Lemma 4.15 and Exercise 4.18.)

Exercise 4.29. Let $\mathscr{P}$ be a projective plane. If α is a Baer involution with Baer subplane $\mathscr{P}_0$ and if β is an involutory (V,l)-elation such that l is a line of $\mathscr{P}_0$ but V is not in $\mathscr{P}_0$, show that $\alpha\beta$ has order 4. Show also that $|\langle\alpha,\beta\rangle|=8$, that $\langle\alpha,\beta\rangle$ contains exactly two Baer involutions and that $\langle\alpha,\beta\rangle$ has a normal subgroup of order 4 consisting of elations with axis l.

References

Too much work has been done on the Lenz-Barlotti classification to allow us to include a comprehensive discussion of it in this book. The classification is listed in [1] (pp. 123–126) and [2] is an excellent survey article on recent research on this topic. Exercise 4.17 is really a result in group theory; a proof is given in [3] (p. 73).

1. Dembowski, H. P.: Finite geometries. Berlin-Heidelberg-New York: Springer 1968.
2. Yaqub, J. C. D. S.: The Lenz-Barlotti classification. Proceedings of Projective Geometry Conference, University of Illinois, Chicago 1967.
3. Hall, M.: The theory of groups. London-New York: MacMillan 1959.

V. Coordinatization

1. Introduction

In this chapter we give a method for coordinatizing a projective plane with a planar ternary ring and establish various algebraic properties of the coordinatizing systems. (A few diagrams are included; but the reader is urged to draw additional ones at every opportunity to illustrate the geometric construction behind the algebraic properties of the planar ternary rings.) Latin squares are introduced and their relation with projective planes is discussed. This discussion ends with a set of exercises using latin squares to show, up to isomorphism, the uniqueness of the projective planes of orders 3 and 4.

The method of coordinatiziation given in this chapter is not the only one in common use. Although we shall always use the method given here there is an appendix to discuss other methods.

This chapter is very elementary and requires virtually no previous knowledge. It may, in fact, be read before Chapter IV.

2. Introduction of Coordinates

In order to show that the planes $\mathscr{P}_2(K)$ are not the only examples of projective planes it is necessary to have ways of representing projective planes and of determining whether two given projective planes are isomorphic. By far the most powerful concept in this respect is that of coordinatization.

Let $\mathscr{P}$ be a projective plane of order n and let R be any set of symbols with cardinal n such that $0, 1 \in R$, $0 \neq 1$, but the symbol "∞" is not in R. Choose any line of $\mathscr{P}$ and label this line l_∞ and then pick any other two lines l_1, l_2 with the only restriction that l_1, l_2, l_∞ form the sides of a nondegenerate triangle. Let $X = l_2 l_\infty$, $Y = l_\infty l_1$ and $O = l_1 l_2$. Finally let I be any point not incident with any side of the chosen triangle.

We now use the elements of R plus the extra symbol ∞ to coordinatize $\mathscr{P}$ relative to the quadrangle O, X, Y, I. However we first label three more points; $A = XI \cap l_1$, $B = YI \cap l_2$ and $J = AB \cap l_\infty$.

In order to coordinatize $\mathscr{P}$ we assign the elements of R to the points of $l_1 \backslash Y$ in an arbitrary manner except that 0 is assigned to O and 1 is assigned to A. If $c \in R$ is assigned to the point $C \in l_1$ then we write $(0, c)$ for C.

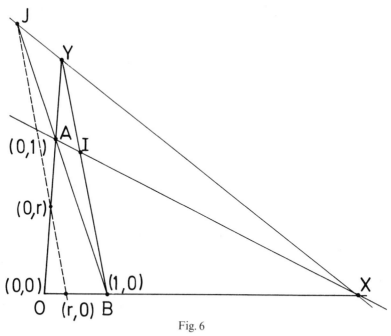

Fig. 6

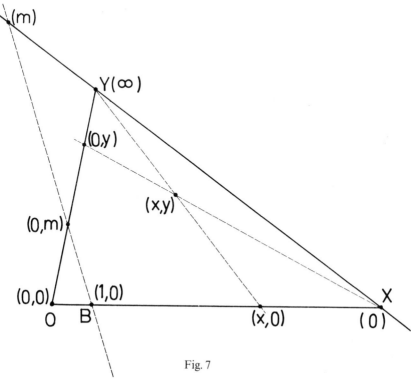

Fig. 7

For any point D on l_2, $D \neq X$, let $D' = JD \cap l_1$ and then, if D' is $(0, d)$, we write $(d, 0)$ for D. Since 0 has been assigned to O, O has been given unique coordinates $(0, 0)$. For any point E not on l_∞, if $XE \cap l_1$ is $(0, g)$ and $YE \cap l_2$ is $(f, 0)$ then E is given the coordinates (f, g). In this manner every point of $\mathscr{P}^{l_\infty}$ has been given unique coordinates (x, y) where $x, y \in R$.

If M is any point of l_∞ other than Y and if the line joining M to $(1, 0)$ meets l_1 in $(0, m)$ then M is given the coordinate (m). Finally we coordinatize Y by (∞) and, in this way, every point of $\mathscr{P}$ has been coordinatized, the coordinatization depending only on the initial choice of O, X, Y, I and the way in which the elements of R are assigned to the points of $l_1 \setminus Y$. (See Figs. 6 and 7.)

We now coordinatize the lines in a way that utilizes the coordinatization of the points. If l is any line not containing Y then, if l meets l_∞ at the point (m) and l_1 at the point $(0, k)$, we give l the coordinates $[m, k]$. If l contains Y but is distinct from l_∞ then we call l the line $[k]$ where k is determined by $l \cap l_2 = (k, 0)$. Finally we call l_∞ the line $[\infty]$ and in this way we have coordinatized every point and line of $\mathscr{P}$. (See Fig. 8.) Having done this we must now build into our coordinatization system a method for determining whether two given elements are incident.

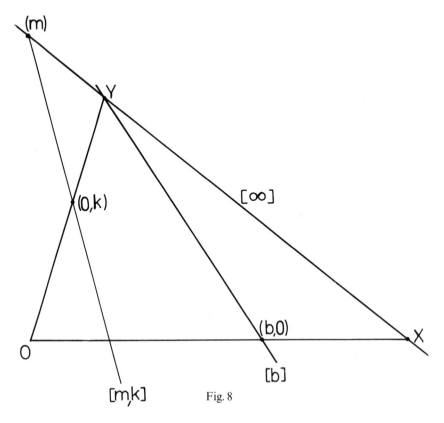

Fig. 8

3. Planar Ternary Rings

If $\mathscr{S}$ is any set then a *ternary operation* T on $\mathscr{S}$ is a rule which assigns to any three ordered elements a, b, c of $\mathscr{S}$ a unique element $T(a, b, c)$ of $\mathscr{S}$. A non-empty set $\mathscr{S}$ with a ternary operation T is called a *ternary ring* which we denote by $(\mathscr{S}, T)$.

If a projective plane $\mathscr{P}$ has been coordinatized by the elements of a set R in the manner described in Section 2, then we use the incidences of $\mathscr{P}$ to define a ternary operation T on R as follows; if $a, b, c \in R$, $T(a, b, c) = k$ if and only if (b, c) is on $[a, k]$. Thus $(0, k)$ is the intersection of l_1 with the line joining (a) to (b, c) so that, given a, b, c, the value of k is uniquely determined. (See Fig. 9.)

Theorem 5.1. *Let $\mathscr{P}$ be a projective plane coordinatized by a set R. If T is defined by $T(a, b, c) = k$ if and only if (b, c) is on $[a, k]$ then the following properties hold.*

(A) $T(a, 0, c) = T(0, b, c) = c$ *for all* $a, b, c \in R$.

(B) $T(a, 1, 0) = T(1, a, 0) = a$ *for all* $a \in R$.

(C) *If* $a, b, c, d \in R$, $a \neq c$, *then there is a unique* $x \in R$ *such that* $T(x, a, b) = T(x, c, d)$.

(D) *If* $a, b, c \in R$ *then there is a unique* $x \in R$ *such that* $T(a, b, x) = c$.

(E) *If* $a, b, c, d \in R$, $a \neq c$, *then there is a unique ordered pair* $x, y \in R$ *such that* $T(a, x, y) = b$ *and* $T(c, x, y) = d$.

Proof. (A) $T(a, 0, c) = k$ implies that the point $(0, c)$ is on the line $[a, k]$, i.e. $(0, c)$ is on the line joining (a) to $(0, k)$. But any line intersects l_1 in a unique point. Thus $(0, c) = (0, k)$, i.e. $c = k$ and $T(a, 0, c) = c$.

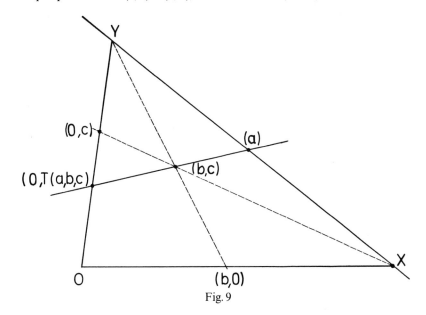

Fig. 9

$T(0, b, c) = k$ implies (b, c) is on the line joining (0) to $(0, k)$. But (b, c) is the intersection of the line joining (0) to $(0, c)$ with the line joining (∞) to (b, c). Thus, since the line joining (b, c) to (0) has a unique intersection with l_1, $c = k$ and $T(0, b, c) = c$.

(B) $T(a, 1, 0) = k$ implies that $(1, 0)$ is on $[a, k]$, i.e. $(1, 0)$ is on the line joining (a) to $(0, k)$. But (a) is the intersection of the line joining $(1, 0)$ to $(0, a)$ with l_∞, and so, since the line joining $(1, 0)$ to (a) intersects l_1 in a unique point, $(0, k) = (0, a)$ or $k = a$. Hence $T(a, 1, 0) = a$.

$T(1, a, 0) = k$ implies $(a, 0)$ is on the line joining (1) to $(0, k)$. But $(a, 0)$ is the intersection of l_2 with the line joining (1) to $(0, a)$. Thus $(0, a) = (0, k)$ and $T(1, a, 0) = a$.

(C) Let $a, b, c, d \in R$, $a \neq c$. Then there is a unique line joining (a, b) to (c, d). Since $a \neq c$ this line does not pass through (∞) and, hence, intersects l_∞ in a unique point (m) where $m \in R$. If this line also intersects l_1 in $(0, k)$ then $T(m, a, b) = T(m, c, d) = k$, and so since (m) is unique, there is a unique $x \in R$ such that $T(x, a, b) = T(x, c, d)$.

(D) $T(a, b, x) = c$ if and only if (b, x) is on the line joining (a) to $(0, c)$. But, for any x, (b, x) is on the line joining (∞) to $(b, 0)$. These two lines intersect in a unique point and so there is a unique x such that $T(a, b, x) = c$.

(E) $T(a, x, y) = b$ if and only if (x, y) is on the line $[a, b]$ and $T(c, x, y) = d$ if and only if (x, y) is on $[c, d]$. But the lines $[a, b]$ and $[c, d]$ intersect in a unique point which, since $a \neq c$, is not on l_∞. Thus there is a unique ordered pair $x, y \in R$ such that $T(a, x, y) = b$ and $T(c, x, y) = d$. $\square$

Properties (A), (B) of Theorem 5.1 reflect the fact that O, the point $(0, 0)$, and I, the point $(1, 1)$, play a special role in the coordinatization of $\mathscr{P}$ and it is not really surprising that $0, 1$ have special properties in the resulting ternary ring. Any ternary ring with two distinct elements $0, 1$ satisfying properties (A)–(E) of Theorem 5.1 is called a *planar ternary ring*, which we shall usually abbreviate to PTR.

Theorem 5.2. *If (R, T) is a PTR then the structure $\mathscr{P}$ defined as follows is a projective plane. The points of $\mathscr{P}$ are ordered pairs (x, y), where $x, y \in R$ together with elements of the form (x), where $x \in R$ and (∞) where ∞ is a symbol not contained in R. Lines are represented by ordered pairs $[m, k]$, where $m, k \in R$ together with elements of the form $[m]$, where $m \in R$ and $[\infty]$. Incidence is defined in the following manner:*

(x, y) *is on* $[m, k] \Leftrightarrow T(m, x, y) = k$,
(x, y) *is on* $[k]$ $\quad \Leftrightarrow x = k$,
$\quad (x)$ *is on* $[m, k] \Leftrightarrow x = m$,
$\quad (x)$ *is on* $[\infty]$ *for all* $x \in R$ *and* (∞) *is on* $[k]$ *for all* $k \in R$. *Finally* (∞) *is on* $[\infty]$.

Proof. We first show that any two points are on a unique line. If $a \neq c$ then, by (C) of Theorem 5.1, for any given $b, d \in R$ there is a unique $m \in R$ such that $T(m, a, b) = T(m, c, d)$. Thus, if $a \neq c$, the points (a, b), (c, d) are on the unique line $[m, T(m, a, b)]$.

Clearly the two points (a, b), (a, d) are on the line $[a]$. If (a, b), (a, d) were also on a line $[m, k]$ then we would have $T(m, a, b) = k = T(m, a, d)$ which contradicts (D) of Theorem 5.1. Thus there is a unique line joining any two distinct points such that neither is on $[\infty]$.

Let (m) be any point not (∞) on $[\infty]$ and (a, b) any point not on $[\infty]$. Any line through (m) is either $[\infty]$ or of the form $[m, k]$ for some k, and the line $[m, k]$ passes through (a, b) if and only if $T(m, a, b) = k$. But, since T is a ternary operation, $T(m, a, b)$ is uniquely determined by m, a and b so that $[m, T(m, a, b)]$ is the unique line of $\mathscr{P}$ containing (m) and (a, b). Since any line not $[\infty]$ which contains (∞) is of the form $[k]$ for some $k \in R$, the unique line of $\mathscr{P}$ joining (∞) to (a, b) is $[a]$. Finally (m_1), (m_2) clearly have $[\infty]$, and no other line, in common so we have shown that any two points of $\mathscr{P}$ are on a unique line of $\mathscr{P}$.

In order to show that any two distinct lines of $\mathscr{P}$ intersect in a unique point of $\mathscr{P}$ it is sufficient, by Exercise 3.2, to show that they have at least one point in common. We first consider two lines $[m_1, k_1]$, $[m_2, k_2]$. If $m_1 \neq m_2$ then, by (E) of Theorem 5.1, there exists a unique ordered pair (a, b) such that $T(m_1, a, b) = k_1$ and $T(m_2, a, b) = k_2$ so that the point (a, b) is on both $[m_1, k_1]$ and $[m_2, k_2]$. If $m_1 = m_2$ then (m_1) is on both lines; hence any pair of distinct lines of the form $[m, k]$ have a common point. Clearly the line $[m, k]$ meets $[\infty]$ at (m) so, in order to show a line of type $[m, k]$ intersects any other line, we need only consider the intersection of $[m, k]$ with $[h]$, $h \neq m$. But these two lines intersect in the point (h, h') where h' is given by $T(m, h, h') = k$, (h' exists by (D) of Theorem 5.1). Since any two lines $[m_1]$, $[m_2]$ intersect at (∞), we have shown that any two lines of $\mathscr{P}$ intersect in a unique point of $\mathscr{P}$.

Finally we must exhibit a quadrangle in $\mathscr{P}$. Let $A = (0)$, $B = (\infty)$, $C = (0, 0)$, $D = (1, 1)$. Then $AB = [\infty]$, $BC = [0]$, $CA = [0, 0]$ and it is clear none of these lines contains either of the other points. Thus A, B, C, D is a quadrangle and $\mathscr{P}$ is a projective plane. $\square$

Note that if we coordinatize $\mathscr{P}$ by choosing (∞) as Y, (0) as X, $(0, 0)$ as O, $(1, 1)$ as I and then assign $b \in R$ to $(0, b)$ on OY we get a PTR identical to (R, T).

If (R, T) and (R', T') are two ternary rings we say that (R, T) is *isomorphic* to (R', T') if there is a one to one mapping α from the elements of R onto R' such that $(T(a, b, c))^\alpha = T'(a^\alpha, b^\alpha, c^\alpha)$ for all $a, b, c \in R$.

Exercise 5.1. Let $\mathscr{P}$ be a projective plane. If a set R is used to coordinatize $\mathscr{P}$ by using a quadrangle O, X, Y, I and if (R, T), (R, T') are two distinct PTR's which arise by assigning the elements of R to the points of OY in different ways, show that $(R, T) \cong (R, T')$.

Exercise 5.2. Let $\mathscr{P}$ be a projective plane. If (R, T) is a coordinatizing PTR relative to O, X, Y, I and if (R', T') coordinatizes $\mathscr{P}$ relative to O', X', Y', I', show that $(R, T) \cong (R', T')$ if and only if there is a collineation $\alpha \in \operatorname{Aut} \mathscr{P}$ with $O^\alpha = O'$, $X^\alpha = X'$, $Y^\alpha = Y'$, $I^\alpha = I'$.

Since (x, y) is on $[k]$ if and only if $x = k$, we might refer to $[k]$ as the line $x = k$.

Before we discuss the algebraic properties of PTR's, it is worth noting that there are many ways of coordinatizing a projective plane even after the coordinatizing set R and the quadrangle O, X, Y, I have been chosen. As well as the one adopted here there are two others in common use, and a brief discussion of these systems is given in the appendix at the end of the chapter.

4. Algebraic Properties of Planar Ternary Rings

A non empty set G with a binary operation $\cdot$ is called a *loop* if:
  (i) $a \cdot x = b$ has a unique solution in x for any $a, b \in G$.
  (ii) $y \cdot a = b$ has a unique solution in y for any $a, b \in G$.
  (iii) G has an element e such that $e \cdot x = x \cdot e = x$ for all $x \in G$ (the element e is called an *identity* of G).

A system which satisfies (i), (ii) is called a *quasigroup* and we exhibit a quasigroup which is not a loop. $G = \{a, b, c\}$ and the operation $\cdot$ is given by the following "operation" table.

	a	b	c
a	a	b	c
b	c	a	b
c	b	c	a

In case any reader has not met such a table before, we also write out the complete effect of $\cdot$ so that the reader may learn how to interpret the table: $a \cdot a = a$, $a \cdot b = b$, $a \cdot c = c$, $b \cdot a = c$, $b \cdot b = a$, $b \cdot c = b$, $c \cdot a = b$, $c \cdot b = c$, $c \cdot c = a$.

An "operation" table for a binary system is often called a *Cayley* table.

Loops have much less structure than groups and very few of the elementary theorems about groups extend to loops. Certainly there is no obvious equivalent to Lagrange's theorem and to illustrate this we exhibit the Cayley table for a loop with five elements in which each element has order 2;

	e	a	b	c	d
e	e	a	b	c	d
a	a	e	c	d	b
b	b	d	e	a	c
c	c	b	d	e	a
d	d	c	a	b	e

Exercise 5.3. Defining isomorphism of loops in the obvious way, show that, up to isomorphism, there is only one loop of order 5 in which each element has order 2.

We now introduce binary operations of addition and multiplication into a PTR (R, T) in such a way that $(R, +)$ and $(R^*, \cdot)$, where R^* is defined as the set of non-zero elements of R, are loops. For any $a, b \in R$ we define $a \cdot b = ab = T(a, b, 0)$ and $a + b = T(1, a, b)$. Since T is a ternary operation, both ab and $a + b$ are uniquely determined by a and b. (See Fig. 10.)

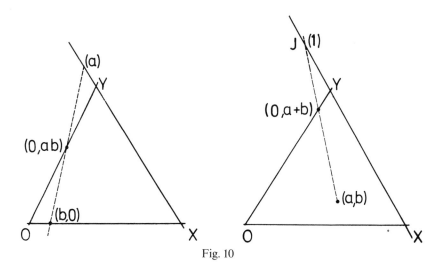

Fig. 10

Theorem 5.3. *If (R, T) is a PTR then $(R, +)$ and $(R^*, \cdot)$ are loops with* 0, 1 *as respective identities.*

Proof. (i) Addition.

For any $a \in R$, $0 + a = T(1, 0, a) = a$ and $a + 0 = T(1, a, 0) = a$ (by (A) and (B) of Theorem 5.1). Thus $(R, +)$ has 0 as an identity.

The equation $a + x = b$ has a unique solution for x if and only if $T(1, a, x) = b$ has a unique solution for x. But this is true by (D) of Theorem 5.1.

The equation $x + a = b$ has a unique solution for x if and only if $T(1, x, a) = b$ has a unique solution for x. By (E) of Theorem 5.1 there exists a unique ordered pair x, y such that $T(1, x, y) = b$ and $T(0, x, y) = a$. But, by (A), $T(0, x, y) = a$ has $y = a$ as a solution and so there exists a unique x such that $T(1, x, a) = b$. Hence, $(R, +)$ is a loop.

(ii) Multiplication.

In this case we first have to show that R^* is closed under multiplication, i.e. that if $x \neq 0$ and $y \neq 0$ then $xy \neq 0$. Suppose $xy = 0$ and $y \neq 0$. Consider the equation $T(u, y, 0) = T(u, 0, 0)$. By (C) there is a unique solution for u, and so, since $u = 0$ is certainly a solution, the only solution to $T(u, y, 0)$ $= T(u, 0, 0)$ is $u = 0$. But since $xy = 0$, $T(x, y, 0) = 0$ and, certainly, $T(x, 0, 0) = 0$. Thus $x = 0$ and R^* is closed under multiplication.

By (B) of Theorem 5.1, $1 \cdot x = T(1, x, 0) = x$ and $x \cdot 1 = T(x, 1, 0) = x$, so 1 is an identity element for $(R^*, \cdot)$.

The equation $ax = b$ has a unique solution for x if and only if $T(a,x,0) = b$ has. By (E) of Theorem 5.1 there is a unique ordered pair (x, y) such that

$$T(a, x, y) = b \tag{1}$$

and

$$T(0, x, y) = 0 . \tag{2}$$

Thus, since (2) implies $y = 0$, $T(a, x, 0) = b$ has a unique solution for x.

The equation $xa = b$ has a unique solution for x if and only if $T(x, a, 0) = b$ has. By (C) of Theorem 5.1, $T(x, a, 0) = T(x, 0, b)$ has a unique solution for x (since $a \neq 0$), and so, since, by (A), $T(x, 0, b) = b$ for all x, $T(x, a, 0) = b$ has a unique solution and $(R^*, \cdot)$ is a loop. □

A PTR was defined as a ternary ring which satisfied conditions (A)–(E) of Theorem 5.1. We now show that if the ternary ring is finite then one of these conditions is a consequence of the other four.

Theorem 5.4. *A finite ternary ring satisfies conditions* (C) *and* (D) *if and only if it satisfies conditions* (D) *and* (E).

Proof. Suppose (R, T) satisfies conditions (C) and (D).

Let $a, c \in R$, $a \neq c$. If for some $b, d \in R$ the equations $T(a, x, y) = b$, $T(c, x, y) = d$ have no solution for x and y then, since R is finite, for some other pair $b_1, d_1 \in R$ the equations $T(a, x, y) = b_1$ and $T(c, x, y) = d_1$ have at least two solutions. Thus, in order to prove that (E) is satisfied, if suffices to show that for given $a, b, c, d \in R(a \neq c)$ the equations $T(a, x, y) = b$ and $T(c, x, y) = d$ have at most one solution for the ordered pair (x, y). Suppose

(1) $T(a, x, y) = b = T(a, u, v)$

and

(2) $T(c, x, y) = d = T(c, u, v)$.

If $x = u$ then $T(c, x, y) = d = T(c, x, v)$ and, by (D), $y = v$. If $x \neq u$, then $T(z, x, y) = T(z, u, v)$ has distinct solutions $z = a$ and $z = c$. Since this contradicts (C), $T(a, x, y) = b$, $T(c, x, y) = d$ have a unique solution for x, y, i.e. (R, T) satisfies condition (E).

Suppose (R, T) satisfies conditions (D) and (E).

Given $a, b, c \in R$ $(a \neq c)$ we define for each $x \in R$ the element xA given by $T(x, a, b) = T(x, c, xA)$. Given x, the element xA is unique by (D), and so, in order to show that (R, T) satisfies condition (C) we have to show that A is a one-to-one mapping from R onto itself. Since R is finite it is sufficient to show that A is one-to-one.

Suppose we have two distinct elements x, y of R such that $xA = yA$. Then $T(x, a, b) = T(x, c, xA) = g$ and $T(y, a, b) = T(y, c, xA) = h$. This means the equations $T(x, u, v) = g$, $T(y, u, v) = h$ with $x \neq y$ have the solutions $u = a$, $v = b$ and $u = c$, $v = xA$. But, by (E), the equations $T(x, u, v) = g$, $T(y, u, v) = h$ have a unique solution for (u, v). Since this contradicts the fact that $a \neq c$, A must be a one-to-one mapping from R onto itself and the theorem is proved. ☐

This theorem now enables us to prove a very strong theorem about finite substructures of planar ternary rings.

Theorem 5.5. *If (R, T) is a PTR and (R_0, T) is a finite non-empty ternary subring of (R, T), then either R_0 consists of 0 alone or (R_0, T) is planar.*

We first prove a lemma.

Lemma 5.6. *If a finite ternary ring (R, T) satisfies* (D) *and if equations of the form $T(x, a, b) = T(x, c, d)$, $a \neq c$, have at most one solution for x, then they have exactly one solution for x.*

Proof of Lemma. Given $x, a, b, c \in R$ $(a \neq c)$ define xA by $T(x, a, b) = T(x, c, xA)$. Then, by (D), xA is uniquely determined. If $x \neq y$ but $xA = yA$, then $T(x, a, b) = T(x, c, xA)$ and $T(y, a, b) = T(y, c, xA)$, which implies that $T(u, a, b) = T(u, c, xA)$ has distinct solutions $u = x$ and $u = y$ and contradicts our assumption. Thus A is one-to-one and, since R is finite, onto. Hence there is a unique x with $xA = d$ and $T(x, a, b) = T(x, c, d)$ has a unique solution. ☐

Proof of Theorem 5.5. For fixed $a, b \in R_0$ consider the mapping S given by $xS = T(a, b, x)$. Since (R_0, T) is a finite ternary ring, R_0 is closed under T, and $xS \in R_0$ for all $x \in R_0$. Furthermore, since (R, T) is planar $xS = yS$ if and only if $x = y$ (by condition (D) of Theorem 5.1). Thus since R_0 is finite, S is a one-to-one mapping from R_0 onto itself and R_0 satisfies (D).

Since every element of R_0 is also an element of R any equation $T(x, a, b) = T(x, c, d)$ $(a, b, c, d \in R_0, a \neq c)$ has at most one solution in R_0. [It has exactly one solution in R, by (C), but this may not lie in R_0.] Hence, by Lemma 5.6, such equations have exactly one solution in R_0. This implies that (R_0, T) satisfies (C) and (D) and thus, since R_0 is finite, (R_0, T) also satisfies (E) (by Theorem 5.4).

We must now show that either $R_0 = \{0\}$ or $0, 1 \in R_0$, thus finishing the proof of the theorem.

Suppose $R_0 = \{a\}$. Then, since R_0 is closed under T, $T(a, a, a) = a$. If $a \neq 0$ the equation $T(x, a, a) = T(x, 0, a)$ has distinct solutions $x = 0$ and $x = a$ in R, since $T(a, 0, a) = T(0, 0, a) = T(0, a, a) = a$ by (A). But this contradicts (C), and so if R_0 has only one element then $R_0 = \{0\}$.

Suppose $a, b \in R_0 (a \neq b)$ and consider the equation $T(x, a, c) = T(x, b, c)$ for any $c \in R_0$. Equations of this type have a unique solution in R which, by application of Lemma 5.6, must lie in R_0. Since, clearly, $x = 0$ is a solution, we have $0 \in R_0$.

Finally since $T(x, a, 0) = T(x, 0, a)$ for $a \in R_0$, $a \neq 0$, has the unique solution 1 in R, $1 \in R_0$. ☐

Exercise 5.4. Let R be the set of rationals and let T be given by $T(a, b, c) = ab + c$. Show that (R, T) is a PTR. If R_0 is the subset of integers show that (R_0, T) is a ternary subring which is not planar.

As an example we now find a PTR to coordinatize the plane $\mathscr{P}_2(K)$ for any skewfield K. Changing the notation of Chapter II a little we recall that $\mathscr{P}_2(K)$ was defined as follows.

Points of $\mathscr{P}_2(K)$ are ordered triples $(x, y, z) \neq (0, 0, 0)$ where $x, y, z \in K$ with $(x, y, z) = (xk, yk, zk), k \neq 0$; lines of $\mathscr{P}_2(K)$ are ordered triples $[l, m, n]$ $\neq [0, 0, 0]$ where $l, m, n \in K$ with $[l, m, n] = [kl, km, kn], k \neq 0$; incidence is given by (x, y, z) is on $[l, m, n]$ if, and only if, $lx + my + nz = 0$.

Choose as $[\infty]$ the line $[0, 0, 1]$. Choose $(0, 1, 0)$ as (∞), $(1, 0, 0)$ as (0), $(0, 0, 1)$ as $(0, 0)$ and $(1, 1, 1)$ as $(1, 1)$. Take as the elements of the PTR the elements of K and assign $k \in K$ to the point $(0, k, 1)$.

The point $(1, 0)$ is the intersection of the line joining $(0, 0)$ to (0) with the line joining (∞) to $(1, 1)$. Thus it is the intersection of $[0, 1, 0]$ with $[-1, 0, 1]$ which is the point $(1, 0, 1)$. Thus (1), which is the intersection of $[\infty]$ with the line joining $(1, 0)$ to $(0, 1)$, is the intersection of $[0, 0, 1]$ with $[-1, -1, 1]$, i.e. $(1, -1, 0)$. Since $(a, 0)$ is collinear with $(0, a)$ and (1), similar considerations show that $(a, 0)$ is $(a, 0, 1)$ and it is now straightforward to show that (a, b) is $(a, b, 1)$. Similarly the point (m) is the point $(1, -m, 0)$.

The dual arguments yield that $[m, k]$ is the line $[m, 1, -k]$ and that $[k]$ is $[1, 0, -k]$.

We now determine the ternary operation T. Let $\oplus$, $\odot$ denote the addition and multiplication of (K, T). For any $a, b \in K$, $a \oplus b = T(1, a, b)$. But $T(1, a, b) = k$ if and only if (a, b) is on $[1, k]$, i.e. if and only if $(a, b, 1)$ is on $[1, 1, -k]$. But this implies $a + b - k = 0$, i.e. that $a + b = k$, and so the addition of the ternary ring (K, T) is the addition of the original skewfield K.

For any $a, b \in K$, $a \odot b = T(a, b, 0)$. But $T(a, b, 0) = k$ if and only if $(b, 0)$ is on $[a, k]$; i.e. if and only if $(b, 0, 1)$ is on $[a, 1 - k]$. Since this implies $ab + 01 - k = 0$, the multiplication of the ternary ring (K, T) is also the multiplication of the skewfield K.

Finally we determine $T(m, x, y)$. $T(m, x, y) = k$ if and only if (x, y) is on $[m, k]$, i.e. if and only if $(x, y, 1)$ is on $[m, 1, -k]$. Since this implies $mx + y - k = 0$ we have $T(m, x, y) = mx + y$, i.e. $T(m, x, y) = m \odot x \oplus y$.

If (R, T) is any PTR with $T(a, b, c) = ab + c$ for all $a, b, c \in R$ then we say that (R, T) is *linear*. We have just shown that $\mathscr{P}_2(K)$ may be coordinatized by a linear PTR (K, T) which is isomorphic to K. Most of the PTR's which we meet in this book will be linear.

5. Latin Squares

A *latin square* of order n is an $n \times n$ matrix with entries from a set R of n distinct elements, which we shall usually take to be $0, 1, 2, \ldots, n-1$, such that each row and column contains every element of R exactly once. Thus the Cayley table for any quasigroup provides an example of a latin square.

Two latin squares are said to be *orthogonal* if the n^2 ordered pairs (i, j), where i and j are the entries from the same position in the respective squares, exhaust the n^2 possibilities.

We now show how orthogonal latin squares and PTR's are closely related.

If (R, T) is a PTR of order n with $R = \{0, 1, \ldots, n-1\}$ then for any $x \in R, x \neq 0$, we define an $n \times n$ matrix $\{x\}$ as follows; in row i and column j of $\{x\}$ the entry is $T(x, i, j)$. Note that the top row of $\{x\}$ is row 0, etc.

Lemma 5.7. *If $\{x\}$ is defined as above then*
(1) *$\{x\}$ is a latin square.*
(2) *$\{x\}$ is orthogonal to $\{y\}$ if $x \neq y$.*

Proof. (1) For fixed i, $T(x, i, j) = T(x, i, k)$ if and only if $j = k$ (by (D) of Theorem 5.1).

For fixed j, there is, by (C), a unique x satisfying $T(x, i, j) = T(x, k, j)$, $i \neq k$. But if $x = 0$ then $T(0, i, j) = j = T(0, k, j)$ and so, for $x \neq 0$, $T(x, i, j) = T(x, k, j)$ implies $i = k$. Thus, since the square $\{x\}$ is only defined for $x \neq 0$, there are no two identical entries in any one row or any one column in $\{x\}$, and so, since $\{x\}$ is an $n \times n$ matrix with entries from R, with $|R| = n$, $\{x\}$ is latin.

(2) Suppose $\{x\}$ is not orthogonal to $\{y\}$. Then there exist $a, b, c, d \in R$ such that $(T(x, a, b), T(y, a, b)) = (T(x, c, d), T(y, c, d))$. Since $\{x\}$ is latin, $T(x, a, b) = T(x, c, d)$ implies $a \neq c$ and $b \neq d$. But now the equation $T(u, a, b) = T(u, c, d)$ $(a \neq c)$ has distinct solutions $u = x$, $u = y$, which contradicts (C) and proves that $\{x\}$ is orthogonal to $\{y\}$. ☐

As an alternative proof of Lemma 5.7, which has the advantage that it does not depend on the finiteness of R, we suggest the following exercise.

Exercise 5.5. Give an alternative proof of Lemma 5.7 as follows;

(i) Show that, for any given x and a, $T(x, i, j) = a$ has a unique solution for j (given i), and for i, (given j). [This shows that each row and column of $\{x\}$ contains the entry a exactly once for any $a \in R$.]

(ii) Show that given $x, y, u, v, x \neq 0, y \neq 0, x \neq y$ the equations $T(x, a, b) = u$, $T(y, a, b) = v$ have a unique solution for a, b. [This shows that when we take the n^2 possibilities (i, j) for entries in identical position in $\{x\}, \{y\}$, the ordered pair (u, v) occurs exactly once; namely in the (a, b) position.]

Any set of latin squares such that any two are orthogonal is called a set of *mutually orthogonal latin squares*. We shall prove that the maximum number of mutually orthogonal $n \times n$ latin squares is $n - 1$. Anticipating this result, a set of $n-1$ mutually orthogonal $n \times n$ latin squares is said to be *complete*.

If A is any latin square and if α is any permutation of the rows, or columns, of A then, clearly, A^α is also a latin square. If B is a latin square which may be obtained from A by a permutation of the rows and columns of A, we say that B is *equivalent* to A.

We now list some properties of a set of mutually orthogonal latin squares.

Property 1. *Any set of mutually orthogonal latin squares is equivalent to a set in which each square has* $(0, 1, ..., n-1)$ *for its top row.* (Such a set is said to be in normal form.)

First we note that a permutation of the ("names" of) the letters within a latin square leaves it latin; such an operation also leaves it orthogonal to any latin square it was already orthogonal to. Hence any set of mutually orthogonal latin squares can be put in normal form.

Property 2. *Any set of mutually orthogonal $n \times n$ latin squares has at most $n-1$ matrices.*

Proof. By Property 1 we may assume that the given set is in normal form. No two of them may have the same entry, s say, in the $(1, 0)$-position (i.e. the position in the left column, and second row from the top) since then the ordered pair (s, s) occurs in the $(1, 0)$-position and in the $(0, s)$-position. None of the squares may have a zero in the $(1, 0)$-position since then the left column would contain two zeros. Thus there are only $n-1$ possible entries for the $(1, 0)$-position and this gives a bound on the maximum number of mutually orthogonal latin squares. $\quad\square$

Property 3. *Given any set of mutually orthogonal $n \times n$ latin squares in normal form, the first square may be assumed to have* $(0, 1, ..., n-1)'$ *in the column 0.*

Proof. Let $A_1, A_2, ..., A_t$ be a set of mutually orthogonal latin squares and let $A_k = (a_{ij}^{(k)})$. Define a mapping β on the rows of each individual A_k so that if $a_{i,0}^{(1)} = s$ then β maps the ith row of A_k onto the sth row $(i = 1, 2, ..., n-1, s = 1, 2, ..., n-1, k = 1, 2, ..., t)$. But if A and B are any pair of orthogonal latin squares and α is a permutation on the rows of each individual square such that α is the same permutation on the rows of A as on the rows of B, then A^α, B^α are clearly orthogonal. Thus $A_1^\beta, A_2^\beta, ..., A_t^\beta$ are a set of mutually orthogonal squares in normal form such that A_1^β has $(0, 1, ..., n-1)'$ in the first column. $\quad\square$

We are now in a position to prove a theorem which relates the study of latin squares with the study of projective planes.

Theorem 5.8. *There exists a finite projective plane of order n if and only if there exists a complete set of $n-1$ mutually orthogonal $n \times n$ latin squares.*

Proof. If $\mathscr{P}$ is a finite projective plane then $\mathscr{P}$ may be coordinatized by a PTR (R, T). By Lemma 5.7 this PTR may be used to construct a complete set of mutually orthogonal latin squares.

Let $R = \{0, 1, ..., n-1\}$ and suppose $A_1, A_2, ..., A_{n-1}$ is a complete set of mutually orthogonal latin squares in normal form, such that A_1 has $(0, 1, ..., n-1)'$ in its left column. Since no two of the A_i have the same entry in the $(1, 0)$-position we can label the squares $\{x\}$ so that $\{x\}$ is the square with x in the $(1, 0)$-position, $x = 1, 2, ..., n-1$. Defining $T(x, i, j)$ $= (i, j)$th entry in $\{x\}$ for $x = 1, ..., n-1$ and $T(0, i, j) = j$ for all i, j, we now verify that T satisfies (A), (B), (D), (E) of Theorem 5.1, and by Theorem 5.4, this shows that (R, T) is a PTR.

(A) $T(x, 0, c) = c$ since all squares are in normal form.

(B) $T(a, 1, 0) = a$ because $\{a\}$ is the square with a as the $(1, 0)$ position. $T(1, a, 0) = a$ since $\{1\}$ has $(0, 1, ..., n-1)'$ as its first column.

(D) Every row of the square $\{a\}$ contains each element of R exactly once. Thus given a, b, c $(a \neq 0)$ there is a unique x such that $T(a, b, x) = c$. If $a = 0$ then $T(0, b, x) = c$ if and only if $x = c$. So, in either case, (D) is satisfied.

(E) For distinct squares $\{a\}, \{c\}$ the n^2 pairs (i, j), where i and j are from the same position in the respective squares, exhaust all n^2 possibilities. Thus given b, d there is a unique ordered pair x, y such that $T(a, x, y) = b$, $T(c, x, y) = d$ provided $a \neq c$ and $a \neq 0$, $c \neq 0$. Suppose $a = 0$, then $T(0, x, y) = b$ if and only if $y = b$. But then $T(c, x, b) = d$ has a unique solution for x, so that (E) is satisfied. $\square$

As a consequence of this theorem one could try to establish the existence of a finite projective plane of a given order n by finding a set of $n-1$ mutually orthogonal latin squares. Many attempts have been made for $n = 10$ but so far, while there are many examples of two orthogonal latin squares, no-one has found a set of three mutually orthogonal 10×10 latin squares.

If a PTR is used to construct a complete set of mutually orthogonal latin squares, or vice versa, in the above way, we shall say that the PTR and the set of orthogonal squares *correspond* to each other.

Theorem 5.9. *A finite PTR (R, T) is linear if and only if in the corresponding complete set of mutually orthogonal latin squares the rows of any square (as vectors) are the same as the rows of any other.*

Proof. Suppose (R, T) is linear.

In position (a, b) of $\{x\}$ we have $T(x, a, b) = xa + b = T(1, xa, b)$ which is in position (xa, b) of $\{1\}$. Thus row a of $\{x\}$ is row xa of $\{1\}$ and the claim of the theorem is proved.

Suppose the rows of $\{1\}$ are the same as the rows of $\{x\}$ and let A_x be the permutation of the rows of $\{x\}$ such that the (a, b) entry of $\{x\}$ is in position (aA_x, b) of $\{1\}$. Then $T(x, a, b) = T(1, aA_x, b) = aA_x + b$. Putting

$b = 0$ gives $T(x, a, 0) = xa = aA_x$, so that $T(x, a, b) = xa + b$ and (R, T) is linear. ☐

To illustrate this discussion we now give a number of exercises which use latin squares to show that, up to isomorphism, the only projective planes of order 3 or 4 are $\mathscr{P}_2(3)$ and $\mathscr{P}_2(4)$.

Exercise 5.6. Show that the only loops of order 4 are the two groups of order 4.

Exercise 5.7. Show that there is no latin square orthogonal to

$$\begin{pmatrix} 0 & 1 & 2 & 3 \\ 1 & 2 & 3 & 0 \\ 2 & 3 & 0 & 1 \\ 3 & 0 & 1 & 2 \end{pmatrix}.$$

Hint: try to construct one starting with

$$\begin{pmatrix} 0 & 1 & 2 & 3 \\ 2 & & & \end{pmatrix}.$$

Since the matrix in Exercise 5.7 is the addition table for $(Z_4, +)$ this shows that if (R, T) is a PTR of order 4 then $(R, +)$ must be elementary abelian.

Exercise 5.8. Verify that

$$\{1\} = \begin{pmatrix} 0 & 1 & a & b \\ 1 & 0 & b & a \\ a & b & 0 & 1 \\ b & a & 1 & 0 \end{pmatrix} \quad \{a\} = \begin{pmatrix} 0 & 1 & a & b \\ a & b & 0 & 1 \\ b & a & 1 & 0 \\ 1 & 0 & b & a \end{pmatrix} \quad \{b\} = \begin{pmatrix} 0 & 1 & a & b \\ b & a & 1 & 0 \\ 1 & 0 & b & a \\ a & b & 0 & 1 \end{pmatrix}$$

is a set of 3 mutually orthogonal latin squares in normal form and that, given $\{1\}$, $\{a\}$ is uniquely determined once a is entered in the $(1, 0)$ position.

Exercise 5.9. Since multiplication in (R, T) is determined by $xy = T(x, y, 0)$, use the squares of Exercise 5.8 to show $(R^*, \cdot) \cong C_3$. Show that (R, T) is linear and that if $\mathscr{P}$ is a projective plane coordinatized by (R, T) then $\mathscr{P} \cong \mathscr{P}_2(4)$.

Exercise 5.10. Show, by using latin squares, that if $\mathscr{P}$ is a projective plane of order 3 then $\mathscr{P} \cong \mathscr{P}_2(3)$.

Appendix

When we coordinatized a projective plane $\mathscr{P}$ we remarked that there were other ways in which this could be done. In this appendix we discuss in some detail a method which was first introduced by Hall and mention briefly a third method due to Pickert.

We shall see, however, that if a given plane when coordinatized by one method gives a linear PTR then the PTR obtained by either of the other systems, using the same basic quadrangle, is also linear. We shall always use the method which we introduced in this chapter.

The Coordinatization Method of Hall

Let O, X, Y, I be a quadrangle of a projective plane of order n and let R be a set of n symbols with two distinct elements which we label 0, 1,
We label the line XY as $[\infty]$ and Y as (∞) where $\infty \notin R$. The elements of R are assigned to the points of OI not on $[\infty]$ in such a way that 0 is assigned to O, and 1 is assigned to I. If a point A on OI has been assigned $a \in R$ we coordinatize A as (a, a). If a point B on OY is such that $BX \cap OI = (b, b)$, then B is given the coordinates $(0, b)$ and if C is on OX such that $CY \cap OI = (c, c)$, then C is $(c, 0)$. For any point E not on any side of the triangle OXY, if $XE \cap OY = (0, y)$ and $YE \cap OX = (x, 0)$ then E is given the coordinates (x, y). In this way we have assigned coordinates to all points not on $[\infty]$. If M is on $[\infty]$, $M \neq Y$, and if M is on the line joining O to $(1, m)$ then M is coordinatized by (m).

The lines of $\mathscr{P}$ are coordinatized as follows; for any $m \in R$, $[m, k]$ is the line joining (m) to $(0, k)$, the line joining (∞) to $(a, 0)$ is $[a]$ and XY is $[\infty]$.

The basic difference so far between this method and the one adopted by us is that Hall's method insures that the set of points of the form (a, a) (as a varies over R) are collinear.

We now introduce a ternary operation T on R by $T(x, m, k) = y$ if and only if (x, y) is on $[m, k]$. This appears to be very different from the operation introduced in Chapter V; however the following exercise shows that this is not so.

Exercise 5.11. Let $\mathscr{P}$ be a projective plane coordinatized by a set R in the above way. If $T(x, m, k) = y$ if and only if (x, y) is on $[m, k]$ show

(A) $T(x, 0, b) = T(0, x, b) = b$ for all $x, b \in R$.
(B) $T(x, 1, 0) = T(1, x, 0) = x$ for all $x \in R$.
(C) Given $x, y, u, v \in R$, $x \neq u$, there is a unique ordered pair $a, b \in R$ such that $T(x, a, b) = y$, $T(u, a, b) = v$.
(D) Given $x, y, a \in R$, there is a unique $b \in R$ with $T(x, a, b) = y$.
(E) Given $a, b, c, d \in R(a \neq c)$ there is a unique $x \in R$ such that $T(x, a, b) = T(x, c, d)$.

Note that (A)–(E) of Exercise 5.11 are (A)–(E) of Theorem 5.1. In the Hall method $a + b$ is defined as $T(a, 1, b)$ and ab as $T(a, b, 0)$ which are slightly different to our definitions. A PTR with addition and multiplication defined in this way will, in this appendix, be called a Hall PTR.

Exercise 5.12. If (R, T) is a Hall PTR show that $(R, +)$, $(R^*, \cdot)$ are loops with respective identities 0 and 1.

Linearity in a Hall PTR is defined by $T(x, m, k) = xm + k$ and once the quadrangle $OXYI$ is chosen, a given plane $\mathscr{P}$ is coordinatizable by a linear Hall PTR if and only if the PTR given by our method is linear.

A slight modification of the Hall method, which is essentially due to Pickert, is to coordinatize $\mathscr{P}$ in the same way but to define a ternary operation S by $S(m, x, k) = y$ if and only if (x, y) is on $[m, k]$. Addition is given by $a + b = S(1, a, b)$, multiplication by $ab = S(a, b, 0)$ while S is linear if $S(m, x, k) = mx + k$. The reader should formulate the properties corresponding to (A)–(E) of Theorem 5.1 for these ternary rings.

The results in the remaining chapters of this book will be totally independent of the method of coordinatization and we shall keep to our definition.

References

These other methods of coordinatizing are found in

1. Hall, M.: Projective planes. Trans. Am. Math. Soc. 54, 229–277 (1943).
2. Pickert, G.: Projektive Ebenen. Berlin-Göttingen-Heidelberg: Springer 1955.

VI. Algebraic Properties of Planar Ternary Rings

1. Introduction

In this chapter we discuss the relation between the geometric structure of a projective plane and the algebraic properties of a planar ternary ring coordinatizing it. Usually the geometric property which interests us is that of being (P, l)-desarguesian which, as we have seen, is equivalent to being (P, l)-transitive. The proofs of the first fifteen theorems are geometric and are elementary in the sense that they follow simply from the earlier chapters. But then we give proofs of two very famous, but fairly difficult, theorems. They are the Skornyakov-San Soucie Theorem (6.16) and the Artin-Zorn Theorem (6.20). The proofs of both of these theorems are very long and involve very detailed algebra. Since these two theorems are so crucial to the whole subject of projective planes, we feel that it is important to give proofs. However, we strongly advise the student to skip these two proofs on his first reading of the book. Both of these proofs are taken from [1] and we wish to thank Professor Heinz Lüneburg for giving us permission to reproduce them here.

We have only included one diagram to illustrate Theorem 6.1 but we strongly urge the reader to draw his own for each proof.

2. The Condition for Linearity

In Chapter V we saw that if (R, T) is a PTR coordinatizing $\mathscr{P}_2(K)$ then T is linear and $R \cong K$. In this chapter we study in detail the relation between the algebraic structure of (R, T) and the geometric structure of the projective plane which it coordinatizes. More precisely we usually study the relation between the algebraic structure of (R, T) and the groups of perspectivities admitted by the corresponding plane. However, by Theorem 4.29, those groups closely reflect the geometric structure of the plane.

Throughout this chapter $\mathscr{P}$ is assumed to be a given projective plane and (R, T) a given PTR coordinatizing $\mathscr{P}$.

We begin by giving necessary and sufficient conditions for (R, T) to be linear.

Theorem 6.1. (R, T) *is linear if and only if any two triangles which are in perspective from* (∞) *in such a way that each has a vertex on* $[0]$ *and such* *that the two pairs of sides containing points of* $[0]$ *meet on* $[\infty]$, *have the property that one of the third sides passes through* (0) *if and only if the final side does.* (See Fig. 11.)

Proof. Let the two triangles be $A_1 B_1 C_1$, $A_2 B_2 C_2$ such that $A_1 A_2$ is the line $x = 0$, $B_1 B_2$ is $x = u$ and $C_1 C_2$ is $x = v$. Let $C_3 = A_1 B_1 \cap A_2 B_2$ be (m), let $B_3 = C_1 A_1 \cap C_2 A_2$ be (n) and let $B_1 C_1 \cap [\infty]$ be (0). Further $A_1 = (0, a)$, $A_2 = (0, d)$, $B_1 = (u, b)$, $B_2 = (u, c)$. Since C_1 is the intersection of $x = v$ (the line $C_1 C_2$) with $y = b$ (the line $X B_1$), C_1 is the point (v, b). If we let C_2 be (v, f) then the configurational property of the theorem implies that $f = c$, and conversely.

(i) Suppose (R, T) is linear.

Since A_1, B_1, C_3 are collinear, $mu + b = a$ and, since A_1, C_1, B_3 are also collinear, $nv + b = a$. However, $(R, +)$ is a loop so that the equation $x + b = a$ has a unique solution; hence $mu = nv$.

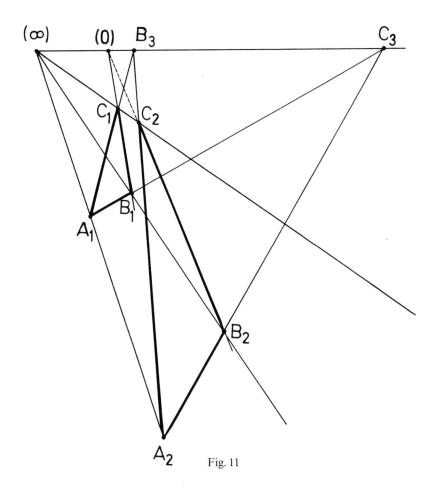

Fig. 11

Similarly the collinearity of A_2, B_2, C_3 gives $mu + c = nv + f = d$. Putting $nv = mu$ this gives $mu + c = mu + f = d$ which, again since $(R, +)$ is a loop, implies $c = f$ as required.

(ii) Suppose $c = f$.

Using the same set of collinear points as in (1) we have

$$T(m, u, b) = T(n, v, b) = a \tag{1}$$

and

$$T(m, u, c) = T(n, v, c) = d. \tag{2}$$

Equations (1) and (2) must hold for arbitrary m, u, b, n, c. (Draw a picture to see what this means!) If we let $c = 0$ and $n = 1$, then m, u, b are still arbitrary. Putting $c = 0$, $n = 1$ in (2) we have $mu = v = d$ and substituting these equalities in (1) we get $T(m, u, b) = T(1, v, b) = v + b = mu + b$, i.e. (R, T) is linear. □

If (R, T) is known to be linear then we shall often refer to R as a planar ternary ring without specifically mentioning T.

3. Additive Properties of (R, T)

In this section we relate the additive structure of (R, T) to the existence in Aut$\mathscr{P}$ of elations with axis $[\infty]$.

Theorem 6.2. (R, T) is linear with associative addition if and only if $\mathscr{P}$ is $((\infty), [\infty])$-transitive.

Proof. (i) Suppose (R, T) is linear with associative addition. If A, B are any distinct points which are collinear with (∞) but not incident with $[\infty]$ we must exhibit a $((\infty), [\infty])$-elation mapping A onto B. In view of Exercise 4.1 it is sufficient to exhibit a collineation of $\mathscr{P}^{[\infty]}$ fixing all lines through (∞) but no points of $\mathscr{P}^{[\infty]}$ and mapping A onto B. (This will extend to a unique collineation of $\mathscr{P}$ which fixes (∞) linewise and, since it fixes no point not on $[\infty]$, must fix $[\infty]$ pointwise by Theorem 4.9.)

If $A = (u, v)$ then, by the collinearity of (∞), A and B, $B = (u, w)$ for some $w \in R$. However, since $(R, +)$ is a loop, there is a unique $a \in R$ such that $w = v + a$ and $B = (u, v + a)$. We now define a mapping ϕ_a of the elements of $\mathscr{P}^{[\infty]}$ by;

$$\phi_a : (x, y) \to (x, y + a), \quad [m, k] \to [m, k + a], \quad [k] \to [k].$$

Clearly ϕ_a is a one-to-one mapping of the points (lines) of $\mathscr{P}^{[\infty]}$ onto the points (lines) of $\mathscr{P}^{[\infty]}$. Furthermore ϕ_a fixes every line of the form $[k]$ and sends A onto B. Thus, if ϕ_a is a collineation of $\mathscr{P}^{[\infty]}$, it certainly extends to the desired collineation of $\mathscr{P}$. To prove the theorem, therefore, we have only to check that ϕ_a is a collineation, i.e. that ϕ_a preserves incidence.

The condition that (x, y) is on $[m, k]$ is $mx + y = k$. But $(x, y)^{\phi_a}$ is on $[m, k]^{\phi_a}$ if and only if $mx + (y + a) = k + a$. However, since addition is associative $mx + (y + a) = (mx + y) + a$ and so $mx + (y + a) = k + a$ if and only if $mx + y = k$.

The point (x, y) is on $[k]$ if and only if $x = k$ which is precisely the condition that $(x, y)^{\phi_a}$ is on $[k]^{\phi_a}$. Thus ϕ_a preserves incidence and half of the theorem is proved.

(ii) Suppose $\mathscr{P}$ is $((\infty), [\infty])$-transitive.

By Theorem 4.29 $\mathscr{P}$ is $((\infty), [\infty])$-desarguesian which implies, by Theorem 6.1, (R, T) is linear. For any point $(0, a)$ on $[0]$ there is a $((\infty), [\infty])$-elation mapping $(0, 0)$ onto $(0, a)$. If we call this elation ϕ_a then, for any point (x, y), $(x, y)^{\phi_a} = (x, u)$ where u depends only on y and a. Thus we may write $(x, y)^{\phi_a} = (x, y^{\alpha_a})$ where α_a is a one-to-one mapping from R onto R such that $0^{\alpha_a} = a$.

The action of ϕ_a on the points of $\mathscr{P}$ is now completely determined and is given by $\phi_a : (x, y) \to (x, y^{\alpha_a})$, $(m) \to (m)$, $(\infty) \to (\infty)$. Since $(m)^{\phi_a} = (m)$ and $(0, k)^{\alpha_a} = (0, k^{\alpha_a})$, the action of ϕ_a on the lines of $\mathscr{P}$ is given by $\phi_a : [m, k] \to [m, k^{\alpha_a}]$, $[k] \to [k]$, $[\infty] \to [\infty]$.

Using the facts that (R, T) is linear and that ϕ_a preserves incidence we have $mx + y = k$ if and only if $mx + y^{\alpha_a} = k^{\alpha_a}$, i.e.

$$mx + y^{\alpha_a} = (mx + y)^{\alpha_a} \quad \text{for all} \quad m, x, y \in R . \tag{1}$$

Putting $y = 0$ and $m = 1$ in (1) gives

$$x + 0^{\alpha_a} = x^{\alpha_a}$$

or

$$x^{\alpha_a} = x + a . \tag{2}$$

Substituting $x^{\alpha_a} = x + a$ in (1) and putting $m = 1$ we get $x + (y + a) = (x + y) + a$ for all $x, y, a \in R$. Hence (R, T) is linear with associative addition. $\square$

If a linear PTR (R, T) has associative addition then (R, T) is called a *Cartesian group*.

Theorem 6.3. *If (R, T) is a Cartesian group then (R, T) satisfies the left distributive law, $a(b + c) = ab + ac$, for all $a, b, c \in R$, if and only if $\mathscr{P}$ is $((0), [\infty])$-transitive.*

Proof. (i) Suppose (R, T) satisfies the left distributive law.

For any $a \in R$ we must exhibit a $((0), [\infty])$-elation mapping $(0, 0)$ onto $(a, 0)$. However, as in the proof of Theorem 6.2, it is sufficient to exhibit the appropriate collineation of $\mathscr{P}^{[\infty]}$. For any $a \in R$ we define θ_a by $\theta_a : (x, y) \to (a + x, y)$, $[m, k] \to [m, ma + k]$, $[k] \to [a + k]$.

Clearly θ_a fixes no point of $\mathscr{P}^{[\infty]}$ but fixes every line of $\mathscr{P}$ through (0) so, in order to prove the theorem, we only have to show that θ preserves incidence.

Since (R, T) is a Cartesian group, it is linear and (x, y) is on $[m, k]$ if and only if $mx + y = k$. The point $(x, y)^{\theta_a}$ is on $[m, k]^{\theta_a}$ if and only if

$m(a + x) + y = ma + k$. But, by the left distributive law $m(a + x) = ma + mx$, so that $m(a + x) + y = (ma + mx) + y$. However, in a cartesian group, addition is associative so that $m(a + x) + y = (ma + mx) + y = ma + (mx + y)$ and this is equal to $ma + k$ if and only if $mx + y = k$. Hence θ_a preserves incidence.

(ii) Suppose $\mathscr{P}$ is $((0), [\infty])$-transitive.

If we define θ_a to be the $((0), [\infty])$-elation which maps $(0, 0)$ onto $(a, 0)$ then, as in the proof of Theorem 6.2, $(x, y)^{\theta_a} = (x^{\beta_a}, y)$ where β_a is a one-to-one mapping of R onto R satisfying $0^{\beta_a} = a$. This completely determines the action of θ_a on the points of $\mathscr{P}^{[\infty]}$. If we let $[m, k]^{\theta_a} = [m, h]$ then, in order to know the action of θ_a on the lines of $\mathscr{P}^{[\infty]}$, we must find h in terms of m, k, a. Since $(0, k)$ is on $[m, k]$, $(0, k)^{\theta_a}$ is on $[m, k]^{\theta_a}$ (i.e. (a, k) is on $[m, h]$) and so, by the linearity of (R, T), $ma + k = h$; hence $[m, k]^{\theta_a} = [m, ma + k]$.

The point (x, y) is on $[m, k]$ if and only if $(x, y)^{\theta_a}$ is on $[m, k]^{\theta_a}$ which, by the linearity of (R, T), gives $mx + y = k$ if and only if $mx^{\beta_a} + y = ma + k$. Thus $mx^{\beta_a} + y = ma + (mx + y)$ which, since $(R, +)$ is associative, is equal to $(ma + mx) + y$. Hence

$$mx^{\beta_a} + y = (ma + mx) + y \quad \text{for all} \quad m, x, y, a \in R. \tag{1}$$

Putting $m = 1$, $y = 0$ in (1) gives

$$x^{\beta_a} = a + x \tag{2}$$

which, when substituted in (1), means

$$m(a + x) + y = (ma + mx) + y. \tag{3}$$

Finally, since $(R, +)$ is a loop, $t + y = s + y$ implies $t = s$ so we get $m(a + x) = ma + mx$ and (R, T) satisfies the left distributive law. $\square$

If (R, T) is a Cartesian group satisfying the left distributive law then we call (R, T) a *left quasifield* or, more often, just a *quasifield*. As mentioned earlier, since (R, T) is linear, we shall often omit the "T" and say that R is a quasifield.

Combining Theorem 6.3 with Theorem 4.19, we have:

Corollary 1. $\mathscr{P}$ *is coordinatized by a quasifield, with the line k as $[\infty]$, if and only if k is a translation line.*

And this leads to:

Corollary 2. *If $\mathscr{P}$ is coordinatized by a quasifield for a particular choice of the points $X = (0)$ and $Y = (\infty)$, then $\mathscr{P}$ is coordinatized by a quasifield whenever any two points on the line XY are chosen as (0) and (∞).*

Note also that, from Theorem 4.14, since $[\infty]$ is an axis of elations for two distinct centres the group $\Pi_{([\infty],[\infty])}$ of all elations of $\mathscr{P}$ with axis $[\infty]$ is abelian. In particular, then, we have $\Pi_{((\infty),[\infty])}$ is abelian. But any ϕ in $\Pi_{((\infty),[\infty])}$ is one of the mappings ϕ_a in the proof of Theorem 6.2 and so

$\phi_a \phi_b = \phi_b \phi_a$ for any $a, b \in R$. However, the associativity of addition gives $\phi_a \phi_b = \phi_{a+b}$ and we have shown:

Theorem 6.4. *If (R, T) is a quasifield than $(R, +)$ is abelian.* $\square$

4. Multiplicative Properties of (R, T)

The last three theorems have shown a close relation between the additive structure of (R, T) and the existence in $\mathscr{P}$ of elations with axis $[\infty]$. We now prove similar results which relate the multiplicative structure to the existence of homologies in $\mathscr{P}$.

Theorem 6.5. *(R, T) is linear with associative multiplication if and only if $\mathscr{P}$ is $((0), [0])$-transitive.*

Proof. (i) Suppose (R, T) is linear with associative multiplication.

For any $a \in R^*$ we must exhibit a $((0), [0])$-homology sending $(1, 0)$ onto $(a, 0)$. Since the required homology does not have axis $[\infty]$ we will show its existence by defining its action on all of $\mathscr{P}$. We define ϕ_a by $\phi_a : (x, y) \to (ax, y)$, $(m) \to (ma^{-1})$, $(\infty) \to (\infty)$, $[m, k] \to [ma^{-1}, k]$, $[k] \to [ak]$ and $[\infty] \to [\infty]$ where $t = a^{-1}$ is the unique solution of $ta = 1$ in $(R^*, \cdot)$. Clearly if ϕ_a is a collineation then it is a $((0), [0])$-homology mapping $(1, 0)$ onto $(a, 0)$ and so, in order to prove the theorem, we need only show that ϕ_a preserves incidence.

As (R, T) is linear, (x, y) is on $[m, k]$ if and only if $mx + y = k$, whereas $(x, y)^{\phi_a}$ is on $[m, k]^{\phi_a}$ if and only if $(ma^{-1})(ax) + y = k$. However multiplication in R is associative so that $(ma^{-1})(ax) = mx$ and $(x, y)^{\phi_a}$ is on $[m, k]^{\phi_a}$ if and only if $mx + y = k$. Straightforward verification shows that ϕ_a preserves incidences between other types of points and lines and proves the first half of the theorem.

(ii) Suppose $\mathscr{P}$ is $((0), [0])$-transitive.

Let ϕ_a be the $((0), [0])$-homology sending $(1, 0)$ onto $(a, 0)$. The action of ϕ_a on $\mathscr{P}$ is determined by two permutations α_a, β_a of the elements of R as follows; for any point $(m) \in [\infty]$, $(m)^{\phi_a} = (m^{\alpha_a})$ with $0^{\alpha_a} = 0$ and for any point $(x, 0)$, $x \in R$, $(x, 0)^{\phi_a} = (x^{\beta_a}, 0)$ where $0^{\beta_a} = 0$ and $1^{\beta_a} = a$. The action of ϕ_a is now given by $(x, y)^{\phi_a} = (x^{\beta_a}, y)$, $(m)^{\phi_a} = (m^{\alpha_a})$, $(\infty)^{\phi_a} = (\infty)$, $[m, k]^{\phi_a} = [m^{\alpha_a}, k]$, $[k]^{\phi_a} = [k^{\beta_a}]$, $[\infty]^{\phi_a} = [\infty]$.

Thus, for all $m, x, y \in R$, $T(m, x, y) = T(m^{\alpha_a}, x^{\beta_a}, y)$ which, putting $y = 0$, gives

$$mx = m^{\alpha_a} x^{\beta_a} \quad \text{for all} \quad m, x \in R . \tag{1}$$

If $x = 1$, then (1) becomes

$$m = m^{\alpha_a} a . \tag{2}$$

We now introduce a new mapping γ_a of R^* onto itself given by $x^{\gamma_a} = xa$ for all $x \in R^*$; i.e. γ_a is the mapping of multiplication on the right by a. From (2) $\alpha_a = \gamma_a^{-1}$ and so

$$mx = m^{\gamma_a^{-1}} x^{\beta_a}. \tag{3}$$

Putting $m = a$ in (3) and using the obvious fact that $a^{\gamma_a^{-1}} = 1$ we have $ax = x^{\beta_a}$ so that

$$mx = m^{\gamma_a^{-1}}(ax). \tag{4}$$

Since γ_a is a permutation on R^*, for any $m \neq 0$ there is a unique $u \in R$ with $m = ua$. Note that since m was arbitrary then so is u. By substituting $m = ua$ in (4) we get

$$(ua)\, x = (ua)^{\gamma_a^{-1}}(ax) = (u^{\gamma_a})^{\gamma_a^{-1}}(ax) = u(ax) \tag{5}$$

for all $u, a, x \in R^*$, and so R has associative multiplication.

To prove the linearity of (R, T) we note that, for all $m, a, x, y \in R, a \neq 0$, $T(m, x, y) = T(ma^{-1}, ax, y)$. Thus, putting $m = a$ we have $T(a, x, y) = T(1, ax, y) = ax + y$ as required. $\square$

Theorem 6.6. *If (R, T) is linear then (R, T) has associative multiplication and satisfies the left distributive law if and only if $\mathscr{P}$ is $((\infty), [0, 0])$-transitive.*

Proof. (i) Suppose (R, T) is linear with associative multiplication and satisfies the left distributive law.

A repetition of the arguments used in the earlier proofs of this chapter shows that the mapping ϕ_a; $(x, y) \to (x, ay)$, $(m) \to (am)$, $(\infty) \to (\infty)$, $[m, k] \to [am, ak]$, $[k] \to [k]$, $[\infty] \to [\infty]$ is a $((\infty), [0, 0])$-homology mapping $(0, 1)$ onto $(0, a)$ and proves that $\mathscr{P}$ is $((\infty), [0, 0])$-transitive.

(ii) Suppose $\mathscr{P}$ is $((\infty), [0, 0])$-transitive.

If ϕ_a is the $((\infty), [0, 0])$-homology sending $(0, 1)$ onto $(0, a)$ then, as in the proof of Theorem 6.5, there are two permutations α_a, β_a of R such that $(x, y)^{\phi_a} = (x, y^{\beta_a})$, $(m)^{\phi_a} = (m^{\alpha_a})$, $(\infty)^{\phi_a} = (\infty)$, $[m, k]^{\phi_a} = [m^{\alpha_a}, k^{\beta_a}]$, $[k]^{\phi_a} = [k]$, $[\infty]^{\phi_a} = [\infty]$ and $0^{\alpha_a} = 0^{\beta_a} = 0$, $1^{\beta_a} = a$.

Since (R, T) is linear, $mx + y = k$ if and only if $m^{\alpha_a} x + y^{\beta_a} = k^{\beta_a}$ which gives

$$m^{\alpha_a} x + y^{\beta_a} = (mx + y)^{\beta_a} \quad \text{for all} \quad m, x, y \in R. \tag{1}$$

Putting $y = 0$ and $x = 1$ in (1) gives $m^{\alpha_a} = m^{\beta_a}$ for all $m \in R$, i.e. $\alpha_a = \beta_a$. Thus

$$m^{\alpha_a} x + y^{\alpha_a} = (mx + y)^{\alpha_a}. \tag{2}$$

Now putting $y = 0$, $m = 1$ in (2), and using the fact that $1^{\alpha_a} = a$ we have

$$ax = x^{\alpha_a} \tag{3}$$

so that α_a is left multiplication by a. Hence (2) becomes

$$(am)\, x + ay = a(mx + y) \quad \text{for all} \quad a, m, x, y \in R. \tag{4}$$

Putting $y = 0$ in (4) gives associative multiplication and the left distributive law follows from (4) by putting $m = 1$. []
Any quasifield which has associative multiplication is called a *nearfield*.

Exercise 6.1. If (R, T) is a quasifield show that (R, T) is a nearfield if and only if $\mathscr{P}$ is $((0), [0])$-transitive.

5. Coordinatizing the Dual Plane

We now consider briefly the problem of coordinatizing the dual plane $\mathscr{P}^*$ of $\mathscr{P}$. Since $\mathscr{P}$ and $\mathscr{P}^*$ have the same order we may use the same coordinatizing set R. If we denote the coordinates of points and lines in $\mathscr{P}^*$ by $(x, y)'$, $[m, k]'$ etc. then we coordinatize $\mathscr{P}^*$ by choosing $(0, 0)' = [0, 0]$, $(0)' = [0]$, $(\infty)' = [\infty]$, $(1)' = [1]$ and assign the elements of R to the points of $[0, 0]'$ so that $(x, 0)' = [x, 0]$. We find one relationship between the coordinatizing PTR (R, T') of $\mathscr{P}^*$ and (R, T):

Theorem 6.7. *If (R, T) is linear with associative addition then (R, T') is also linear with associative addition.*

Proof. By Theorem 6.2, $\mathscr{P}$ is $((\infty), [\infty])$-transitive. However, if α is any $((\infty), [\infty])$-elation of $\mathscr{P}$ then α fixes every line through (∞) and every point on $[\infty]$ and so, considered as a collineation of $\mathscr{P}^*$, α fixes every point on $[\infty]'$ and every line on $(\infty)'$, i.e. α is a $((\infty)', [\infty]')$-elation on $\mathscr{P}^*$. Hence $\mathscr{P}^*$ is $((\infty)', [\infty]')$-transitive if and only if $\mathscr{P}$ is $((\infty), [\infty])$-transitive and, thus, again using Theorem 6.2, (R, T') is linear with associative addition. []

Exercise 6.2. If (R, T) is a quasifield show that (R, T') is linear with associative addition and satisfies the right distributive law; $(a + b)c = ac + bc$ for all $a, b, c \in R$ (such a system is called a *right quasifield*).

Note that if (R, T) is a quasifield then $\mathscr{P}$ is a translation plane with respect to $[\infty]$ and so $\mathscr{P}^*$ is the dual of a translation plane with respect to $(\infty)'$. Thus $\mathscr{P}^*$ is $((\infty)', [\infty]')$-transitive and $((\infty)', [0]')$-transitive and so we have:

Theorem 6.8. *If (R, T) is linear with associative addition then (R, T) satisfies the right distributive law if and only if $\mathscr{P}$ is $((\infty), [0])$-transitive.*

Corollary. *(R, T) is a right quasifield if and only if (∞) is a translation point.*

A quasifield satisfying the right distributive law is called a *division ring*. As an immediate consequence of Theorems 6.3 and 6.8 we have

Theorem 6.9. *(R, T) is a division ring if and only if $\mathscr{P}$ is $((\infty), [\infty])$, $((0), [\infty])$ and $((\infty), [0])$-transitive.*

Exercise 6.3. Prove Theorem 6.9 directly (i.e. without dualizing).

Note that Theorem 6.9 can be reworded to say that (R, T) is a division ring if and only if $\mathscr{P}$ is a translation plane with respect to $[\infty]$ and the dual of a translation plane with respect to (∞).

6. Division Rings with Inverse Properties

A loop G is said to have the *right inverse property* (RIP) if for each $x \in G$ there is an element $x^{-1} \in G$ such that $(yx) x^{-1} = y$ for all $y \in G$. G is said to have the *left inverse property* (LIP) if for each $x \in G$ there is an element x^{-1} such that $x^{-1}(xy) = y$ for all $y \in G$. If G has RIP and LIP then we say that G has the *inverse property* (IP).

Lemma 6.10. *Let G be a loop with either RIP or LIP. Then, for any $x \in G$, $x^{-1} x = 1 = x x^{-1}$ and, consequently, $(x^{-1})^{-1} = x$.*

Proof. Suppose, for instance, G has RIP. Since $(yx) x^{-1} = y$ for all $y \in G$ we have, putting $y = x^{-1}, (x^{-1} x) x^{-1} = x^{-1}$ and so, since $bx^{-1} = x^{-1}$ has the unique solution $b = 1$ in G, $x^{-1} x = 1$. However, putting $y = 1$ gives $x x^{-1} = 1$ and thus $x^{-1} x = 1 = x x^{-1}$. Clearly, now, $(x^{-1})^{-1} = x$. ☐

If (R, T) is a division ring then T is linear and so as with any other linear PTR we shall often refer to R as a division ring without refering to the ternary operation T. If R is a division ring then R is certainly a quasifield so that $(R, +)$ is a group. Now, clearly, any group has both inverse properties and so any statement of the form "R is a division ring with RIP" is intended to mean that "$(R^*, \cdot)$ has RIP".

Theorem 6.11. *If R is a division ring then R has RIP if and only if $\mathscr{P}$ is $((0, 0), [0])$-transitive.*

Proof. (i) Suppose $\mathscr{P}$ is $((0, 0), [0])$-transitive.

For any $b \in R$ let ϕ_b be the $((0, 0), [0])$-elation sending (0) onto $(b, 0)$, then $[m, 0]^{\phi_b} = [m, 0]$ for all $m \in R$. As the point (m) is the intersection of $[\infty]$ with $[m, 0], (m)^{\phi_b}$ is the intersection of $[\infty]^{\phi_b} = [b]$ with $[m, 0]^{\phi_b} = [m, 0]$ so that $(m)^{\phi_b} = (b, w)$ for some w. However, (b, w) is on $[m, 0]$ and so, since R is linear, $mb + w = 0$ giving

$$(m)^{\phi_b} = (b, -mb). \tag{1}$$

Since ϕ_b has centre $(0, 0)$ the line $y = 0$ is left invariant and, since the axis of ϕ_b is $[0]$, any point with $x = 0$ is left invariant. Let $m, x \in R, x \neq 0, -b$ and let u, n be defined by $(x, 0)^{\phi_b} = (u, 0)$ and $[m, mx]^{\phi_b} = [n, mx]$; then, since (m) and $(x, 0)$ are on $[m, mx], (m)^{\phi_b}$ and $(x, 0)^{\phi_b}$ are on $[m, mx]^{\phi_b}$. Substituting the known values for $(m)^{\phi_b}, (x, 0)^{\phi_b}$ and $[m, mx]^{\phi_b}$ gives

$$nb - mb = mx \tag{2}$$

and

$$nu = mx. \tag{3}$$

Equation (2) determines n for given m, b, x and Eq. (3) then determines u. If, for a given x and b, k is the particular value of n when $m = 1$ then $ku = x$ and $kb = b + x$. Thus if k, b, x, u, are such that $kb = b + x$, $ku = x$, and if $nb = m(b + x)$ for any m, then $nu = mx$. Thus, in the special case when $b = -1$, the conditions $k = 1 - x$, $ku = x$ and $n = m(1 - x)$ must imply that $n = mk$ so that $(mk)u = nu = mx = m(ku)$. Since $\mathscr{P}$ is $((0, 0), [0])$-transitive, as x and b vary k will vary over all possible values in R. Hence, for any m, k if u satisfies $k = 1 - ku$ we have

$$(mk)\, u = m(ku)\,. \tag{4}$$

Let $v = 1 + u$ so that $kv = 1$, then

$$
\begin{aligned}
(mk)\, v = (mk)\,(1 + u) &= mk + mk(u) && \text{(by the left distributive law)}\\
&= mk + m(ku) && \text{(by (4))}\\
&= m\big(k(1 + u)\big)\\
&= m(kv) = m && \text{since } kv = 1\,.
\end{aligned}
$$

But k was defined by $[1, x]^{\phi_b} = [k, x]$ and thus, as b varies over R^*, k takes all possible values in R^* so that R has RIP.

(ii) Suppose R has RIP.

For any $b \in R^*$ define ϕ_b by:

$$(x, y) \to \big((b^{-1} + x^{-1})^{-1}, (yx^{-1})\,(b^{-1} + x^{-1})^{-1}\big) \text{ for } x \neq 0, x \neq -b$$

$$(0, y) \to (0, y)$$

$$(-b, y) \to (yb^{-1})$$

$$(m) \to (b, -mb)$$

$$(\infty) \to (\infty)$$

$$[m, k] \to [m + kb^{-1}, k]$$

$$[k] \to [(b^{-1} + k^{-1})^{-1}] \text{ for } k \neq 0, k \neq -b$$

$$[0] \to [0]$$

$$[-b] \to [\infty]$$

$$[\infty] \to [b]\,.$$

Certainly ϕ_b fixes $(0, 0)$ linewise, $[0]$ pointwise and maps (0) onto $(b, 0)$; so, in order to show that $\mathscr{P}$ is $((0, 0), [0])$-transitive it is sufficient to prove that ϕ_b preserves incidence.

Clearly this involves a detailed case analysis considering incidence between the various "types" of points and lines. We shall consider only incidences between points of type (x, y), $x \neq 0$, $-b$ and lines of type $[m, k]$. The point (x, y) is on $[m, k]$ if and only if $mx + y = k$ and $(x, y)^{\phi_a}$ is on $[m, k]^{\phi_a}$ if and only if $(m + kb^{-1})\,(b^{-1} + x^{-1})^{-1} + (yx^{-1})\,(b^{-1} + x^{-1})^{-1} = k$.

Suppose $(m + kb^{-1})(b^{-1} + x^{-1})^{-1} + (yx^{-1})(b^{-1} + x^{-1})^{-1} = k$ then

$$[(m + kb^{-1})(b^{-1} + x^{-1})^{-1} + (yx^{-1})(b^{-1} + x^{-1})^{-1}](b^{-1} + x^{-1})$$
$$= k(b^{-1} + x^{-1}). \tag{1}$$

But since R is a division ring it satisfies the right distributive law. Using the fact that R has RIP and that, by Lemma 6.10, $(b^{-1} + x^{-1})^{-1}(b^{-1} + x^{-1}) = 1$ we get

$$(m + kb^{-1}) + yx^{-1} = k(b^{-1} + x^{-1}). \tag{2}$$

Now R satisfies the associative law for addition and the left distributive law so that (2) simplifies to

$$m + kb^{-1} + yx^{-1} = kb^{-1} + kx^{-1} \tag{3}$$

Since R under addition is an abelian group we can cancel kb^{-1} from each side to give

$$m + yx^{-1} = kx^{-1} \tag{4}$$

and now multiplying both sides on the right by x and using the right distributive law, RIP, and Lemma 6.10 we get $mx + y = k$ as required.

We leave the rest of the proof as a long, but not difficult, exercise. Note that we have already used each of the properties given to R to prove incidence is preserved between the types of points and lines considered. □

A $((0, 0), [0])$-elation of $\mathscr{P}$ fixes (∞) but moves $[\infty]$ so that $\mathscr{P}$ has at least two translation lines through (∞). But, by Theorem 4.20 (a), this implies that every line through (∞) is a translation line and Theorem 6.11 may be restated as

Theorem 6.12. *R is a division ring with RIP if and only if every line through (∞) is a translation line of $\mathscr{P}$.*

Lemma 6.13. *If R is a RIP division ring then R also satisfies $x((yz) y) = ((xy) z) y$. In particular R satisfies the right alternative law, $x(y^2) = (xy) y$.*

Proof. For any appropriate $y, z \in R$, put $t = [(yz) y + y] \, [y^{-1} - (y + z^{-1})^{-1}]$. Then

$$t(y + z^{-1}) = [yz + 1 - ((yz) y)(y + z^{-1})^{-1} - y(y + z^{-1})^{-1}](y + z^{-1})$$
$$= (yz + 1)(y + z^{-1}) - ((yz) y) - y$$
$$= (yz) y + y + y + z^{-1} - (yz) y - y$$
$$= y + z^{-1}.$$

Thus $t = 1$ and $[(yz) y + y]^{-1} = y^{-1} - (y + z^{-1})^{-1}$. Hence

$$x = xt = x([(yz) y + y] \, [y^{-1} - (y + z^{-1})^{-1}])$$
$$= (x[(yz) y + y]) [y^{-1} - (y + z^{-1})^{-1}] \text{ by RIP of } R$$
$$= (x((yz) y) + xy)(y^{-1} - (y + z^{-1})^{-1}).$$

Let $w = (((xy)z)y + xy)(y^{-1} - (y + z^{-1})^{-1})$, then we must show that $w = x$.

$$
\begin{aligned}
w(y + z^{-1}) &= [(xy)z + x - ((xy)z\,y)(y + z^{-1})^{-1} - xy(y + z^{-1})^{-1}](y + z^{-1}) \\
&= ((xy)z + x)(y + z^{-1}) - ((xy)z)y - xy \\
&= ((xy)z)y + xy + xy + xz^{-1} - ((xy)z)y - xy \\
&= xy + xz^{-1} = x(y + z^{-1}).
\end{aligned}
$$

Hence $w(y + z)^{-1}) = x(y + z^{-1})$ and so, since $(R^*, \cdot)$ is a loop, $w = x$. ☐

Theorem 6.14. *If R is a division ring with RIP then R has LIP if and only if $\mathscr{P}$ is $((0, 0), [0, 0])$-transitive.*

Proof. (i) Suppose $\mathscr{P}$ is $((0, 0), [0, 0])$-transitive.

Let α be the $((0, 0), [0, 0])$-elation mapping (∞) onto $(0, -1)$. For any $m \in R$, $m \neq 0$, $(m)^\alpha$ is on $[m, 0]^\alpha = [m, 0]$ and $[\infty]^\alpha = [0, -1]$, so if $(m)^\alpha = (t, -1)$ we have $mt - 1 = 0$, i.e.

$$(m)^\alpha = (m^{-1}, -1). \tag{1}$$

Since $(0, y)^\alpha$ is on $[0]$, $(0, y)^\alpha$ is either $(0, v)$ for some $v \in R$ or (∞); but for any $y \neq 0$, $(0, y)^\alpha$ is also on the line joing $(y)^\alpha = (y^{-1}, -1)$ to $(1, 0)^\alpha = (1, 0)$, so that, for $y \neq 1$,

$$(0, y)^\alpha = (0, (y^{-1} - 1)^{-1}). \tag{2}$$

For any non-zero ab, $(ab)^{-1} - 1 = (ab)^{-1}(1 - ab)$, so that

$$
\begin{aligned}
[(ab)^{-1} - 1][(1 - ab)^{-1} - 1] &= [(ab)^{-1}(1 - ab)](1 - ab)^{-1} - [(ab)^{-1} - 1] \\
&= (ab)^{-1} - [(ab)^{-1} - 1] \\
&= 1.
\end{aligned}
$$

Thus we have

$$[(1 - ab)^{-1} - 1]^{-1} = (ab)^{-1} - 1. \tag{3}$$

Hence, putting $y = 1 - ab$ in (2) gives

$$(0, 1 - ab)^\alpha = (0, (ab)^{-1} - 1). \tag{4}$$

The line $[b]^\alpha$ is the join of $(b, 0)^\alpha = (b, 0)$ to $(\infty)^\alpha = (0, -1)$, thus

$$[b]^\alpha = [-b^{-1}, -1]. \tag{5}$$

Similarly $(b, 1 - ab)^\alpha$ is on both $[b]^\alpha = [-b^{-1}, -1]$ and $[a - b^{-1}, 0]^\alpha = [a - b^{-1}, 0]$. But, by inspection, $(a^{-1}, b^{-1}a^{-1} - 1)$ is on both these lines so

$$(b, 1 - ab)^\alpha = (a^{-1}, b^{-1}a^{-1} - 1). \tag{6}$$

However, from (4), $(0, 1 - ab)^\alpha = (0, (ab)^{-1} - 1)$ so that

$$(b, 1 - ab)^\alpha = (c, (ab)^{-1} - 1) \quad \text{for some} \quad c. \tag{7}$$

Thus, from (6) and (7), $(ab)^{-1} = b^{-1}a^{-1}$ or $b^{-1} = (ab)^{-1}a$. Hence $b = (b^{-1})^{-1} = [(ab)^{-1}a]^{-1} = a^{-1}(ab)$ and R has LIP.

(ii) Suppose R has LIP.

The above mapping α defined by (1), (2), (5), (6) is a $((0, 0), [0, 0])$-elation mapping (∞) onto $(0, -1)$, and thus sending $[\infty]$ onto $[0, -1]$. Thus, by Exercise 4.13, $[0, -1]$ is a translation line and, using Theorem 4.20, every line through (0) is a translation line. Hence, in particular, $[0, 0]$ is a translation line and $\mathscr{P}$ is $((0, 0), [0, 0])$-transitive. ☐

Since R has RIP, every line through (∞) is a translation line and so, if R also has LIP, $\mathscr{P}$ has three non-concurrent translation lines which, by Theorem 4.20, means that every line of $\mathscr{P}$ is a translation line. Theorem 6.14 may now be restated as:

Theorem 6.15. *If R is a division ring with RIP then R has LIP if and only if $\mathscr{P}$ is a Moufang plane.*

Exercise 6.4. Show that R is a skewfield if and only if $\mathscr{P}$ is (V, l)-transitive for all choices of V and l.

If a division ring D satisfies both alternative laws, (i) $x(xy) = x^2y$ and (ii) $(xy)y = xy^2$, for all $x, y \in D$, then we call D an *alternative division ring* (in the literature, sometimes called an alternative field). We have seen that any division ring with IP is alternative (Lemma 6.13), so a Moufang plane can be coordinatized infact by an alternative division ring. Now in fact it is also true that any alternative division ring has IP, so the two classes of rings are identical. But historically they have been sometimes separated, and there has been much algebraic study of various systems satisfying the alternative laws, or perhaps one of them only (e.g., a *right* or *left* alternative division ring), or even weak forms of them. But from a geometric point of view, we are really only interested in RIP division rings and in IP division rings, and we want to prove that in this context, RIP implies IP (and for the experts, we mention that right alternative does *not* imply alternative, but geometrically this fact has no interest). This will have the powerful corollary that the presence of two distinct translation lines implies that all lines are translation lines, and dually, the presence of two distinct translation points implies that all points are translation points. Note that if all lines are translation lines, then all points are translation points, and conversely.

In fact, all alternative division rings have been classified; this result, the Bruck-Kleinfeld Theorem, is completely algebraic and is not included in the book. But the special case of finite alternative division rings is included: the Artin-Zorn Theorem (really a special case of the Bruck-Kleinfeld Theorem) asserts that a finite alternative division ring is a field. This implies that the presence of two translation lines (or points) in a finite plane forces the plane to be a $\mathscr{P}_2(q)$. In Chapter IX, Section 4, we give examples of division rings which include non-associative but alternative examples, and these are in fact the most general possible; but we shall not concern ourselves anymore with the very deep questions connected with alternative division rings and the Bruck-Kleinfeld Theorem.

The rest of this chapter is devoted to proving the results cited above: an RIP division ring is alternative (Theorem 6.16), an alternative division rings is IP (Theorem 6.17), and a finite alternative division ring is a field (Theorem 6.20). Even though the proofs are not at all geometric, and hence ought to be skipped by most readers on their first time through, the study of projective planes cannot go forward without the central corollaries mentioned above (two translation elements of the same kind implies Moufang, and finite Moufang implies Desargues). Hence we have included the proofs for the sake of completeness.

7. The Skornyakov-San Soucie Theorem

In this section we prove the following:

Theorem 6.16 (Skornyakov-San Soucie Theorem). *A division ring D with RIP is alternative.*

Proof.[3]. For any a, b, c in D we define the *associator* $[a, b, c,] = (ab)c - a(bc)$ and the *commutator* $[a, b] = ab - ba$. It is straightforward to verify that each of these is additive in all arguments.

Throughout the section we shall let (RA) denote the right alternative law, $((xy) y = xy^2)$, and let (M) denote the identity $x((yz) y) = ((xy) z) y$. (The M standing for Moufang who first studied division rings with this property.) By Lemma 6.13, since D has RIP, D satisfies both (RA) and (M). We now establish a long series of results which we shall call **(1)**, **(2)**, ...

(1) $[a, b, c] = -[a, c, b]$ for all a, b, c in D.

Proof. By (RA) $[x, y, y] = 0$ for all $x, y \in D$ and we have

$$0 = [a, b + c, b + c] = [a, b, b] + [a, b, c] + [a, c, b] + [a, c, c]$$
$$= [a, b, c] + [a, c, b] . \quad \square$$

(2) $[a, b, bc] = [a, b, c] b$ for all a, b, c in D.

Proof. By (M) $a((bc) b) = ((ab) c) b$ and hence by **(1)**

$$[a, b, bc] = -[a, bc, b] = -((a(bc)) b - a((bc) b))$$
$$= -((a(bc)) b - ((ab) c) b)$$
$$= ((ab) c - a(bc)) b$$
$$= [a, b, c] b . \quad \square$$

---

[3] The proofs of this theorem and of Theorem 6.18 are taken from [1]. We wish to thank Professor H. Lüneburg for giving us permission to reproduce them here.

We define three "functions", f, g, h, each of four variables, on D by

$$f(d, a, b, c) = [da, b, c] - [d, ab, c] + [d, a, bc] - d[a, b, c] - [d, a, b]\,c\,,$$

$$g(a, d, b, c) = [a, d, bc] + [a, b, dc] - [a, d, c]\,b - [a, b, c]\,d\,,$$

$$h(d, a, b, c) = [da, b, c] + [d, a, [b, c]] - d[a, b, c] - [d, b, c]\,a\,.$$

(3) $f(d, a, b, c) = g(a, d, b, c) = h(d, a, b, c) = 0$ for all a, b, c, d in D.

Proof. $f(d, a, b, c) = ((da)\,b)\,c - (da)(bc) - (d(ab))\,c + d((ab)\,c)$
$$+ (da)(bc) - d(a(bc)) - d((ab)\,c) + d(a(bc))$$
$$- ((da)\,b)\,c + (d(ab))\,c = 0\,.$$

Now replace b by $b + d$ in **(2)** to get

$$0 = [a, b + d, (b + d)\,c] - [a, b + d, c]\,(b + d)$$
$$= [a, b, bc] + [a, b, dc] + [a, d, bc] + [a, d, dc]$$
$$- [a, b, c]\,b - [a, b, c]\,d - [a, d, c]\,b - [a, d, c]\,d$$

and by **(2)** the first and fifth terms cancel and so do the fourth and eighth leaving $g(a, d, b, c)$. Now, we have

$$0 = f(d, a, b, c) - g(d, c, a, b) = ([da, b, c] - [d, ab, c]$$
$$+ [d, a, bc] - d[a, b, c] - [d, a, b]\,c) - ([d, c, ab]$$
$$+ [d, a, cb] - [d, c, b]\,a - [d, a, b]\,c)\,.$$

The last terms of each bracket cancel with each other and, by **(1)**, $-[d, ab, c] - [d, c, ab] = 0$ and $[d, c, b]\,a = -[d, b, c]\,a$. Thus, by the additivity property of the commutator, we have

$$0 = [da, b, c] + [d, a, bc - cb] - d[a, b, c] - [d, b, c]\,a$$
$$= h(d, a, b, c). \quad \square$$

Note. (M) is not used in proving $f(d, a, b, c) = 0$.

(4) $[a, b, c^2] = [a, bc + cb, c]$ for all a, b, c, in D.

Proof. $0 = f(a, b, c, c) = [ab, c, c] - [a, bc, c] + [a, b, c^2]$
$$- a[b, c, c] - [a, b, c]\,c\,.$$

By (RA) $[ab, c, c] = a[b, c, c] = 0$ and by **(1)** and **(2)** $-[a, b, c]\,c = [a, c, b]\,c = [a, c, cb] = -[a, cb, c]$ so we have $0 = [a, b, c^2] - [a, bc, c] - [a, cb, c]$ as desired. $\quad \square$

If a, b are any two elements of D we define $A(a, b)$ by

$$A(a, b) = \{x \text{ in } D \,|\, [x, a, b] = x[b, a]\}\,.$$

(5) $A(a, b) = A(b, a)$ is a subgroup of the additive group of D. Furthermore, x is in $A(a, b)$ if and only if $(xa)\,b = x(ba)$.

Proof. Let x be in $A(a, b)$. Then $(xa) b - x(ab) = x(ba) - x(ab)$ by definition of $A(a, b)$ and hence $(xa) b = x(ba)$. The converse is clearly true also. We obviously have $[0, a, b] = 0 = 0[b, a]$ so $0 \in A(a, b)$ and $A(a, b)$ is non empty. Let $x, y \in A(a, b)$. Then $[x - y, a, b] = [x, a, b] - [y, a, b] = x[b, a] - y[b, a] = (x - y) [b, a]$.
Hence $x - y \in A(a, b)$ so $A(a, b)$ is a subgroup as claimed.

Finally, $x \in A(a, b) \Leftrightarrow [x, a, b] = x[b, a]$

$$\Leftrightarrow -[x, b, a] = -x[a, b]$$
$$\Leftrightarrow [x, b, a] = x[a, b]$$
$$\Leftrightarrow x \in A(b, a). \quad \square$$

(6)   If $[a, a, b] \neq 0$, then $A(a, b) \cap A(a, ba) = \{0\}$.

Proof. Let x be in $A(a, b) \cap A(a, ba)$. Since $x \in A(a, b)$, by **(5)** $(xa) b = x(ba)$ and hence by (M)

$$x((ab) a) = ((xa) b) a = (x(ba)) a.$$

By **(5)** since $x \in A(a, ba) = A(ba, a)$, we have

$$(x(ba)) a = x(a(ba))$$

and by the two equations above

$$0 = (x(ba)) a - (x(ba)) a = x((ab) a) - x(a(ba))$$
$$= x[a, b, a] = -x[a, a, b]$$

and since $[a, a, b] \neq 0$ by hypothesis, $x = 0$. $\quad \square$

(7) $[x, a, b], [x, a, b] a \in A(a, b)$ for all $x \in D$.

Proof. By **(2)** $[x, a, ab^2] = [x, a, b^2] a$ and since D is right alternative we have $0 = g(x, a, ab, b) = [x, a, (ab) b] + [x, ab, ab] - [x, a, b] (ab) - [x, ab, b] a = [x, a, b^2] a - [x, a, b] (ab) - [x, ab, b] a$ and applying **(4)** to the first term yields

$$0 = [x, ab + ba, b] a - [x, a, b] (ab) - [x, ab, b] a$$
$$= [x, ba, b] a - [x, a, b] (ab)$$

because of additivity. By **(1)** and **(2)** $[x, ba, b] = -[x, b, ba] = -[x, b, a] b = [x, a, b] b$ and we have

$$0 = ([x, a, b] b) a - [x, a, b] (ab)$$

which means $[x, a, b] \in A(b, a) = A(a, b)$ by **(5)**.

Now by (RA) and **(2)** $[x, ba, ba^2] = [x, ba, (ba) a]$
$$= [x, ba, a] (ba).$$

We also have by (RA) and (M) that

$$(ba) a^2 = ((ba) a) a = b(a^2 a) = ba^3$$

and by **(2)** $[x, b, ba^3] = [x, b, a^3] b$. Using these results we have

$$0 = g(x, ba, b, a^2) = [x, ba, ba^2] + [x, b, (ba) a^2] - [x, ba, a^2] b$$
$$- [x, b, a^2] (ba) = [x, ba, a] (ba) + [x, b, a^3] b - [x, ba, a^2] b$$
$$- [x, b, a^2] (ba).$$

By **(4)** $[x, b, a^2] = [x, ab + ba, a]$ so we have

$$0 = [x, ba, a] (ba) + [x, b, a^3] b - [x, ba, a^2] b - [x, ab, a] (ba)$$
$$- [x, ba, a] (ba).$$

The first and last terms cancel and we get

$$([x, b, a^3] - [x, ba, a^2]) b = [x, ab, a] (ba). \tag{$*$}$$

Now $0 = g(x, b, a^2, a) b = [x, b, a^3] b + [x, a^2, ba] b - ([x, b, a] a^2) b - ([x, a^2, a] b) b = [x, ab, a] (ba) - ([x, b, a] a^2) b - ([x, a^2, a] b) b$ by **(1)** and **(*)**.

We have by **(1)** and **(2)** $[x, a^2, a] = -[x, a, a^2] = -[x, a, a] a = 0$. Thus using (RA) we get

$$[x, ab, a] (ba) = (([x, b, a] a) a) b.$$

Since $[x, ab, a] = -[x, a, ab] = -[x, a, b] a$ by **(2)**, we have $([x, a, b] a)(ba) = (([x, a, b] a) a) b$ and $[x, a, b] a \in A(a, b)$. □

(8)   If $[a, a, b] \neq 0$, then $[a, b] \neq 0$.

Proof. Assume that $[a, b] = 0$. Then $ab = ba$ and $A(a, ab) = A(a, ba)$. By **(6)** $A(a, b) \cap A(a, ba) = \{0\} = A(a, b) \cap A(a, ab)$. By **(2)** and **(7)** $[a, a, b] a = [a, a, ab] \in A(a, b) \cap A(a, ab) = \{0\}$ so that $[a, a, b] a = 0$ which implies $a = 0$ since $[a, a, b] \neq 0$ by hypothesis. But then $[a, a, b] = [0, 0, b] = 0$ contrary to hypothesis. □

(9) $[[a, b], a, b] = 0$ for all $a, b \in D$.

Proof. Let $[a, b] = q$. Then by **(7)** $[q, a, b] \in A(a, b)$ and so by the definition of $A(a, b)$

$$[[q, a, b], a, b] = [q, a, b] [b, a] = -[q, a, b] q.$$

By **(7)**, $[[q, a, b], a, b] \in A(a, b)$, so that $-[q, a, b] q \in A(a, b)$. Also by **(7)** $[q^2, a, b] \in A(a, b)$. Thus since

$$0 = h(q, q, a, b) = [q^2, a, b] + [q, q, q] - q[q, a, b] - [q, a, b] q$$

and since $[q, q, q] = 0$ by (RA), we have

$$q[q, a, b] = [q^2, a, b] - [q, a, b] q$$

and belongs to $A(a, b)$ because $A(a, b)$ is a group under addition by **(5)**. Put $[q, a, b] = s$, then since $qs \in A(a, b)$

$$[qs, a, b] = (qs) [b, a] = -(qs) q.$$

Similarly, since by **(7)** $s = [q, a, b] \in A(a, b)$, $[s, a, b] = -sq$ and $q[s, a, b]$
$= -q(sq)$. Finally, from the above we have

$$0 = h(q, s, a, b) = [qs, a, b] + [q, s, q] - q[s, a, b] - [q, a, b] s$$
$$= -(qs) q + (qs) q - q(sq) + q(sq) - s^2 .$$

Thus, $s^2 = 0$ and we have $0 = s = [[a, b], a, b]$. $\square$

(10) $[[a, b], a, c] = -[[a, c], a, b]$ for all $a, b, c \in D$.

Proof. By **(9)** $0 = [[a, b + c], a, b + c]$; expanding this by linearity and
applying **(9)** gives

$$0 = [[a, b + c], a, b] + [[a, b + c], a, c]$$
$$= [[a, b], a, b] + [[a, c], a, b] + [[a, b], a, c] + [[a, c], a, c]$$
$$= [[a, c], a, b] + [[a, b], a, c] \quad \text{as desired.} \quad \square$$

(11) $[x[a, b], a, b] = [x, a, b] [a, b]$ for all $a, b, x \in D$.

Proof. Put $[a, b] = q$. Then

$$0 = h(x, q, a, b) = [xq, a, b] + [x, q, q] - x[q, a, b] - [x, a, b] q$$

and since $[x, q, q] = 0$ by (RA) and $[q, a, b] = 0$ by **(9)** the result is proved. $\square$

(12) $[(xa) b - x(ba), a, b] = 0$ for all $a, b, x \in D$.

Proof. Put $[a, b] = q$. By **(7)** and the definition of $A(a, b)$ we get
$[[x, a, b], a, b] = [x, a, b] [b, a] = -[x, a, b] q$. Moreover, $xq + [x, a, b]$
$= x(ab) - x(ba) + (xa) b - x(ab) = (xa) b - x(ba)$.
Hence by the above remarks and **(11)** we have

$$0 = -[[x, a, b], a, b] + [[x, a, b], a, b]$$
$$= [x, a, b] q + [[x, a, b], a, b]$$
$$= [xq, a, b] + [[x, a, b], a, b] = [xq + [x, a, b], a, b]$$
$$= [(xa) b - x(ba), a, b] . \quad \square$$

(13) $[[a, a, b]^2, a, [a, b]] = [[a, a, b]^2, a, ab] = [[a, a, b]^2, a, ba] = 0$
for all $a, b \in D$.

Proof. Let $p = [a, a, b]$ and $q = [a, b]$. By **(7)** $pa \in A(a, b)$ and, by the
definition of $A(a, b)$, this yields $[pa, a, b] = -(pa) q$. Also by **(7)** $p \in A(a, b)$
so $[p, a, b] = -pq$ and $[p, a, b] a = -(pq) a$. But

$$0 = h(p, a, a, b) = [pa, a, b] + [p, a, q] - p[a, a, b] - [p, a, b] a$$
$$= -(pa) q + [p, a, q] - p^2 + (pq) a$$
$$= -(pa) q + (pa) q - p(aq) - p^2 + (pq) a .$$

This gives $p^2 = (pq)\, a - p(aq)$ and so using **(12)**

$$[p^2, a, q] = [(pq)\, a - p(aq), a, q] = - [(pq)\, a - p(aq), q, a] = 0\,.$$

Also $p^2 = p[a, a, b] = - p[a, b, a] = -p((ab)a) + p(a(ba))$; but by **(M)** $p((ab)\, a) = ((pa)\, b)\, a$ so that we have $p^2 = - ((pa)\, b)\, a + p(a(ba))$.
However $p \in A(a, b)$, so by **(5)** $(pa)\, b = p(ba)$ which gives

$$p^2 = - (p(ba))\, a + p(a(ba))\,.$$

Thus by **(12)**

$$[p^2, a, ba] = - [p^2, ba, a] = [(p(ba))\, a - p(a(ba)), ba, a] = 0.$$

Finally, $[p^2, a, ab] = [p^2, a, q] + [p^2, a, ba] = 0 + 0 = 0.$ □

    (14) $[x, y, z]\,(wz) + [x, w, z]\,(yz) = ([x, y, z]\, z)\, w + ([x, w, z]\, z)\, y$ for all $x, y, z, w \in D$.

Proof. By **(7)** $[x, y + w, z] \in A(y + w, z)$ and hence by **(5)**

$$[x, y + w, z]\,(yz + wz) = ([x, y + w, z]\, z)\,(y + w)\,.$$

Thus,

$$[x, y, z]\,(yz) + [x, y, z]\,(wz) + [x, w, z]\,(yz) + [x, w, z]\,(wz)$$
$$= ([x, y, z]\, z)\, y + ([x, y, z]\, z)\, w + ([x, w, z]\, z)\, y + ([x, w, z]\, z)\, w\,.$$

Since $[x, y, z] \in A(y, z)$ and $[x, w, z] \in A(w, z)$ (by **(7)**) the first and last terms of each side of the above equation are equal respectively by **(5)** and can be cancelled leaving the desired result. □

    (15) $[x, y, z]\,(yw) + [x, y, w]\,(yz) = ([x, y, z]\, w)\, y + ([x, y, w]\, z)\, y.$

Proof. By **(7)** $[x, y, z + w] \in A(y, z + w) = A(z + w, y)$. Hence by **(5)** $[x, y, z + w]\,(yz + yw) = ([x, y, z + w]\,(z + w))\, y$ and the proof is completed by proceeding as in **(14)**. □

    (16) $[x, y, z]\,(wu) + [x, y, u]\,(wz) + [x, w, z]\,(yu) + [x, w, u]\,(yz)$
$$= ([x, y, z]\, u)\, w + ([x, y, u]\, z)\, w + ([x, w, z]\, u)\, y + ([x, w, u]\, z)\, y.$$

Proof. We leave this proof as an exercise. (Hint; in **(14)** replace z by $z + u$, expand by additivity and distributivity and use **(14)** again.) □

    (17)   The following identity is true in any non-associative (or associative) ring:

$$[xy, z] - x[y, z] - [x, z]\, y = [x, y, z] - [x, z, y] + [z, x, y]\,.$$

Proof. $[xy, z] - x[y, z] - [x, z]\, y = (xy)z - z(xy) - x(yz) + x(zy) - (xz)y + (zx)y$ and $[x, y, z] - [x, z, y] + [z, x, y] = (xy)z - x(yz) - (xz)y + x(zy) + (zx)\, y - z(xy)$. □

Proof of Theorem 6.16. Assume D is a counterexample. Then there are elements a, b in D such that $[a, a, b] \ne 0$ since RA implies that $[b, a, a] = 0$ for all $a, b \in D$. By **(8)** $[a, b] \ne 0$. Put $p = [a, a, b]$, $q = [a, b]$.

Step (A). $[q, a, q] = 0 = [q, a, ba]$:

Proof. By **(2)** and **(9)** $[q, a, ab] = [q, a, b] a = 0$ so that $[q, a, q]$ $= [q, a, ab - ba] = -[q, a, ba]$ and by **(7)** $[q, a, ba] \in A(a, ba)$. Also, by **(10)** $[q, a, ba] = [[a, b], a, ba] = -[[a, ba], a, b]$ which by **(7)** is in $A(a, b)$. Hence, $[q, a, ba] \in A(a, b) \cap A(a, ba) = \{0\}$ by **(6)** since $p = [a, a, b] \neq 0$. Thus $[q, a, ba] = 0$ and $[q, a, q] = 0$. ☐

Step (B). $x + x = 0$ for all $x \in D$:

Proof. $0 = h(q, a, a, b) = [qa, a, b] + [q, a, q] - q[a, a, b] - [q, a, b] a$. By Step (A) $[q, a, q] = 0$ and by **(9)** $[q, a, b] = 0$ so we have $0 = [qa, a, b] - qp$ or equivalently $qp = [qa, a, b]$. Furthermore, by **(11)** $[aq, a, b] = [a, a, b] q$ $= pq$ and thus

$$pq - qp = [aq - qa, a, b] = [[a, q], a, b]$$
$$= -[[a, b], a, q] \quad \text{by (10)}$$
$$= -[q, a, q] = 0 \quad \text{by Step (A)} .$$

Therefore, $pq = qp$.

By **(2)** and **(13)** $[p^2, a, b] a = [p^2, a, ab] = 0$ and since $q = [a, b] \neq 0$, $a \neq 0$ so that $[p^2, a, b] = 0$. Now, since $p = [a, b] \in A(a,b)$, $[p, a, b] = pq$; thus

$$0 = h(p, p, a, b) = [p^2, a, b] + [p, p, q] - p[p, a, b] - [p, a, b] p$$
$$= [p, p, q] + p(pq) + (pq) p$$
$$= -[p, q, p] + p(pq) + (pq) p$$
$$= -(pq) p + p(qp) + p(pq) + (pq) p .$$

Thus, $0 = p(qp + pq)$ and since $p \neq 0$, $qp + pq = 0$ but $pq = qp$ and we have $(p + p) q = 0$ which, since $q \neq 0$, yields $0 = p + p = (1 + 1) p$. However, $p \neq 0$ so we have $1 + 1 = 0$ and hence $x + x = (1 + 1) x = 0$ as claimed. ☐

[Note: this result that any counterexample must have characteristic two is the result of Skornyakov.] Since D has characteristic two, it is not necessary to distinguish between $+$ and $-$ and this will not always be done in what follows.

Step (C). $[q, x, y] = 0$ for all x, y in D:

Proof. By **(9)** $[q, a, b] = 0$. If in **(15)** we replace x, y, z, w by q, a, b, x respectively, then we have

$$[q, a, b] (ax) + [q, a, x] (ab) = ([q, a, b] x) a + ([q, a, x] b) a ,$$

which implies $[q, a, x] (ab) = ([q, a, x] b) a$ so that $[q, a, x] \in A(a, b)$. Now in **(15)** replacing x, y, z, w by q, a, ba, x gives $[q, a, ba] (ax) + [q, a, x] (a(ba))$ $= ([q, a, ba] x) a + ([q, a, x] (ba)) a$. By Step (A) $[q, a, ba] = 0$ and we have $[q, a, x] (a(ba)) = ([q, a, x] (ba)) a$ and hence $[q, a, x] \in A(a, ba) \cap A(a, b) = \{0\}$ since $[a, a, b] \neq 0$. That is $[q, a, x] = 0$ for all $x \in D$. Also, by **(9)** $[q, b, a]$ $= -[q, a, b] = 0$ and $[q, ba, a] = -[q, a, ba] = 0$. Hence, using **(15)** as above we find $[q, b, y] \in A(a, b)$ and $[q, ba, y] \in A(a, ba)$ for all $y \in D$. Now $0 = g(q, x, b, a) = [q, x, ba] + [q, b, xa] - [q, x, a] b - [q, b, a] x$. The last two terms become 0 on transposition of the second and third entries in each of the associators. Hence $[q, x, ba] = - [q, b, xa] \in A(a, b) \cap A(a, ba)$ by the above remarks. Therefore, we have

$$[q, b, xa] = 0 \quad \text{for all} \quad x \in D. \tag{$**$}$$

Since $a \neq 0$, if $y \in D$, there exists $x \in D$ such that $y = xa$ and thus

$$[q, b, y] = 0 \quad \text{for all} \quad y \in D.$$

In (16) replacing x, y, z, w, u by q, a, b, x, y gives $[q, a, b] (xy) + [q, a, y] (xb) + [q, x, b] (ay) + [q, x, y] (ab) = ([q, a, b] y) x + ([q, a, y] b) x + ([q, x, b] y) a + ([q, x, y] b) a$. By the above remarks, transposing second and third entries in the associators when necessary, all terms but the last on each side of the equation are zero. Hence $[q, x, y] \in A(a, b)$. Repeating this process, replacing x, y, z, w, u in (16) by q, a, ba, x, y and using $(**)$ together with the above remarks gives $[q, x, y] \in A(a, b) \cap A(a, ba)$ which yields $[q, x, y] = 0$ by (6). $\square$

If $A = A(a, b)$ and $N = \{x \in D \,|\, [x, a, b] = 0\}$ then

Step (D). $D = A + N$ and $NA = A$.

Proof. By (7) $[x, a, b] \in A$ and by (9) $q \in N$. Also, by (11) $[xq, a, b] = [x, a, b] q$ and hence we have

$$0 = [1, a, b] = [q^{-1}q, a, b] = [q^{-1}, a, b] q,$$

which, since $q \neq 0$, means $[q^{-1}, a, b] = 0$.

If we put $d = [x, a, b] q^{-1}$, then by RIP

$$[d, a, b] = [[x, a, b] q^{-1}, a, b] = [[(xq^{-1}) q, a, b] q^{-1}, a, b]$$
$$= [([xq^{-1}, a, b] q) q^{-1}, a, b] \qquad \text{by (11)}$$
$$= [[xq^{-1}, a, b], a, b] \qquad \text{by RIP}$$

Since $[xq^{-1}, a, b] \in A = A(a, b)$ by (7) we have

$$[[xq^{-1}, a, b], a, b] = - [xq^{-1}, a, b] q = [xq^{-1}, a, b] q$$

by Step (B). Hence using RIP and (11)

$$[d, a, b] = [xq^{-1}, a, b] q = [(xq^{-1}) q, a, b] = [x, a, b] = [x, a, b] (q^{-1}q) = dq.$$

Thus

$$[d, a, b] = dq \quad \text{and} \quad d \in A. \tag{$\dagger$}$$

Now put $x + d = y$ or equivalently $x = y + d$ by Step (B). Then using $(\dagger)$ and Step (B) $[y, a, b] = [x + d, a, b] = [x, a, b] + [d, a, b] = 2[x, a, b] = 0$ so that $y \in N$. Thus we have, for any $x \in D$,

$$x = y + d = d + y \in A + N.$$

Now, if n, s are arbitrary elements of N and A respectively, then

$$0 = h(n, s, a, b) = [ns, a, b] + [n, s, q] + n[s, a, b] + [n, a, b] s.$$

Since $n \in N$, $[n, a, b] = 0$ and since $s \in A(a, b)$, $[s, a, b] = -sq$, thus

$$0 = [ns, a, b] + [n, s, q] - n(sq)$$
$$= [ns, a, b] + (ns) q + n(sq) - n(sq).$$

Hence, by Step (B), $[ns, a, b] = (ns) q$ which implies $ns \in A$ and, consequently that $NA \subseteq A$. But since $1 \in N$, $A = 1A \subseteq NA \subseteq A$ which gives $NA = A$. $\square$

Step (E). $[x, q] = 0$ for all $x \in D$ (i.e., q is in the centre of D).

Proof. $0 = h(x, y, a, b) = [xy, a, b] + [x, y, q] + x[y, a, b] + [x, a, b] y$ and hence

$$[xy, a, b] = [x, y, q] + x[y, a, b] + [x, a, b] y. \tag{††}$$

For any $s \in A$, from (††) we have

$$[p(qs), a, b] = [p, qs, q] + p[qs, a, b] + [p, a, b] (qs).$$

Since $qs \in NA = A$ by Step (D), we have $[qs, a, b] = (qs) q$ and $[p, qs, q] = [p, q, qs] = [p, q, s] q = [p, s, q] q$. Also, $p \in A$ and hence $[p, a, b] = pq$. Thus

$$[p(qs), a, b] = [p, s, q] q + p((qs) q) + (pq) (qs)$$
$$= ((ps) q) q + (p(sq)) q + p((qs) q) + (pq) (qs).$$

Also, by (††),

$$[p(sq), a, b] = [p, sq, q] + p[sq, a, b] + [p, a, b] (sq)$$
$$= [p, sq, q] + p([s, a, b] q) + [p, a, b] (sq) \quad \text{by (11)}$$
$$= [p, sq, q] + p((sq) q) + (pq) (sq)$$

(since $[s, a, b] = sq$ because $s \in A$). Hence by the two preceding equations

$$pq(qs + sq) = [p(qs), a, b] + [p(sq), a, b] + ((ps) q) q + (p(sq)) q$$
$$+ p((qs) q) + (p(sq)) q + p((sq) q) + p((sq) q).$$

However $(p(sq)) q + (p(sq)) q = 0 = p((sq) q) + p((sq) q)$ by Step (B) and by (M) $p((qs) q) = ((pq) s) q$, so we have for $s \in A$

$$pq(qs + sq) = [p(qs), a, b] + [p(sq), a, b] + ((ps) q) q + ((pq) s). \tag{$***$}$$

By (††) we have

$$[ps, a, b] = p[s, a, b] + [p, a, b] s + [p, s, q]$$
$$= p(sq) + (pq) s + [p, s, q] \quad \text{since } p, s \in A$$
$$= p(sq) + (pq) s + [p, q, s] \quad \text{by Step (B) and (1)}$$
$$= p(sq) + (pq) s + (pq) s + p(qs) = p(sq) + p(qs).$$

Since $[ps, a, b] \in A$ by **(7)** we then have

$$(p(sq))\,q + (p(qs))\,q = [ps, a, b]\,q = [[ps, a, b], a, b] = [p(sq) + p(qs), a, b]$$
$$= [p(sq), a, b] + [p(qs), a, b]\,.$$

Substituting this into the first two terms of (∗∗∗) gives

$$pq(qs + sq) = (p(sq))\,q + (p(qs))\,q + ((ps)\,q)\,q + ((pq)\,s)\,q\,.$$

Adding together the first and third, and second and fourth terms gives:

$$pq(qs + sq) = [p, s, q]\,q + [p, q, s]\,q$$
$$= [p, q, s]\,q + [p, q, s]\,q = 0 \quad \text{(by (1) and Step (B))}.$$

Hence, since $p \neq 0$ implies $pq \neq 0$, $qs + sq = 0$ which is the same as $[s, q] = 0$ for all $s \in A$. If $n \in N$, then $np \in A$ by Step (D) and thus $[np, q] = 0$. If in **(17)** we put $x = n$, $y = p$, $z = q$ we get $[np, q] + n[p, q] + [n, q]\,p = [n, p, q] + [n, q, p] + [q, n, p]$. Since $[np, q] = 0 = n[p, q]$, $(p \in A)$, and $[n, p, q] = -[n, q, p]$ and $[q, n, p] = 0$ by Step (C), we have $[n, q]\,p = 0$. However, $p \neq 0$, so we have $[n, q] = 0$ for all $n \in N$. Now, if $x \in D$, by Step (D), $x = n + s$ where $n \in N$ and $s \in A$ and we have

$$[x, q] = [n, q] + [s, q] = 0$$

proving Step (E). □
   Finally, by Step (E) $[a, q] = 0$, so by **(8)** $[a, a, q] = 0$. Thus

$$[a, a, ab] = [a, a, ba] \in A(a, ba) \quad \text{by (7)}.$$

Also by **(2)**, $[a, a, ab] = [a, a, b]\,a \in A(a, b)$ by **(7)**. Thus by **(6)**, $[a, a, b]\,a = 0$ since $[a, a, b] = p \neq 0$ by hypothesis. Thus $pa = 0$ and, since $p \neq 0$, $a = 0$. But if $a = 0$, $p = [a, a, b] = [0, 0, b] = 0$ contrary to hypothesis. Therefore, Theorem 6.16 is proved. □

8. The Artin-Zorn Theorem

We continue with our series of numbered results.

(18) If D is an alternative division ring and α is a permutation of $x, y, z \in D$, then

$$[x, y, z] = (\operatorname{sgn}\alpha)\,[x^\alpha, y^\alpha, z^\alpha]\,.$$

(Here $\operatorname{sgn}\alpha = 1$ or -1 according as α is an even or odd permutation.)

Proof. Consider the permutation α defined by $x^\alpha = x$, $y^\alpha = z$, $z^\alpha = y$. Then since D is, in particular, a right alternative division ring, by **(1)**

$$[x, y, z] = -[x, z, y] = \operatorname{sgn}\alpha\,[x^\alpha, y^\alpha, z^\alpha]\,.$$

Now consider the permutation β given by $x^\beta = y$, $y^\beta = x$, $z^\beta = z$. Then since D is alternative

$$0 = [x + y, x + y, z] = [x, x, z] + [x, y, z] + [y, x, z] + [y, y, z]$$

and since the first and last terms on the right are zero,

$$[x, y, z] = -[y, x, z] = \operatorname{sgn}\beta[x^\beta, y^\beta, z^\beta] .$$

Since α and β generate the full symmetric group on $\{x, y, z\}$, the proof is complete. □

We now define a function k by

$$k(w, x, y, z) = [wx, y, z] - [x, y, z] w - x[w, y, z] ;$$

k is clearly additive in all arguments for $w, x, y, z \in D$, an alternative division ring.

(19) If α is a permutation of w,x,y,z, then $k(w,x,y,z) = (\operatorname{sgn}\alpha)k(w^\alpha, x^\alpha, y^\alpha, z^\alpha)$.

Proof. $k(w, x, y, z) - k(x, y, z, w) + k(y, z, w, x)$

$$= [wx, y, z] - [x, y, z] w - x[w, y, z] - [xy, z, w] + [y, z, w] x$$
$$+ y[x, z, w] + [yz, w, x] - [z, w, x] y - z[y, w, x] .$$

By **(3)** the function f is identically zero in any non-associative ring (see the note following the proof of **(3)**) and incorporating the first, fourth and seventh terms into $f(w, x, y, z)$, adding the two extra terms and rearranging by **(18)** where necessary, we find that the above expression can be reduced to:

$$f(w, x, y, z) + w[x, y, z] + [w, x, y] z - [x, y, z] w - x[w, y, z]$$
$$+ [y, z, w] x + y[x, z, w] - [z, w, x] y - z[y, w, x] .$$

By **(18)** $[w, y, z] = [y, z, w]$, $[x, z, w] = [z, w, x]$ and $[y, w, x] = [w, x, y]$. Thus we find

$$k(w, x, y, z) - k(x, y, z, w) + k(y, z, w, x)$$
$$= [w, [x, y, z]] + [[w, x, y], z] + [[y, z, w], x] + [y, [z, w, x]] .$$

Similarly,

$$k(x, y, z, w) - k(y, z, w, x) + k(z, w, x, y)$$
$$= [x, [y, z, w]] + [[x, y, z], w] + [[z, w, x], y] + [z, [w, x, y]] .$$

Since $[a, b] = -[b, a]$, adding these gives

$$k(w, x, y, z) + k(z, w, x, y) = 0 .$$

Thus if we define β by $x^\beta = w$, $y^\beta = x$, $z^\beta = y$, $w^\beta = z$, then since β is an odd permutation, $\operatorname{sgn}\beta = -1$ and we have shown

$$k(w, x, y, z) = (\operatorname{sgn}\beta) k(w^\beta, x^\beta, y^\beta z^\beta) .$$

Also by **(18)**

$$k(w, x, y, z) = [wx, y, z] - [x, y, z] \, w - x[w, y, z]$$
$$= -[wx, z, y] + [x, z, y] \, w + x[w, z, y]$$
$$= -k(w, x, z, y)$$

and if $\gamma = (yz)$, then $\operatorname{sgn} \gamma = -1$ and we have

$$k(w, x, y, z) = (\operatorname{sgn} \gamma) \, k(w^\gamma, x^\gamma, y^\gamma, z^\gamma) \, .$$

However, $\langle \beta, \gamma \rangle = S_4$ the full symmetric group on $\{x, y, z, w\}$ and the proof is complete. []

In Lemma 6.13 we showed that RIP implies (RA). We now prove the converse.

Theorem 6.17. *Every alternative division ring D has* IP.

Proof. By definition of k, $k(w, x, y, y) = 0$ for w, x, y in an alternative division ring. Hence by **(19)** $k(w, x, y, z) = 0$ if two of the arguments are equal. Thus,

$$0 = k(z, x, y, z) = [zx, y, z] - [x, y, z] \, z - x[z, y, z]$$

and by **(18)**, $[z, y, z] = -[z, z, y] = 0$ in an alternative division ring; we then have $[zx, y, z] = [x, y, z] \, z$. Let $aa' = 1$. Then

$$0 = [1, a, b] = [aa', a, b] = [a', a, b] \, a \, .$$

Since $a \neq 0$, this means $[a', a, b] = 0$. Moreover, $[a', a, b] = -[b, a, a']$ by **(18)** so we have $[b, a, a'] = 0$ and hence $(ba) \, a' = b(aa') = b$ for all b so that D has RIP. Setting $b = a'$ gives $(a' a) \, a' = a'$ which implies $a' a = 1$ and since $0 = [a', a, b]$, D also has LIP; i.e., D has IP. []

Theorems 6.16 and 6.17 show that RIP implies IP and thus, by Theorems 6.12 and 6.15, we have

Theorem 6.18. *Let $\mathscr{P}$ be a projective plane. Then $\mathscr{P}$ is a Moufang plane if and only if $\mathscr{P}$ has two distinct translation lines.*

Lemma 6.19. *If a and b are two elements of the alternative division ring D, then a and b are contained in an associative subring of D.*

Proof. Put $E = \{a, b\}$. If $x, y, z \in E$, then $[x, y, z] = 0$ since at least two of x, y, z are equal and D is alternative. If we put

$$X = \{x \in D \,|\, [e, f, x] = 0 \text{ for all } e, f \in E\} \, ,$$

then $E \subseteq X$. Let $e_1, e_2, e_3 \in E$ and $x \in X$. Then

$$[e_1 x_1, e_2, e_3] = k(e_1, x, e_2, e_3) + [x_1, e_2, e_3] + x[e_1, e_2, e_3] = 0$$

(since the first and third terms must have at least two equal arguments and the second one is 0 by definition of X and **(18)**). Hence by **(18)**

$$[e_2, e_3, e_1 x] = 0 \text{ so that } e_1 x \in X, \text{ or } EX \subseteq X.$$

Now, define Y by $Y = \{y \in D \mid [e, x, y] = 0 \text{ for all } e \in E, x \in X\}$. Since $E \subseteq X$, we have that $Y \subseteq X$. From the definition of X, we have $E \subseteq Y$. For $x \in X$, $y \in Y$ and $e, f \in E$ consider

$$[e, f, yx] = [yx, e, f] = k(y, x, e, f) + [x, e, f] y + x[y, e, f].$$

The last two terms on the right are zero by the definitions of X and Y and **(18)**. Also $k(y, x, e, f) = -k(e, x, y, f)$ by **(19)** which gives

$$[e, f, yx] = -k(e, x, y, f) = -[ex, y, f] + [x, y, f] e + x[e, y, f] = 0$$

since $EX, E \subseteq X$ and then by definition of Y all the terms are zero. Thus $yx \in X$ and hence $YX \subseteq X$. Now define R by

$$R = \{r \in D \mid [x, y, r] = 0 \text{ for all } x \in X, y \in Y\}.$$

By definition of X and Y, $E \subseteq R$. Since $E \subseteq Y$, $[x, e, r] = 0$ for all $r \in R$ which gives $R \subseteq Y$. Thus $E \subseteq R \subseteq Y \subseteq X$ and hence $[r, s, t] = 0$ for all $r, s, t \in R$; so R is associative. For r and $s \in R$, we have for all $x \in X$ and $y \in Y$

$$[x, y, r - s] = [x, y, r] - [x, y, s] = 0$$

so R is an additive subgroup of D. Also, $k(y, x, r, s) = [yx, r, s] - [x, r, s] y - x[y, r, s] = 0$ since $YX \subseteq X$ and $R \subseteq Y$. It follows from this and **(19)** that $k(r, s, y, x) = 0$. Thus

$$0 = [rs, y, x] - [s, y, x] r - s[r, y, x] = [rs, y, x].$$

Hence, $rs \in R$ and R is an associative subring as desired. □

Exercise 6.4. Show that if D is any division ring, then the *centre Z*, defined to be the set of elements of D which *commute* with every element of D and which *associate* with every pair of elements of D, in any order, is non-empty and is a subfield of D. Hence any division ring contains an associative sub-division ring.

Theorem 6.20 (The Artin-Zorn Theorem). *Every finite alternative division ring is a field.*

Proof. Let D be a counter-example. Let C be a maximal associative subring of D. Then $C \neq D$ since an associative alternative division ring is a skewfield. Since D has no zero-divisors, C can have none. So C is a skewfield and hence by finiteness (Result 1.13), a field. As a result, the multiplicative group C^* of C is cyclic. Let C^* be generated by a. Since $C \neq D$, there exists $b \in D \setminus C$. By Lemma 6.19 there is an associative subring R containing a and b. Then, since C^* is cyclic, $C \cup \{b\} \subseteq R$. Since $b \notin C$, $R \neq C$

and $R \neq D$ because D is not associative; thus we have contradicted the maximality of C and the theorem is proved. □

Corollary. *A finite Moufang plane is pappian (and desarguesian).*

9. Summary of Results from Previous Sections

It is very important that the student has a clear grasp of the results of this chapter, especially of the relationships between *algebraic* properties of the coordinatizing PTR and the *geometric* properties of the associated projective plane; these latter properties are often expressed in terms of the presence of certain (V, k)-transitivities. In the table reproduced below, we have summarized some of the most important of these relations. It is important to note that these conditions are equivalences: the algebraic condition is equivalent to the geometric condition. Also, we have sometimes written the geometric condition in a number of equivalent forms (and hence these conditions imply each other). And finally, remember that in the finite case there are additional simplifications: in particular, the first three types collapse into one. The table has no redundancies otherwise; that is, for each type there are non-trivial examples, even finite ones. The relation between the types is given in the diagram:

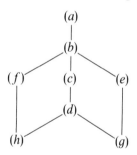

Some remarks on the table are in order: planes of type (a), (b) or (c) have the property that all coordinatizing PTR's have the same structure (in the case of types (a) and (b), that is, pappian and desarguesian planes respectively, all PTR's are even isomorphic). But for all the proper members of the remaining types (that is, planes which are not also of "higher" types), there are elements which play a special role in the coordinatization, and if some other element is chosen in this role, the PTR may have no "nice" properties at all. Thus for a proper translation plane there is a unique line k which must play the role of $[\infty]$ in order to obtain a quasifield, and so on.

The dual of a plane of type (a), (b), (c) or (d) is again of the same type (indeed, any pappian plane is even self-dual), while the duals of planes of type (g) are of type (h), and the duals of (e) are (f). We shall show later

Algebraic properties of PTR	Geometric properties of plane	Name
(a) Field	(V, k)-transitivity, all V and k, plus the "Configuration of Pappus" (see Theorem 2.6).	Pappian plane
(b) Skewfield	(V, k)-transitivity, all V and k; or (V, k)-transitivity, all V and k such that V is on k, plus (V, k)-transitivity for one non-incident pair V and k.	Desarguesian plane
(c) Alternative (or IP) division ring	(V, k)-transitivity, all incident pairs V and k; or (k, k)-transitivity, all k; or (V, V)-transitivity, all V; or (k, k)-transitivity, two distinct k; or (V, V)-transitivity, two distinct V.	Moufang plane
(d) Division ring	(k, k)-transitivity and (V, V)-transitivity, for one incident pair k and V (then $V = (\infty)$ and $k = [\infty]$).	Division ring plane
(e) Nearfield	(k, k)-transitivity and (V, m)-transitivity, with V on k but not on m (then $k = [\infty]$ and either $V = (0)$ and m contains (∞), or $V = (\infty)$ and m contains (0)).	Nearfield plane
(f) Right nearfield	(V, V)-transitivity and (W, k)-transitivity, with V but not W on k (then $V = (\infty)$ and either $k = [\infty]$ and W lies on $[0]$, or $k = [0]$ and W lies on $[\infty]$).	Dual nearfield plane
(g) Quasifield	(k, k)-transitivity, for one k; or (A, k)-transitivity, for two distinct points A on the line k (in either case, then $k = [\infty]$).	Translation plane
(h) Right quasifield	(V, V)-transitivity, for one V; or (V, k)-transitivity, for two distinct lines k on the point V (in either case, $V = (\infty)$).	Dual translation plane

in this book that there exist proper examples of each type (there also exist many planes which are of none of these types, and we shall exhibit examples of those as well).

10. Homomorphisms

In this brief section we utilize Theorem 5.5 and our information about planar ternary rings gained in this chapter to prove or sketch a few results on homomorphisms of projective planes. In fact this is a subject about which not very much is known at the present time, and our results are very elementary. Clearly, a *homomorphism* of the projective plane $\mathscr{P}_1$ onto the projective plane $\mathscr{P}_2$ will mean a mapping ϕ which sends the points and lines, respectively, of $\mathscr{P}_1$ onto the points and lines of $\mathscr{P}_2$, and which preserves incidence. We have first:

Theorem 6.21. *Let ϕ be a homomorphism from the projective plane $\mathscr{P}_1$ onto the projective plane $\mathscr{P}_2$. Then either ϕ is an isomorphism, or the pre-image of any element of $\mathscr{P}_2$ consists of infinitely many elements.*

Proof. Let X_2, Y_2, O_2, I_2 be a coordinatizing quadrangle for $\mathscr{P}_2$ (see Chapter V), and (R_2, T_2) a PTR which corresponds. Choose X_1, Y_1, O_1, I_1 in $\mathscr{P}_1$ so that $X_1^\phi = X_2$, $Y_1^\phi = Y_2$, $O_1^\phi = O_2$, $I_1^\phi = I_2$, and let (R_1, T_1) be a PTR for $\mathscr{P}_1$ corresponding to this quadrangle choice. We shall use subscripts to distinguish coordinates in the two planes; e.g., $(x, y)_1$ is a point of $\mathscr{P}_1$, while $(u, v)_2$ is a point of $\mathscr{P}_2$.

Let Z be the set of all x in R_1 such that $(x, 0)_1^\phi = (0, 0)_2$. It is clear that $(1)_1^\phi = (1)_2$, so $(0, y)_1^\phi = (0, 0)_2$ if and only if the line on $(1)_1$, $(0, y)_1$, and $(y, 0)_1$ goes to the line on $(1)_2$ and $(0, 0)_2$. Thus $(0, y)_1^\phi = (0, 0)_2$ if and only if y is in Z. From this it is immediate that $(x, y)_1^\phi = (0, 0)_2$ if and only if x and y are in Z. Finally $(m)_1$, $(0, m)_1$ and $(1, 0)_1$ are collinear, so $(m)_1^\phi = (0)_2$ if and only if m is in Z.

Now choose m, x, y in Z. The line on $(m)_1$ and $(x, y)_1$ passes through $(0, T_1(m, x, y))_1$, but also the line on the images $(m)_1^\phi = (0)_2$, $(x, y)_1^\phi = (0, 0)_2$ passes through $(0, 0)_2$: that is, $T_1(m, x, y)$ is in Z. So (Z, T_1) is a ternary ring. Thus from Theorem 5.5 (Z, T_1) is either infinite, consists of 0 alone, or is planar. But if it were planar, then it would have to have a (multiplicative) identity, and this clearly must be the element 1 of R_1; since 1 is definitely not in Z, it follows that Z is either infinite or consists of 0 alone. If Z consists of 0 alone, then the pre-image of $(0, 0)_2$ is just $(0, 0)_1$; if Z is infinite, then there are infinitely many points of $\mathscr{P}_1$ which map onto $(0, 0)_2$.

Now it is straightforward to show that the pre-image of the point $(x, y)_2$ consists exactly of all the points $(u + z_1, v + z_2)_1$, where $(u, v)_1$ is some point such that $(u, v)_1^\phi = (x, y)_2$ and z_1 and z_2 are in Z; similarly, the pre-image of $(m)_2$ contains at least all the points $(k + z)_1$, where $(k)_1^\phi = (m)_2$ and z is in Z, while the pre-image of $(\infty)_2$ contains all the points $(w)_1$, where $wz = 1$ for some z in Z. So if the pre-image of one point (that is, $(0, 0)_2$) is infinite, then the pre-image of every point is infinite. ☐

Corollary. *If $\mathscr{P}_1$, $\mathscr{P}_2$ are finite projective planes and ϕ is a homomorphism from $\mathscr{P}_1$ onto $\mathscr{P}_2$, then ϕ is an isomorphism.*

Now in fact we can analyze homomorphisms in even greater detail, using planar ternary rings. Using the techniques of the proof of Theorem 6.21, it is easy enough to define ϕ completely in terms of Z. Let J be the set of elements x in R_1 such that $xz = 1$ for some element z in Z, and let $\bar{R}_1$ be the set of elements of R which are not in J. Then we can prove:

(i) $(\bar{R}_1, T_1)$ is a ternary ring.

In addition, we can prove that ϕ induces a mapping α of $\bar{R}_1$ onto R_2 such that:

(ii) $[T_1(x, y, z)]^\alpha = T_2(x^\alpha, y^\alpha, z^\alpha)$, for all x, y, z in $\bar{R}_1$;

(iii) $(x, y)_1^\phi = (x^\alpha, y^\alpha)_2$ if both x and y are in $\bar{R}_1$

$(m)_1^\phi = (m^\alpha)_2$ if m is in $\bar{R}_1$.

It is possible to analyze ϕ and Z and $\bar{R}_1$ further, but the details are laborious (though interesting).

*Exercise 6.5.** Prove (i), (ii) and (iii) above.

*Exercise 6.6.** Let (R_1, T_1) be the planar ternary ring of rational numbers, and let (R_2, T_2) be the planar ternary ring $GF(p)$, where p is a prime; let $\mathscr{P}_1$ and $\mathscr{P}_2$ be the projective planes coordinatized by the two planar ternary rings, respectively. Then show that there is a homomorphism of $\mathscr{P}_1$ onto $\mathscr{P}_2$, and the subsets Z, J and $\bar{R}_1$ of the discussion above are given by:

$Z = \{a/b$ in R_1 such that $(a, b) = 1$ and p divides $a\}$,

$J = \{a/b$ in R_1 such that $(a, b) = 1$ and p divides $b\}$,

$\bar{R}_1 = \{a/b$ in R_1 such that $(p, b) = 1\}$.

It is possible, using the approach above, to show that (P, k)-transitivity must be preserved by homomorphism, and hence that the homomorphic image of a translation plane is a translation plane, and the homomorphic image of a desarguesian plane must be desarguesian. (However it can happen that (P, k)-transitivity with P not on k might induce (P^ϕ, k^ϕ)-transitivity where P^ϕ is incident with k^ϕ. So homologies might become elations.)

References

This chapter is self contained. However, as we mentioned earlier, the proofs of the Skornyakov-San Soucie Theorem and the Artin-Zorn Theorem are taken from the excellent set of lecture notes [1].

1. Lüneburg, H.: Lectures on projective planes. University of Illinois at Chicago Circle 1969.

VII. Quasifields and Translation Planes

1. Introduction

Among the many kinds of planar ternary rings that we have met (at least potentially) up to now, a certain family plays a very special role: those which are linear, with associative addition and the left distributive law. They correspond to projective planes such that all possible elations with $[\infty]$ as axis actually exist, i.e., translation planes. Desarguesian planes are of this sort, but in fact there are many others, and in preparation for Chapter IX, where examples will be given, we study these planes in more detail. Although a given quasifield determines a unique (up to isomorphism) projective plane, it often occurs that a given projective plane may be coordinatized by non-isomorphic quasifields. We define the kernel of a quasifield and the major part of the chapter is devoted to proving that if two non-isomorphic quasifields coordinatize the same translation plane then their kernels are isomorphic. To do this we adopt the approach of André and introduce the concept of a congruence partition.

At the end of the chapter there is a brief discussion of spreads and a few exercises to give examples of quasifields which are not fields; thus establishing for the first time in this book the actual existence of non-desarguesian translation planes.

We want to emphasize that almost all known finite projective planes or their duals are in the class considered here (some important exceptions will be dealt with in Chapters IX and X).

2. Algebraic Definitions and Basic Properties

In Chapter VI a quasifield was defined as a PTR obtained by coordinatizing a projective plane $\mathscr{P}$ which had $[\infty]$ as a translation line. We now give a purely algebraic definition of a quasifield and leave as an exercise the proof that the two definitions are equivalent. Throughout this chapter when we regard a quasifield Q as a PTR, the ternary operation is understood to be the natural linear one, i.e., $T(m, x, y) = mx + y$, and we shall simply say that Q is a PTR.

A *quasifield* Q is a set with two binary operations $+, \cdot$ satisfying:
(1) $(Q, +)$ is a group;
(2) $(Q^*, \cdot)$ is a loop;
(3) $x(y + z) = xy + xz$ for all $x, y, z \in Q$;
(4) the additive identity 0 satisfies $0x = 0$ for all $x \in Q$;
(5) $ax = bx + c$ has a unique solution for x, given $a, b, c \in Q, a \neq b$.

Exercise 7.1. Prove that a "quasifield" as defined above is a planar ternary ring and is a "quasifield" by the definition in Chapter VI.

Exercise 7.1 shows that (by Theorem 6.4) $(Q, +)$ is in fact an abelian group.

**Exercise 7.2. Deduce from axioms (1)–(5) in the algebraic definition of a quasifield Q that $(Q, +)$ is abelian.

Exercise 7.3. (a) Prove that axioms (1), (2), (3), (5) in the definition of a quasifield imply (4).
(b) If F is any field, define a new multiplication by $x \cdot y = xy$ if $x \neq 0$ and $0 \cdot y = y$. Show $(F, +, \cdot)$ satisfies axioms (1), (2), (3) for a quasifield, but does not satisfy (4).

For reasons which will soon become apparent (see Theorem 7.3) we now define a slightly more general class of algebraic systems which, however, includes all quasifields.

A *weak quasifield* W is a set with two binary operations $+, \cdot$ satisfying:
(1) $(W, +)$ is an abelian group;
(2) $(W^*, \cdot)$ is a loop;
(3) $x(y + z) = xy + xz$ for all $x, y, z \in W$;
(4) the additive identity 0 satisfies $0 \cdot x = 0$ for all $x \in W$.

Since, as we have already observed, addition in a quasifield is abelian, any quasifield is a weak quasifield.

Exercise 7.4. Show that a finite weak quasifield is a quasifield. (Hint: see Theorem 5.5.)

Weak quasifields have some basic algebraic properties which we now establish. We shall always write 0 for the additive identity in a weak quasifield.

Lemma 7.1. *If W is a weak quasifield then*
(a) $a \cdot 0 = 0$ *for all $a \in W$;*
(b) $a(-b) = -(ab)$ *for all $a, b \in W$.*

Proof. (a) Clearly $1 + 0 = 1$ and so, by (3), $a = a \cdot 1 = a(1 + 0) = a \cdot 1 + a \cdot 0 = a + a \cdot 0$. Thus, since $(W, +)$ is a group, $a \cdot 0 = 0$.
(b) Using (a), $0 = a \cdot 0 = a(b + (-b))$ for any $a, b \in W$. Thus $ab + a(-b) = 0$ and, since $(W, +)$ is a group, $-(ab) = a(-b)$. $\square$

In what follows, we write $-ab$ for $-(ab)$.

If W is any weak quasifield we define the *kernel*, K, of W as the set of all elements $k \in W$ such that

(i) $(x + y)k = xk + yk$ and
(ii) $(xy)k = x(yk)$

for all $x, y \in W$.

Theorem 7.2. *The kernel K of a weak quasifield W is a skewfield. Furthermore W may be regarded as a right vector space over K, in the natural way.*

Proof. Let a, b be any two elements of W and h, k be any elements of K.

$(a + b)(h - k) = (a + b)h + (a + b)(-k)$	(since W satisfies left distributive law)
$\quad = (a + b)h - (a + b)k$	(by Lemma 7.1)
$\quad = ah + bh - ak - bk$	(since h, k distribute on the right)
$\quad = ah - ak + bh - bk$	(since addition is abelian)
$\quad = a(h - k) + b(h - k)$	(using left distributive law and Lemma 7.1)
$(ab)(h - k) = (ab)h - (ab)k$	(by Lemma 7.1 and the left distributive law)
$\quad = a(bh) - a(bk)$	(since h, k associate on the right)
$\quad = a(bh - bk)$	(using left distributive law)
$\quad = a(b(h - k))$	(by Lemma 7.1 and the left distributive law).

Hence K is closed under subtraction and is non-empty (0 and 1 are in K), so $(K, +)$ is an (abelian) subgroup of $(W, +)$.

We must now show that $(K^*, \cdot)$ is a group. Clearly, from the definition of K, multiplication in K is associative and $1 \in K$, so we must show that K is closed under multiplication and that every element of K has a (left) inverse in K.

For any $a, b \in W$, $l, k \in K$, $(a + b)(kl) = [(a + b)k]l = (ak + bk)l = (ak)l + (bk)l = a(kl) + b(kl)$: also $(ab)(kl) = ((ab)k)l = (a(bk))l = a((bk)l) = a(b(kl))$ and so $kl \in K$.

If $k \in K$, $k \neq 0$, then, since $(W^*, \cdot)$ is a loop, there is a unique element $k' \in W$ such that $k'k = 1$. We wish to show $k' \in K$. For any $a, b \in W$, $((a + b)k')k = (a + b)(k'k) = (a + b) \cdot 1 = a \cdot 1 + b \cdot 1 = a(k'k) + b(k'k) = (ak')k + (bk')k = (ak' + bk')k$ and so, since $k \neq 0$, $(a + b)k' = ak' + bk'$.

Similarly $((ab)k')k = (ab)(k'k) = (ab) \cdot 1 = a \cdot (b \cdot 1) = a \cdot (b(k'k)) = a((bk')k) = (a(bk'))k$ and so, since $k \neq 0$, $a(bk') = (ab)k'$. Hence $k' \in K$ and K^* is a multiplicative group. Since both distributive laws obviously hold in K, K is a skewfield.

Using the addition of W as vector addition it is clear that we can now regard W as a right vector space over its kernel K, (in fact over any sub-skewfield of K). $\square$

This last observation will be used frequently in later chapters. We now show that if W is finite dimensional over K then, in fact, W is a quasifield. This is an exceptionally useful result in showing that a given system is a quasifield.

Theorem 7.3. *Let W be a weak quasifield with kernel K. If W is a finite dimensional vector space over any sub-skewfield H of K then W is a quasifield.*

Proof. We must show that given any $a, b, c \in W$, $a \neq b$ the equation $ax = bx + c$ has a unique solution for x.

For any $a \in W$ define a mapping L_a of W onto itself by $xL_a = ax$; i.e., L_a is left multiplication by a. Since W satisfies the left distributive law, $(x + y)L_a = a(x + y) = ax + ay = xL_a + yL_a$. Furthermore, for any $h \in H$, $(xh)L_a = a(xh) = (ax)h = (xL_a)h$. Thus L_a is a linear transformation of W over H and, since W^* is a multiplicative loop, is non-singular unless $a = 0$.

We must show that $xL_a = xL_b + c$ has a unique solution for x given any a, b, c with $a \neq b$; this is equivalent to showing $x(L_a - L_b) = c$ has a unique solution for x. Clearly, since W has finite dimension over H, $x(L_a - L_b) = c$ has a unique solution if and only if $L_a - L_b$ is non-singular. If $L_a - L_b$ is singular then there exists $w \neq 0$ such that $w(L_a - L_b) = 0$. But this implies $aw - bw = 0$ or $aw = bw$, and thus $a = b$; a contradiction and so $L_a - L_b$ is non-singular.

Hence $x(L_a - L_b) = c$ has a unique solution for x and W is a quasifield. □

Corollary. *A finite weak quasifield is a quasifield.* (Compare Exercise 7.4.)

3. Congruence Partitions

If a quasifield Q coordinatizes a projective plane $\mathscr{P}$ then, by Exercise 7.1, $\mathscr{P}$ is a translation plane with respect to $[\infty]$. Under these circumstances we call $\mathscr{P}$ a *translation plane;* i.e., if no translation line is mentioned it is to be assumed to be $[\infty]$. The elations with axis $[\infty]$ will be called *translations.* Similarly an affine plane $\mathscr{A}$ will be called an *affine translation plane* if $\mathscr{A} = \mathscr{P}^{[\infty]}$ for a translation plane $\mathscr{P}$; the collineations of $\mathscr{A}$ induced by translations of $\mathscr{P}$ will be called *translations of A.* Since any affine plane is associated with a unique projective plane no confusion will arise if we identify an affine translation plane $\mathscr{A}$ with the translation plane $\mathscr{P}$ with which it is associated.

Exercise 7.5. Show that a collineation $\alpha \neq 1$ of an affine translation plane $\mathscr{A}$ is a translation if and only if $X^\alpha \neq X$ for all $X \in \mathscr{A}$ and l^α is parallel to l for all $l \in \mathscr{A}$.

We recall some of the properties of the translation group of a translation plane.

Lemma 7.4. *Let $\mathscr{P}$ be a translation plane and let $\Pi = \operatorname{Aut}\mathscr{P}$. Then*

(i) $\Pi_{([\infty],[\infty])} = \bigcup\limits_{X \in [\infty]} \Pi_{(X,[\infty])}$,

(ii) $\Pi_{(A,[\infty])} \cap \Pi_{(B,[\infty])} = 1$ *for all A, B on $[\infty]$, $A \neq B$,*

(iii) $\Pi_{([\infty],[\infty])} = \Pi_{(A,[\infty])} \cdot \Pi_{(B,[\infty])}$ *for any pair A, B on $[\infty]$, $A \neq B$.*

Proof. Parts (i) and (ii) are obvious since they are true for all groups of elations with a common axis. Part (iii) is Exercise 4.25 but we shall give a proof for completeness.

Let π be any element of $\Pi_{([\infty],[\infty])}$, then $\pi \in \Pi_{(C,[\infty])}$ for some point $C \in [\infty]$. If $C = A$ or B then, clearly, $\pi \in \Pi_{(A,[\infty])} \cdot \Pi_{(B,[B,[\infty])}$, so assume $C \neq A$, $C \neq B$. By Theorem 4.7 π is uniquely determined by the image of a single point T not on $[\infty]$. Since $C \neq A$ or B, $AT \neq A\,T^{\pi}$ and $BT \neq BT^{\pi}$ so if $S = AT \cap BT^{\pi}$ then $S \neq T$, T^{π}. Since $\mathscr{P}$ is a translation plane there exist collineations $\alpha \in \Pi_{(A,[\infty])}$, $\beta \in \Pi_{(B,[\infty])}$ such that $T^{\alpha} = S$ and $S^{\beta} = T^{\pi}$. Hence $\alpha\beta \in \Pi_{([\infty],[\infty])}$ with $T^{\alpha\beta} = T^{\pi}$ and thus $\alpha\beta = \pi$. This implies $\Pi_{([\infty],[\infty])} \subseteq \Pi_{(A,[\infty])} \cdot \Pi_{(B,[\infty])} \subseteq \Pi_{([\infty],[\infty])}$ and the theorem is proved. $\square$

Lemma 7.4 leads us to consider the following: let G be a group and let $\mathscr{S}$ be a set of subgroups of G. Then $\mathscr{S}$ is called a *partition of G* if (a) $X \neq 1$ for all $X \in \mathscr{S}$, (b) $G = \bigcup\limits_{X \in \mathscr{S}} X$ and (c) $X \cap Y = 1$ for all X, $Y \in \mathscr{S}$, $X \neq Y$. Any subgroup in $\mathscr{S}$ is called a *component* of $\mathscr{S}$. One possible partition of G is $\mathscr{S} = \{G\}$; any partition with at least two components is called *non-trivial.*

A non-trivial partition $\mathscr{S}$ of G is called a *congruence partition* if $G = X \cdot Y$ for any X, $Y \in \mathscr{S}$ with $X \neq Y$. Lemma 7.4 proves

Theorem 7.5. *If $\mathscr{P}$ is a translation plane with $\Pi = \operatorname{Aut}\mathscr{P}$ then $\mathscr{S} = \{\Pi_{(X,[\infty])} \mid X \in [\infty]\}$ is a congruence partition of $\Pi_{([\infty],[\infty])}$.*

We now show that any group which admits a congruence partition is in fact the translation group of a translation plane. The following discussion of the relation between the properties of congruence partitions and translation planes is based on the original work of André.

If G is any group with a partition $\mathscr{S}$, define $\mathscr{A}(G, \mathscr{S})$ by

(i) the points of $\mathscr{A}(G, \mathscr{S})$ are the elements of G,

(ii) the lines of $\mathscr{A}(G, \mathscr{S})$ are the right cosets Xg for all $X \in \mathscr{S}$, $g \in G$,

(iii) the point g is on the line Xh if and only if $g \in Xh$.

Theorem 7.6. *If $\mathscr{S}$ is a congruence partition of G then*

(a) $\mathscr{A}(G, \mathscr{S})$ *is an affine translation plane;*

(b) $G \cong \Pi_{([\infty],[\infty])}$ *where $\Pi = \operatorname{Aut}(\mathscr{A}(G, \mathscr{S}))$;*

(c) G *is abelian.*

Proof. Let $\mathscr{A} = \mathscr{A}(G, \mathscr{S})$. We first show that $\mathscr{A}$ is an affine plane. Let x, y be any two distinct points of G, whence $xy^{-1} \neq 1$. (Throughout this proof 1 will denote the unit element of G. As we do not need to mention the identity collineation there is no danger of confusing the unit elements of the groups G and Π.) Since $G = \bigcup_{X \in \mathscr{S}} X$ there is a $U \in \mathscr{S}$ such that $xy^{-1} \in U$ so that $x \in Uy$ and there is at least one line containing both x and y. Suppose $x \in Uy$ and $x \in Vy$, then $xy^{-1} \in U \cap V$ and thus, since $x \neq y$, $U \cap V \neq 1$. Since $\mathscr{S}$ is a congruence partition this implies $U = V$ and so the line joining x and y is unique.

If $a \notin Xb$ we must show there is a unique $Y \in \mathscr{S}$ such that $Ya \cap Xb = \phi$. (Clearly if $a \in Yc$ then $Ya = Yc$.) Since $a \notin Xb$, $Xa \cap Xb = \phi$ and so we must show that if $Ya \cap Xb = \phi$ then $Y = X$. Suppose $Y \neq X$ then, as $\mathscr{S}$ is a congruence partition, $G = Y \cdot X$, and there exist $x \in X$, $y \in Y$ such that $ab^{-1} = yx$ or $a = yxb$. Thus $Ya = Yyxb = Yxb$ and so $xb \in Ya \cap Xb$ which contradicts $Ya \cap Xb = \phi$, and shows that Xa is the only line through a not intersecting Xb. Note that this implies that any parallel class of $\mathscr{A}$ is exactly the set of cosets of an element of $\mathscr{S}$.

Since $|\mathscr{S}| \geq 2$ there exist distinct subgroups $U, V \in \mathscr{S}$. If $u \in U$, $v \in V$, $u \neq 1 \neq v$, then $1, u, v$ are three non-collinear points and $\mathscr{A}$ is an affine plane.

In order to prove that $\mathscr{A}$ is a translation plane we shall show (i) that $G \cong \Pi_{([\infty],[\infty])}$ where $\Pi = \mathrm{Aut}\,\mathscr{A}$ and (ii) $\Pi_{([\infty],[\infty])}$ is transitive on the points of $\mathscr{A}$.

For any $g \in G$ define a mapping α_g on the points and lines of $\mathscr{A}$ by $x^{\alpha_g} = xg$ and $(Ux)^{\alpha_g} = Uxg$. Clearly $x \in Uy$ if and only if $x^{\alpha_g} \in (Uy)^{\alpha_g}$ so that $\alpha_g \in \Pi$. In fact, since $y^{\alpha_g} \neq y$ if $g \neq 1$ for any $y \in G$ and $(Ux)^{\alpha_g} = Uxg$ is parallel to Ux, $\alpha_g \in \Pi_{([\infty],[\infty])}$ (by Exercise 7.5). The proof that $\mathscr{A}$ is a translation plane is completed by proving that the mapping $\theta : g \to \alpha_g$ is an isomorphism from G onto $\Pi_{([\infty],[\infty])}$. We leave this as an exercise.

Finally G is abelian by Theorem 4.14. ☐

As a direct corollary of the last two theorems we have:

Theorem 7.7. *Let $\mathscr{P}$ be a translation plane and let $\Pi = \mathrm{Aut}\,\mathscr{P}$. If $\mathscr{S} = \{\Pi_{(X,[\infty])} | X \in [\infty]\}$ then $\mathscr{P}^{[\infty]} \cong \mathscr{A}(\Pi_{([\infty],[\infty])}, \mathscr{S})$.*

Exercise 7.6. If $\mathscr{S}$ is a congruence partition of a group G and if $\mathscr{S}'$ is a congruence partition of G' show that $\mathscr{A}(G, \mathscr{S}) \cong \mathscr{A}(G', \mathscr{S}')$ if and only if there is an isomorphism θ from G onto G' such that $\mathscr{S}^\theta = \mathscr{S}'$. (Meaning that the subgroups of $\mathscr{S}$ are mapped by θ onto the subgroups of $\mathscr{S}'$.)

As we have just seen, any group G which admits a congruence partition $\mathscr{S}$ is abelian and thus the set of all endomorphisms of G form a ring. We define the *kernel* K of $\mathscr{S}$ as the set of all endomorphisms μ of G such that $X^\mu \subseteq X$ for all $X \in \mathscr{S}$.

Theorem 7.8. *The kernel K of a congruence partition $\mathscr{S}$ of G is a skewfield whose multiplicative group is isomorphic to $\Pi_{(1,[\infty])}$, where $\Pi = \mathrm{Aut}\,\mathscr{A}(G, \mathscr{S})$.*

Proof. We first note that the points of $\mathscr{A}(G, \mathscr{S})$ are the elements of G so that $\Pi_{(1,[\infty])}$ is the group of homologies with centre 1 and special axis (where 1 is the unit element of G). Let U be an element of $\mathscr{S}$.

Let e denote the identity isomorphism of G and let 0 be the endomorphism $g^0 = 1$ for all $g \in G$; then clearly, $0, e \in K$. For any $\alpha, \beta \in K$, $U^{\alpha\beta} \subseteq U^{\beta} \subseteq U$ and so $\alpha\beta \in K$. Also, for any $u \in U, u^{\alpha \pm \beta} = u^{\alpha} u^{\pm \beta} \in U^{\alpha} U^{\beta} \subseteq UU \subseteq U$ so that $\alpha \pm \beta \in K$ and, hence, K is a ring. Since the product of mappings is always associative, K is an associative ring and so, in order to show that K is a skewfield, we must show that for any $\gamma \in K, \gamma \neq 0$, there is a mapping $\gamma^{-1} \in K$ with $\gamma\gamma^{-1} = e$.

We first show that if $\alpha \in K$, $\alpha \neq 0$, then α is a monomorphism of G into G and we do this by showing that if $\beta \in K$ such that $g^{\beta} = 1$ for any $g \in G$, $g \neq 1$ then $\beta = 0$. Let A be the (necessarily unique) subgroup of $\mathscr{S}$ containing g and let h be any other element of G such that $h \notin A$. Since $\mathscr{S}$ is a partition there is a unique $B \in \mathscr{S}$ containing h. The element gh cannot belong to A, since h does not, nor can it belong to B and so there is a subgroup $C \in \mathscr{S}$ such that $gh \in C$ and, since $C \neq B$, $B \cap C = 1$. Now $h^{\beta} = 1 h^{\beta} = g^{\beta} h^{\beta} = (gh)^{\beta}$, and so as $h^{\beta} \in B^{\beta}$ and $(gh)^{\beta} \in C^{\beta}$, $h^{\beta} \in B^{\beta} \cap C^{\beta}$. However $B^{\beta} \subseteq B$, $C^{\beta} \subseteq C$ and $B \cap C = 1$, so that $h^{\beta} = 1$ for all $h \notin A$. Repeating the argument with any other subgroup of $\mathscr{S}$ playing the role of A we see that $\beta = 0$.

We must now show that every non-zero element of K is an isomorphism of G onto G; i.e., for any $g' \in G$ and $\alpha \in K$, $\alpha \neq 0$, we must find $g \in G$ such that $g^{\alpha} = g'$. Let U be the subgroup in $\mathscr{S}$ containing g' then there is a component V of $\mathscr{S}$ with $U \neq V$ and if $h \in V$, $h \neq 1$, then, by the above discussion, $h^{\alpha} \neq 1$.

If $h^{\alpha} g' = 1$ then $h^{\alpha} = g'^{-1}$ and so, since $h^{\alpha} \in V^{\alpha} \subseteq V, h^{\alpha} = g'^{-1} \in U \cap V = 1$, which is not so. Thus $h^{\alpha} g' \neq 1$, and there is a unique $W \in \mathscr{S}$ such that $h^{\alpha} g' \in W$. Furthermore $W \neq U$ since $W = U$ would imply $h^{\alpha} g' \in U$ so that $h^{\alpha} \in U$ which is not true. Similarly $W \neq V$.

The line Uh is parallel to U and passes through h. However, because $U \neq W$, Uh is not parallel to W and there is a unique point $x \in Uh \cap W$. Since $x \in Uh$ there is an element $g \in U$ such that $x = gh = hg$ (G is abelian) and so $x^{\alpha} \in (Uh \cap W)^{\alpha} = U^{\alpha} h^{\alpha} \cap W^{\alpha} \subseteq Uh^{\alpha} \cap W$. But Uh^{α} and W are two lines of $\mathscr{A}(G, \mathscr{S})$ which intersect in a unique point and so, since $g' h^{\alpha} \in Uh^{\alpha}$ and $h^{\alpha} g'$ was chosen to belong to W, we have $x^{\alpha} = g' h^{\alpha} = h^{\alpha} g'$. (Again because G is abelian.) Thus $h^{\alpha} g' = x^{\alpha} = (hg)^{\alpha} = h^{\alpha} g^{\alpha}$ and $g^{\alpha} = g'$ as required.

We have now shown that α is an automorphism of G which means that α^{-1} exists and so, in order to show that K is a skewfield, it is now sufficient to show $\alpha^{-1} \in K$ i.e., that $U^{\alpha^{-1}} \subseteq U$ for all $U \in \mathscr{A}$. But this is trivially true and so K is a skewfield.

If for any $\alpha \in K$, $\alpha \neq 0$, we define $(Ug)^{\alpha} = Ug^{\alpha}$ for all $U \in \mathscr{S}$, $g \in G$ we see that α induces a collineation of $\mathscr{A}(G, \mathscr{S})$ which, since $U^{\alpha} = U$, fixes every line which is represented by an element of $\mathscr{S}$. If we denote this collineation by α^* and put $\Pi = \operatorname{Aut}(\mathscr{A}(G, \mathscr{S}))$, then this gives $\alpha^* \in \Pi_{(1, [\infty])}$.

Conversely if $\theta \in \Pi_{(1, [\infty])}$ then θ induces a one-to-one mapping β of G onto itself such that $1^{\beta} = 1$ and $U^{\beta} = U$ for all $U \in \mathscr{S}$. In order to show that $\beta \in K$ we must show that β is an endomorphism of G; i.e., that $(gh)^\beta = g^\beta h^\beta$ for all $g, h \in G$.

If, as before, we let α_g be the collineation of $\mathscr{A}(G, \mathscr{S})$ sending x onto xg and Ux to Uxg, then, noting that $\alpha_g \alpha_h = \alpha_{gh}$, we have $(gh)^\theta = (1^{\alpha_{gh}})^\theta = (1^\theta)^{\theta^{-1} \alpha_{gh} \theta} = 1^{\theta^{-1} \alpha_g \alpha_h \theta} = 1^{(\theta^{-1} \alpha_g \theta)(\theta^{-1} \alpha_h \theta)}$. But, by Lemma 4.11, since α_g is a translation, $\theta^{-1} \alpha_g \theta$ is a translation and so, by Theorem 4.7, is uniquely determined by its action on one element of $\mathscr{A}(G, \mathscr{S})$. However $1^{\theta^{-1} \alpha_g \theta} = 1^{\alpha_g \theta} = g^\theta = g^\beta = t$ say, and α_t is also a translation with $1^{\alpha_t} = t$ so that $\theta^{-1} \alpha_g \theta = \alpha_t$ where $t = g^\beta$. Hence if $h^\beta = h^\theta = s$ we have $(gh)^\beta = (gh)^\theta = (1^{\alpha_t})^{\alpha_s} = ts = g^\beta h^\beta$ and $\beta \in K$.

Clearly $\beta^* = \theta$ so that the mapping $\alpha \to \alpha^*$ for all $\alpha \in K^*$ is a mapping of the multiplicative group of K^* onto $\Pi_{(1, [\infty])}$.

Let $\alpha, \beta \in K^*$, then for all $g \in G$, $g^{\alpha^*} = g^\alpha$ and for all $U \in \mathscr{S}$, $(Ug)^{\alpha^*} = Ug^\alpha$ and so, since $x^{(\alpha\beta)} = x^{\alpha\beta} = x^{\alpha^* \beta^*}$ and $(Ug)^{(\alpha\beta)^*} = Ug^{(\alpha\beta)} = (Ug^\alpha)^{\beta^*} = Ug^{\alpha^* \beta^*}$, the mapping $\gamma \to \gamma^*$ is a homomorphism from K^* onto $\Pi_{(1, [\infty])}$. Finally suppose δ is any element of K^* such that δ^* is the identity collineation; then $x = x^{\delta^*} = x^\delta$ for all $x \in G$, so that $\delta = e$ and the mapping $\gamma \to \gamma^*$ is an isomorphism. □

If $\mathscr{P}$ is a given translation plane then, provided X and Y are chosen as points on $[\infty]$, we have seen that $\mathscr{P}$ is always coordinatized by a quasifield no matter which points are chosen for O and I. However, as we shall see in Chapter IX, different choices of X, Y, O, I, provided X and Y are on $[\infty]$, may lead to different, that is non-isomorphic, quasifields. Each of these quasifields has a kernel but, as yet, no relation has been established between $\mathscr{P}$ and the kernels of the quasifields which coordinatize $\mathscr{P}$.

By Theorem 7.7 $\mathscr{P}^{[\infty]} \cong \mathscr{A}(\Pi_{([\infty], [\infty])}, \mathscr{S})$ where $\Pi = \operatorname{Aut} \mathscr{P}^{[\infty]}$ and $\mathscr{S} = \{\Pi_{(X, [\infty])} | X \in [\infty]\}$. The congruence partition $\mathscr{S}$ also has a kernel which is a skewfield whose multiplicative group is isomorphic to $\Pi_{(1, [\infty])}$. However any point P of $\mathscr{P}^{[\infty]}$ is an element, g say, of $\Pi_{([\infty], \infty])}$ so that $\Pi_{(P, [\infty])} = \alpha_g^{-1} \Pi_{(1, [\infty])} \alpha_g \cong \Pi_{(1, [\infty])}$, and, thus, this kernel is independent of the point assigned to the identity element of $\Pi_{([\infty], [\infty])}$; i.e., the kernel of $\mathscr{S}$ is an invariant of $\mathscr{P}^{[\infty]}$ independent of any coordinate system. In this sense we may refer to it as the kernel of $\mathscr{P}^{[\infty]}$.

4. Properties of the Kernel

We shall now show that the kernel of a quasifield Q is isomorphic to the kernel of the translation plane which it coordinatizes. This will justify our two different uses of the same word and provide an algebraic invariant of all quasifields coordinatizing a given translation plane.

We first show that any two quasifields which coordinatize the same plane have kernels with isomorphic multiplicative groups.

Lemma 7.9. *Let Q be a quasifield and let $\Pi = \operatorname{Aut} \mathscr{P}(Q)$. Then the multiplicative group of $K(Q)$ is isomorphic to $\Pi_{((0, 0), [\infty])}$.*

Proof. Let δ be any $((0,0), [\infty])$-homology of $\mathscr{P}(Q)$ and let $(1,0)^\delta = (k,0)$. Then, as in the proof of Theorem 6.5, δ induces a one-to-one mapping λ of Q onto itself such that $(x, 0)^\delta = (x^\lambda, 0)$ with $0^\lambda = 0$ and $1^\lambda = k$. Clearly, since $(1)^\delta = (1)$, $(0, y)^\delta = (0, y^\lambda)$ and so $[m, k]^\delta = [m, k^\lambda]$. The incidence preserving property of δ gives us that (x, y) is on $[m, k]$ if and only if (x^λ, y^λ) is on $[m, k^\lambda]$ which leads to

$$mx^\lambda + y^\lambda = (mx + y)^\lambda \quad \text{for all} \quad m, x, y \text{ in } Q. \tag{1}$$

Putting $y = 0$ and $x = 1$ in (1) gives

$$mk = m^\lambda. \tag{2}$$

Thus λ is right multiplication by k. Furthermore first by putting $m = 1$ in (1) and then by putting $y = 0$, again in (1), we obtain $xk + yk = (x + y)k$ for all x, y in Q and $m(xk) = (mx)k$ for all m, x in Q. Thus $k \in K(Q)$ and any $((0, 0), [\infty])$-homology of $\mathscr{P}(Q)$ is of the form $(x, y) \to (xk, yk)$ for some $k \in K(Q)$.

Clearly for any k in $K(Q)$ the collineation $(x, y) \to (xk, yk)$ is a $((0, 0), [\infty])$-homology of $\mathscr{P}(Q)$ and the mapping which sends this homology onto k is the required isomorphism between the group of $((0, 0), [\infty])$-homologies of $\mathscr{P}(Q)$ and the multiplicative group of $K(Q)$. □

Corollary. *If K is the kernel of $\mathscr{P}(Q)$ then $(K^*, \cdot) \cong (K^*(Q), \cdot)$.*

Proof. By Theorem 7.8 and Lemma 7.9 both these multiplicative groups are isomorphic to the group of $(A, [\infty])$-homologies for any affine point A of $\mathscr{P}(Q)$. □

Since a finite (skew) field is uniquely determined by its multiplication, this corollary proves that K and $K(Q)$ are isomorphic fields. However there exist non-isomorphic infinite skewfields with isomorphic multiplicative groups, so we have more to do in the infinite case.

Lemma 7.10. *Let Q be any quasifield and let $\Pi = \text{Aut}\,\mathscr{P}(Q)$. Then $\Pi_{([\infty], [\infty])} \cong (Q, +) \oplus (Q, +)$.*

Proof. Since $\Pi_{([\infty], [\infty])} = \Pi_{((\infty), [\infty])} \cdot \Pi_{((0), [\infty])}$, for any γ in $\Pi_{([\infty], [\infty])}$ there is a unique $\beta \in \Pi_{((\infty), [\infty])}$ and a unique $\alpha \in \Pi_{((0), [\infty])}$ such that $\gamma = \beta\alpha$. But, as we saw in the proofs of Theorems 6.2 and 6.3, for any β in $\Pi_{((\infty), [\infty])}$ there is a unique b in Q such that $(x, y)^\beta = (x, y + b)$ and for any α in $\Pi_{((0), [\infty])}$ there is a unique a in Q with $(x, y)^\alpha = (x + a, y)$. Thus if we write $\tau(a, b)$ for the translation given by $(x, y) \to (x + a, y + b)$, we see that $\gamma = \tau(a, b)$. But clearly the group of all $\tau(a, b)$ is isomorphic to the direct sum $(Q, +) \oplus (Q, +)$. □

Also the correspondence $(a, b) \Leftrightarrow \tau(a, b)$ is a one-to-one mapping between the points of $\mathscr{P}(Q)$ and the elements of $\Pi_{([\infty], [\infty])}$. Having made this "identification" it is clear that the elements of the congruence partition of $\Pi_{([\infty],[\infty])}$ are the lines of $\mathscr{P}(Q)$ through the origin. We shall use this identification to prove the following.

Theorem 7.11. *Let $\mathscr{P}^{[\infty]}$ be a translation plane with kernel K and let Q be any quasifield coordinatizing $\mathscr{P}^{[\infty]}$. If $K(Q)$ denotes the kernel of Q then $K \cong K(Q)$.*

Proof. From Theorem 7.8 and Lemma 7.9, we know $(K^*, \cdot) \cong (K^*(Q), \cdot)$ $\cong \Pi_{(A,[\infty])}$ for any affine point A of $\mathscr{P}^{[\infty]}$. If θ is any $((0, 0), [\infty])$-homology then, from the proof of Lemma 7.9, there exists an element k in $K(Q)$ such that θ is given by $(x, y) \to (xk, yk)$. Furthermore any $((0, 0), [\infty])$-homology is of this form for some k in $K^*(Q)$.

Given any k in $K^*(Q)$ we can now define a similar mapping η_k on the translations of $\mathscr{P}^{[\infty]}$ given by $\tau(a, b)^{\eta_k} = \tau(ak, bk)$. We shall show that η_k is in K, that every element of K is of this form and, finally, that the mapping which sends k onto η_k for each k in $K(Q)$ is an isomorphism from $K(Q)$ onto K.

Since k is in $K^*(Q)$, $(a+c)k = ak + ck$ for all a, c in Q so that $\tau(a+c, b+d)^{\eta_k}$ $= \tau((a + c) k, (b + d) k) = \tau(ak + ck, bk + dk) = \tau(ak, bk) + \tau(ck, dk)$ $= \tau(a, b)^{\eta_k} + \tau(c, d)^{\eta_k}$ for all a, b, c, d in Q. Clearly η_k is a one-to-one mapping and, hence η_k is in K for all k in $K^*(Q)$. (Here we emphasize that $\tau(a, b)$ is the translation sending $(x, y) \to (x + a, y + b)$ for all (x, y) in $\mathscr{P}^{[\infty]}$.)

If α is any element of K then α is an automorphism of $\Pi_{([\infty],[\infty])}$ fixing each component of the congruence. Thus if we make our identification of the elements of $\Pi_{([\infty],[\infty])}$ with the points of $\mathscr{P}^{[\infty]}$, α becomes a collineation of $\mathscr{P}^{[\infty]}$ fixing every line through $(0, 0)$, i.e., a $((0, 0), [\infty])$-homology. Hence $\alpha = \eta_k$ for some k in $K^*(Q)$.

Let ϕ be the mapping from $K(Q)$ onto K given by $k^\phi = \eta_k$ for all $k \neq 0$, and let $0^\phi = 0$. For any k_1, k_2 in $K^*(Q)$, $\eta_{k_1 + k_2}$ is given by $\tau(a, b)^{\eta_{k_1 + k_2}}$ $= \tau(a(k_1 + k_2), b(k_1 + k_2)) = \tau(ak_1 + ak_2, bk_1 + bk_2) = \tau(ak_1, bk_1) + \tau(ak_2, bk_2)$ $= \tau(a, b)^{\eta_{k_1}} + \tau(a, b)^{\eta_{k_2}}$, so that $\eta_{k_1 + k_2} = \eta_{k_1} + \eta_{k_2}$. Similarly, since k_1, k_2 are in $K^*(Q)$, $\eta_{k_1 k_2}$ is given by $\tau(a, b)^{\eta_{k_1 k_2}} = \tau(a(k_1 k_2), b(k_1 k_2)) = \tau((ak_1)k_2, (bk_1)k_2)$ $= \tau(ak_1, bk_1)^{\eta_{k_2}} = \tau(a, b)^{\eta_{k_1} \eta_{k_2}}$, so that $\eta_{k_1 k_2} = \eta_{k_1} \cdot \eta_{k_2}$. Thus ϕ is an isomorphism and the theorem is proved. ☐

5. Spreads

Exercise 7.7. If V is a vector space of dimension $2d$ over $GF(q)$ a *spread* of V is a set of $q^d + 1$ d-dimensional subspaces $W_1, W_2, \ldots, W_{q^d+1}$ of V such that $W_i \cap W_j = \{0\}$ for $i \neq j$. Show how any spread of V may be used to construct a translation plane. (Hint: take as points the elements of V and the components of the spread as the lines through the point represented by 0.)

Exercise 7.8. Let W be a vector space of dimension d over $GF(q)$ and let $V = W \oplus W = \{(x, y) | x, y \in W\}$. Then V is a vector space of dimension

$2d$ and, clearly, any vector space of even dimension can be written in this form. Define $W_1 = \{(x, 0) \mid x \in W\}$ $W_2 = \{(0, y) \mid y \in W\}$ and, for $i = 3, 4, \ldots,$ $q^d + 1$, $W_i = \{(x, x\, T_i) \mid x \in W$ and T_i a non-singular linear transformation of $W\}$. Show (a) that $W_1, W_2, \ldots, W_{q^d+1}$ form a spread if and only if $T_i T_j^{-1}$ is fixed point free for $i \neq j$ and (b) that they form a spread if and only if $T_i - T_j$ is non-singular for $i \neq j$.

In Exercise 7.8 the translation plane of Exercise 7.7 has, essentially, been coordinatized by the elements of W. This means that multiplication has been introduced on the elements in such a way that, with the natural vector space addition, W is a quasifield. What is this multiplication? Clearly it must depend on the choice of the T_i; but how? We strongly urge the interested reader to look into this problem. Furthermore he should try to find a suitable set of linear transformations for a given vector space (possibly for one of small dimension over a small field), and try to find a set that will give a desarguesian plane. There are many obvious questions. For instance (a) what restriction on the T_i will force the plane to be desarguesian? (b) what is the kernel of W? – Must it always contain $GF(q)$ or could it be smaller? Note, by the way, that if $\{T_3, T_4, \ldots, T_{q^d+1}\}$ is a set of linear transformations such that $T_i T_j^{-1}$ is fixed point free for $i \neq j$ then so is $T_3 S, T_4 S, \ldots, T_{q^d+1} S$ for any non-singular S. Thus, without any loss of generality, we may always take one of the T_i to be the identity.

Exercise 7.9. Show that the plane constructed from the spread of Exercise 7.8 is a nearfield plane (see Theorem 6.6 for the definition) if and only if the set of T_i forms a multiplicative group.

Clearly if we regard V as an elementary abelian additive group then any spread is a congruence partition. Thus the above exercises give a possible construction for translation planes. Furthermore the results of this chapter show that any translation plane may be constructed in this way. A brief discussion of techniques for constructing these planes is contained in Chapter X.

As we saw in Chapter II any $2d$-dimensional vector space V over $GF(q)$ coordinatizes a projective geometry $\mathscr{P}_{2d-1}(q)$. A spread of V then becomes a set of mutually disjoint $(d-1)$-dimensional complete subgeometries of $\mathscr{P}_{2d-1}(q)$ such that each point is contained in exactly one of these subgeometries. This observation gives a geometrical way of interpreting this construction, and thus gives a way of "embedding" any translation plane in a projective space of higher dimension. It is a matter of choice whether one prefers to adopt the vector space or projective geometry interpretation.

In Chapter IX we shall exhibit many families of quasifields and establish their algebraic properties. However, the reader might like to try the following exercises.

Exercise 7.10. Let F be the field $GF(9)$. Define a new "multiplication" $\circ$ on F by $x \circ y = xy$ if x is a square and $x \circ y = xy^3$ if x is not a square.

Show that F with its original addition and new multiplication is a quasifield but not a field.

Exercise 7.11. Find the kernel of the quasifield of Exercise 7.10. Is the quasifield of Exercise 7.10 a nearfield?

Exercise 7.12. Repeat the last two exercises with $F = GF(q^2)$ and $\circ$ given by $x \circ y = xy$ if x is a square and $x \circ y = xy^q$ if x is a non-square for any prime power q, $q \neq 2^t$.

Exercise 7.13. What happens in Exercise 7.12 when $q = 2^t$?

Exercise 7.14. Show that a quasifield of prime order must be a field and deduce that a translation plane of prime order is desarguesian.

References

Finite nearfields have been completely classified and the interested reader should consult [1], [2] or [3].

1. Hall, M., Jr.: The theory of groups. New York: MacMillan 1960.
2. Passman, D. S.: Permutation groups. New York: W. A. Benjamin, Inc. 1968.
3. Zassenhaus, H.: Über endlicher Fastkörper. Abhandl. Math. Sem. Univ. Hamburg **11**, 187–220 (1935).

VIII. Division Rings and their Planes

1. Introduction

This chapter is really a specialization of the preceding one, in that we want, algebraically, to add the other distributive law to a quasifield; geometrically, this means that we want to consider planes which are both translation planes and dual translation planes. It is easy enough to see that the axis (from the first point of view) must lie on the centre (from the second point of view) to give us anything more interesting than an alternative division ring. The fact is that division rings have (not surprisingly) more structure than quasifields and yet in some curious way, they seem to be less complicated. Thus it is well known that many finite non-desarguesian translation planes have non-soluble collineation groups, while all the known finite non-desarguesian division ring planes have soluble groups: in Section 6 we shall prove that this must be so in certain cases, e.g. when the order of the plane is not a square.

Since a division ring is both a right and left quasifield, it will possess a right and left kernel, with all that implies; it also possesses in fact a "middle" kernel. These subsets are however called *semi-nuclei* in this context, and the word kernel is not used; we may consider dimension in each case. As in Chapter VII, we will be laying a ground-work for, among other things, the examples that will be given in Chapter IX, where we will use the results of this chapter without comment. As in the case of quasifields, isomorphic planes may be coordinatized by non-isomorphic division rings. In order to examine the relation between division rings coordinatizing isomorphic planes the concept of isotopism is introduced and we then prove that two division rings coordinatize isomorphic planes if and only if they are isotopic.

2. The Semi-nuclei

As we have seen in Chapter VI a projective plane $\mathscr{P}$ may be coordinatized by a division ring if and only if $\mathscr{P}$ is a translation plane with respect to $[\infty]$ and the dual of a translation plane with respect to (∞). It was also shown that a division ring is a quasifield with the extra property that it satisfies the right distributive law. Thus we have: —

Lemma 8.1. *An associative division ring is a skewfield.*

One of the most helpful notions in the study of division ring planes, and particularly in determining their collineation groups, is that of the three semi-nuclei of a division ring. The *right nucleus* N_r of a division ring D is the set of all elements d in D such that $(xy)d = x(yd)$ for all x, y in D. The *left nucleus* N_l and *middle nucleus* N_m are defined analogously; N_l, N_m, N_r are the *semi-nuclei* of D. The intersection N of the three *semi-nuclei* of D is called the *nucleus* while the set of all n in N such that $dn = nd$ for all d in D is called the *centre* of D and is denoted by Z.

Exercise 8.1. If D is any division ring show that N_l, N_m, N_r, N and Z are all skewfields.

Exercise 8.1 is easy as it is almost a direct consequence of the definitions. However, it has many important corollaries as it allows us to consider D as a vector space over a sub-skewfield in many different ways.

Exercise 8.2. If D is any division ring show that D may be considered as a left vector space over N_l, N_m, N or Z and as a right vector space over N_m, N_r, N or Z.

Exercise 8.2 is, again, very easy but will prove to be very useful. It is easily seen that, in fact, D is an algebra over each of these skewfields.

Exercise 8.3. Show that a finite dimensional algebra D with identity over a skewfield F is a division ring if and only if it has no zero divisors.

Since a division ring D satisfies both distributive laws, every element of its right nucleus N_r associates and distributes on the right. Thus, if we regard D as a quasifield, N_r is the kernel of D and, by Lemma 7.9 the multiplicative group of N_r^* is isomorphic to the group of $((0, 0), [\infty])$-homologies of $\mathscr{P}(D)$. The other two semi-nuclei have similar geometrical interpretations.

Theorem 8.2. *Let D be a division ring with semi-nuclei N_r, N_m, N_l and let $\mathscr{P}(D)$ be the projective plane coordinatized by D. Then*

(a) *the multiplicative group of N_r^* is isomorphic to the group of $((0, 0), [\infty])$-homologies of $\mathscr{P}(D)$.*

(b) *the multiplicative group of N_l^* is isomorphic to the group of $((\infty), [0, 0])$-homologies of $\mathscr{P}(D)$.*

(c) *the multiplicative group of N_m^* is isomorphic to the group of $((0), [0])$-homologies of $\mathscr{P}(D)$.*

Proof. As we remarked earlier, since N_r is the kernel of D regarded as a quasifield, (a) is proved by Lemma 7.9.

If α is a $((\infty), [0, 0])$-homology then α induces a permutation on the points of $[\infty]$ and on the points of $[0]$. Let $(m)^\alpha = (m^\beta)$ and $(0, b)^\alpha = (0, b^\gamma)$, so that $(x, y)^\alpha = (x, y^\gamma)$ and $[m, k]^\alpha = [m^\beta, k^\gamma]$. Thus, since α preserves incidence, we have

$$mx + y = k \Leftrightarrow m^\beta x + y^\gamma = k^\gamma \tag{1}$$

or

$$m^\beta x + y^\gamma = (mx + y)^\gamma \quad \text{for all} \quad m, x, y \text{ in } D. \tag{2}$$

Since α fixes the point $(0, 0)$, $0^\gamma = 0$ and putting $y = 0$ and $x = 1$ in (2) this gives $m^\beta = m^\gamma$ for all m in D. Thus

$$\beta = \gamma. \tag{3}$$

Suppose $1^\beta = a$ then if we put $m = 1$ and $y = 0$ in (2) we get $ax = x^\gamma$ for all x in D. Thus, if L_a denotes left multiplication by a, we have

$$\beta = \gamma = L_a. \tag{4}$$

Substituting L_a for β and γ in (2) and putting $y = 0$, we obtain $(am) x = a(mx)$ for all m, x in D which implies that a is in N_l. But, clearly, if, for any n in N_l, we put $\beta = \gamma = L_n$ then the mapping α given by $(m)^\alpha = (m^\beta)$, $(x, y)^\alpha = (x, y^\gamma)$, $[m, k]^\alpha = [m^\beta, k^\gamma]$ and $[x]^\alpha = [x]$ is a $((\infty), [0, 0])$-homology of $\mathcal{P}(D)$. If α_1 is given by L_{a_1} and α_2 is given by L_{a_2} then direct computation immediately shows that $\alpha_1 \alpha_2$ is given by $L_{a_2 a_1}$. Thus the multiplicative group of N_l^* is anti-isomorphic to the group of $((\infty), [0, 0])$-homologies of $\mathcal{P}(D)$ and hence, since any group is anti-isomorphic to itself, (b) is proved.

The proof of (c) is very similar to that of (b) but is slightly more complicated. We suggest that the reader attempts to construct his own proof before reading the one given here.

Let θ be a $((0), [0])$-homology and let the permutations which it defines on the points of $[\infty]$ and $[0, 0]$ be given by $(m)^\theta = (m^\lambda)$ and $(x, 0)^\theta = (x^\mu, 0)$ respectively. Then $(x, y)^\theta = (x^\mu, y)$ and $[m, k]^\theta = [m^\lambda, k]$ and so, since θ preserves incidence, we have

$$m^\lambda x^\mu + y = mx + y \quad \text{for all} \quad m, x, y \text{ in } D \tag{5}$$

which simplifies to

$$m^\lambda x^\mu = mx \quad \text{for all} \quad m, x \text{ in } D. \tag{6}$$

If we define a and b by $1^\lambda = a$ and $1^\mu = b$ then (6) gives $ab = 1$. If we now put $m = 1$ and $x = a$ in (6) we get $a \cdot a^\mu = a$, i.e., $a^\mu = 1$. Similarly $b^\lambda = 1$ and now putting $m = b$, $x = a$ in (6) we obtain $ba = 1$.

Putting $m = b$ in (6) gives

$$x^\mu = bx \quad \text{for all} \quad x \text{ in } D, \tag{7}$$

i.e., $\mu = L_b$ while putting $x = a$ gives

$$\lambda = R_a, \tag{8}$$

where R_a is right multiplication by a.

Thus, for any m, x in D, we have

$$(ma)(bx) = mx . \tag{9}$$

If we first put $m = 1$ in (9), and use the fact that $ab = 1$ we get

$$a(bx) = x = (ab)x \quad \text{for all} \quad x \text{ in } D . \tag{10}$$

Similarly, putting $x = 1$ gives

$$(ma)b = m(ab) \quad \text{for all} \quad m \text{ in } D . \tag{11}$$

Combining (9), (10), and (11) we have

$$(ma)(bx) = mx = m((ab)x) = m(a(bx)) \quad \text{for all} \quad m, x \text{ in } D . \tag{12}$$

Finally, writing $bx = z$ in (12) and observing that, since D^* is a multiplicative loop, as x varies over D then so does z, we have

$$(ma)z = m(az) \quad \text{for all} \quad m, z \text{ in } D, \text{ i.e., } a \in N_m^* . \tag{13}$$

Similarly $b \in N_m^*$ and so, since a and b uniquely determine each other, θ determines a unique element of N_m^*.

Clearly for any n in N_m^* we can define a $((0), [0])$-homology θ_n by $(x, y)^{\theta_n} = (x^\mu, y), [m, k]^{\theta_n} = [m^\lambda, k], (m)^{\theta_n} = (m^\lambda)$ and $[x]^{\theta_n} = [x^\mu]$ where $\lambda = R_n$ and $\mu = L_{n-1}$. The mapping sending θ_n onto n is now the required isomorphism between the two groups. $\square$

Exercise 8.4. Prove (b) of Theorem 8.2 by using the fact that $\mathscr{P}(D)$ is the dual of a translation plane with respect to (∞) and dualizing (a).

Exercise 8.5. If D is a division ring use Theorem 8.2 to show that D is a nearfield if and only if D is a skewfield.

3. Collineation Groups of Division Ring Planes

As we saw in Chapter VI, a projective plane is a division ring plane if it is a translation plane with respect to a line and the dual of a translation plane with respect to a point. Thus any Moufang plane or desarguesian plane is a division ring plane. This, of course, we already knew. To avoid confusion we shall call a division ring *proper* if it is not an alternative (or IP) division ring and shall call a division ring plane *proper* if it is not a Moufang plane.

Lemma 8.3. *If D is a proper division ring then $\mathscr{P}(D)$ is a proper division ring plane and* $\operatorname{Aut}\mathscr{P}(D)$ *fixes (∞) and $[\infty]$.*

Proof. Since any Moufang plane may be coordinatized by an IP division ring it is sufficient to show that if $\operatorname{Aut}\mathscr{P}(D)$ moves either (∞) or $[\infty]$ then $\mathscr{P}(D)$ is a Moufang plane.

If $\text{Aut}\,\mathscr{P}(D)$ has a collineation, α say, such that $[\infty]^\alpha \neq [\infty]$ then, by Exercise 4.13, $[\infty]^\alpha$ is also a translation line for $\mathscr{P}(D)$ and so, by Theorem 6.17, $\mathscr{P}(D)$ is a Moufang plane. Dually if $\text{Aut}\,\mathscr{P}(D)$ has a collineation moving (∞), $\mathscr{P}(D)$ is the dual of a translation plane with respect to two distinct points and again by Chapter VI, is a Moufang plane. ∐

The collineation group of a proper division ring plane has many interesting properties which we shall now discuss in some detail.

If $\mathscr{P}(D)$ is a proper division ring plane then any elation with centre (∞) and affine axis is called a *shear*. Using the notation of Chapter IV if Π denotes $\text{Aut}\,\mathscr{P}(D)$ then the group of shears with axis $[0]$ is $\Pi_{((\infty),[0])}$.

Lemma 8.4. *Let* $\mathscr{P}(D)$ *be a division ring plane and let* $\Pi = \text{Aut}\,\mathscr{P}(D)$. *If* $\Sigma = \Pi_{([\infty],[\infty])} \cdot \Pi_{((\infty),[0])}$ *then*
(i) Σ *is a group*
(ii) $\Pi_{([\infty],[\infty])} \lhd \Sigma$ *and* $\Pi_{([\infty],[\infty])} \cap \Pi_{((\infty),[0])} = 1$ *so that* Σ *is metabelian.*

Proof. If H, K are subgroups of a given group G then the complex HK is a subgroup if either H or K is normal in G. Thus if we show that $\Pi_{([\infty],[\infty])} \lhd \Pi$ we will have proved (i) and the first part of (ii). By Corollary 2 of Lemma 4.11, for any α in Π we have $\alpha^{-1}\Pi_{([\infty],[\infty])}\alpha = \Pi_{([\infty]^\alpha,[\infty]^\alpha)}$ and so, by Lemma 8.3 $\Pi_{([\infty],[\infty])} \lhd \Pi$. From the definition of the two groups it is clear that $\Pi_{([\infty],[\infty])} \cap \Pi_{((\infty),[0])} = 1$, and the fact that Σ is metabelian is now apparent since, by Theorem 4.14, $\Pi_{([\infty],[\infty])}$ and $\Pi_{((\infty),[0])}$ are abelian. ∐

Exercise 8.6. Let $\mathscr{P}(D)$ be a division ring plane with $\Pi = \text{Aut}\,\mathscr{P}(D)$. If X, Y are any distinct points on $[\infty]\setminus(\infty)$, and l, m are any distinct lines on $(\infty)\setminus[\infty]$, show that there is a collineation σ in Σ with $X^\sigma = Y$ and $l^\sigma = m$. Show further that if A, B are any distinct affine points then there is a collineation μ in Σ with $X^\mu = Y$ and $A^\mu = B$.

Lemma 8.5. *With the notation of Lemma 8.4, if* D *is a proper division ring then* $\Sigma \lhd \Pi$.

Proof. By the dual of Theorem 7.4, Σ contains all the elations with centre (∞). Moreover, by Lemma 8.3, Π fixes (∞) and $[\infty]$ so that every elation of $\mathscr{P}(D)$ either has centre (∞) or axis $[\infty]$. Thus Σ contains all the elations of $\mathscr{P}(D)$ and in fact, since Σ is generated by the shears and translations, is generated by them. But, by Lemma 4.11, the conjugate of any elation is again an elation and so, since conjugation in Π is merely a permutation of a set of generators of Σ, Σ is normal in Π. ∐

As we shall now show, not only is Σ normal in Π but Π even splits over Σ. For any proper division ring plane $\mathscr{P}(D)$ the group Λ of collineations fixing any triangle ABC with $A = (\infty)$, B ($\neq A$) on $[\infty]$ and C not on $[\infty]$ is called the *autotopism group* of $\mathscr{P}(D)$ with respect to the triangle ABC. The triangle ABC is called the *autotopism triangle* of Λ. By Exercise 8.6, the full collineation group of $\mathscr{P}(D)$ is transitive on all triangles which are candidates to be autotopism triangles (i.e., triangles having one vertex as (∞), a second vertex on $[\infty]$ and the third, therefore, not on $[\infty]$). Thus all of the autotopism groups of $\mathscr{P}(D)$ are conjugate in $\operatorname{Aut}\mathscr{P}(D)$. For this reason we are justified in talking about *the* autotopism group of $\mathscr{P}(D)$ and, unless otherwise stated, we shall take (∞), (0), $(0,0)$ as the autotopism triangle.

Theorem 8.6. *Let* $\mathscr{P}(D)$ *be a proper division ring plane and let* $\Pi = \operatorname{Aut}\mathscr{P}(D)$. *If* $\Sigma = \Pi_{([\infty],[\infty])} \cdot \Pi_{((\infty),[0])}$ *and if* Λ *is the autotopism group of* $\mathscr{P}(D)$ *then*

(i) $\Pi = \Sigma \Lambda$,
(ii) $\Sigma \cap \Lambda = 1$.

Proof. Let π be any collineation in Π and suppose $(m)^\pi = (0)$ and $(a,b)^\pi = (0,0)$. Since $\mathscr{P}(D)$ is a translation plane there is a translation α with $(a,b)^\alpha = (0,0)$ while the $((\infty),[0])$-transitivity of $\mathscr{P}(D)$ implies the existence of a $((\infty),[0])$-elation β with $(m)^\beta = (0)$. The collineation $\alpha\beta$ is in Σ and $(\alpha\beta)^{-1}\pi$ is in Λ, so that $\Pi = \Sigma \cdot \Lambda$.

Suppose γ is in $\Sigma \cap \Lambda$. Then there exists a translation λ and a shear μ such that $\gamma = \lambda\mu$. But $(0)^\lambda = (0)$, (since (0) is on the axis of λ), so that $(0) = (0)^\gamma = (0)^\mu$. Thus, since the identity is the only shear fixing (0), $\mu = 1$. Similarly $\lambda = 1$ and the theorem is proved. $\square$

The structure of Σ is well known so that, in some sense the structure of Π is known once the structure of Λ is known. For instance, in any division ring plane $\mathscr{P}(D)$, the group Σ is always soluble so that the full collineation group is soluble if and only if the autotopism group is. Consequently the autotopism group plays a crucial role in the study of collineation groups of division ring planes.

4. Autotopisms

Let $\mathscr{P}$ be a division ring plane coordinatized by a division ring D and let α be an autotopism of $\mathscr{P}$. Since, by definition, α fixes the three points (∞), (0), $(0,0)$, α must also fix the lines $[\infty]$, $[0,0]$ and $[0]$ and must, therefore, induce a permutation on the points of each of these lines. Thus α defines three permutations P, Q, R of D given by

$$(m)^\alpha = (mP),$$
$$(a,0)^\alpha = (aQ,0),$$
$$(0,b)^\alpha = (0,bR) \quad \text{where} \quad 0P = 0Q = 0R = 0.$$

Since α is a collineation of $\mathscr{P}$, $(x,y)^\alpha = (xQ, yR)$ for all x, y in D and $[m,k]^\alpha = [mP, kR]$ for all m, k. Thus, using the fact that (x,y) is on $[m,k]$ if and only if (xQ, yR) is on $[mP, kR]$ and that the condition for (x,y) to be on $[m, k]$ is $mx + y = k$, we have

$$mP \cdot xQ + yR = (mx + y) R \quad \text{for all} \quad m, x, y \text{ in } D.$$ (1)

Putting $y = 0$ in (1) gives

$$mP \cdot xQ = (mx) R$$ (2)

which, on putting $m = 1$, gives

$$1P \cdot xQ = xR.$$ (3)

Substituting from Eq. (3) in Eq. (1) with $m = 1$ gives

$$xR + yR = (x + y) R \quad \text{for all} \quad x, y \text{ in } R.$$ (4)

Hence R is an additive mapping.

Now put $y = 0$ in (1) and write $m = a + b$. This gives

$$
\begin{aligned}
(a + b) P \cdot xQ &= ((a + b) x) R \quad \text{for all} \quad a, b, x \text{ in } D \\
&= (ax + bx) R \\
&= (ax) R + (bx) R \quad \text{since } R \text{ is additive} \\
&= aP \cdot xQ + bP \cdot xQ \quad \text{by (2)} \\
&= (aP + bP) \cdot (xQ).
\end{aligned}
$$

Thus $(a + b) P = aP + bP$ so that P is additive. An almost identical argument proves that Q is also additive.

The above discussion shows that any collineation α in the autotopism group of a division ring plane $\mathscr{P}(D)$ gives rise to a triple $\alpha(P, Q, R)$ of non-singular additive mappings of D onto itself satisfying $(xP)(yQ) = (xy) R$ for all x, y in D.

Conversely it is easily seen that given three non-singular additive mappings P, Q, R of D onto D satisfying $xP \cdot yQ = (xy) R$ then an auto-topism α of $\mathscr{P}(D)$ may be defined by $(m)^\alpha = (mP)$, $(a, b)^\alpha = (aQ, bR)$.

Exercise 8.7. Show that the mapping α defined above is an autotopism of $\mathscr{P}(D)$.

Any triple $\alpha(P, Q, R)$ of non-singular additive maps of a division ring D onto itself satisfying $(xP)(yQ) = (xy) R$ for all x, y in D is called an *autotopism* of D. Having used the same word in two apparently different definitions we now justify this usage by showing that the set of all auto-topisms of a given division ring D form a group which is isomorphic to the autotopism group of $\mathscr{P}(D)$.

Lemma 8.7. *The autotopisms of a division ring D form a group under the composition* $\alpha(P, Q, R) \cdot \alpha(P_1, Q_1, R_1) = \alpha(PP_1, QQ_1, RR_1)$.

Proof. Let $\varLambda$ denote the set of all autotopisms of D. We must first show that $\varLambda$ is closed under the given rule for composition. Since PP_1, QQ_1, RR_1, are all clearly non-singular additive mappings, we have to show that $(xPP_1)(yQQ_1) = (xy)RR_1$ for all x, y in D. But, since $\alpha(P_1, Q_1, R_1)$ and $\alpha(P, Q, R)$ are autotopisms, $(xPP_1)(yQQ_1) = [(xP)(yQ)]R_1 = (xy)RR_1$.

The given rule is obviously associative so that, in order to prove the lemma, we need only show that each element of Λ has an inverse in Λ. If $\alpha(P, Q, R)$ is in Λ then we must find $\alpha(P', Q', R')$ in Λ such that $\alpha(P, Q, R) \cdot \alpha(P', Q', R') = \alpha(1, 1, 1)$. By the rule of composition, if such an element exists it must be $\alpha(P^{-1}, Q^{-1}, R^{-1})$; thus we must show that $\alpha(P^{-1}, Q^{-1}, R^{-1})$ is in Λ. But $[(xP^{-1})(yQ^{-1})]R = (xP^{-1})P \cdot (yQ^{-1})Q$ (since $(ab)R = aP \cdot bQ$ for all a, b in D) and thus $(xP^{-1})(yQ^{-1}) = (xy)R^{-1}$ as required.   □

Lemma 8.8. *Let D be a division ring. Then the autotopism group of D is isomorphic to the autotopism group of $\mathscr{P}(D)$.*

Proof. This is clear if we interpret the additive maps P, Q, R as permutations of the points of the lines $[\infty], [0, 0], [0]$ respectively.   □

Now suppose that a, b are the unique elements of D given by $bP = 1 = aQ$. Then by first putting $x = b$ and then $y = a$ in the identity $(xP)(yQ) = (xy)R$ we have $xQ = (bx)R$ and $xP = (xa)R$ for all x in D. Furthermore, since $(ba)R = 1$, R and a completely determine b so that the autotopism $\alpha(P, Q, R)$ is uniquely determined by R and the element a. However, this is not a completely unexpected result, as is shown by the following exercise.

Exercise 8.8. Let $\mathscr{P}$ be any projective plane. If α and β are collineations of $\mathscr{P}$ such that $A^\alpha = A^\beta$ for all points A on a given line l and if $X^\alpha = X^\beta$, $Y^\alpha = Y^\beta$ for some pair of distinct points X, Y not on l, show that $\alpha = \beta$.

The above discussion is summarized by the following lemma.

Lemma 8.9. *If $\alpha(P, Q, R)$ is an autotopism of a division ring D then there exist unique elements a, b of D such that $P = R_a R$ and $Q = L_b R$.*

Clearly if any two of the mappings P, Q, R are the identity then so is the third. When one of the mappings is the identity then the collineations induced on $\mathscr{P}(D)$ are merely the homologies which were discussed in the proof of Theorem 8.2. We note that if $\alpha = \alpha(P, Q, R)$ and $\beta = \beta(A, B, C)$ are two autotopisms of a division ring plane then (P, Q, R) and (A, B, C) agree in one of the three mappings if and only if $\alpha\beta^{-1}$ is a homology. If the triples agree in two places then the collineations α and β are equal.

5. Isotopism

As we saw in Exercise 8.6 the full collineation group Π of a division ring plane $\mathscr{P}$ is transitive on triangles having (∞) as one vertex and any other special point as (0). Thus, by Exercise 5.2, if D_1 and D_2 are any pair of division rings which may be used to coordinatize $\mathscr{P}$, then each one may be used to coordinatize $\mathscr{P}$ with the same choice of points for (∞), (0) and $(0, 0)$. If S is the point which when chosen as $(1, 1)$ gives D_1 as coordinatizing division ring, and if T gives rise to D_2 in a similar manner then, again using Exercise 5.2, D_1 is isomorphic to D_2 if and only if there is a collineation θ of $\mathscr{P}$ such that $(\infty)^\theta = (\infty)$, $(0)^\theta = (0)$, $(0, 0)^\theta = (0, 0)$ and $S^\theta = T$. But θ is an autotopism of $\mathscr{P}$ and thus D_1 is isomorphic to D_2 if and only if S and T are in the same orbit of the autotopism group of $\mathscr{P}$. Thus we have established:

Theorem 8.10. *The non-isomorphic division rings coordinatizing a given division ring plane $\mathscr{P}$ are in one-to-one correspondence with the orbits of the autotopism group of the points not incident with any side of the autotopism triangle.*

The only known planes for which this number is one are the desarguesian planes, and it has been conjectured that these are the only ones.

Theorem 8.10 poses an obvious problem: If two non-isomorphic division rings coordinatize the same plane what can be said about them? Anticipating the answer we define two division rings D_1, D_2 to be *isotopic* if there is a triple (P, Q, R) of non-singular additive mappings from D_1 onto D_2 such that $(xP) \odot (yQ) = (x \cdot y) R$ for all x, y in D_1 (where $\odot$ is the multiplication of D_1 and $\cdot$ is the multiplication of D_2). The triple (P, Q, R) is called an *isotopism* from D_1 onto D_2. The motivation for the definition and the importance of the concept are both apparent from the next theorem.

Theorem 8.11. *Two division rings coordinatize isomorphic planes if and only if they are isotopic.*

Proof. (i) Let D_1, D_2 be two isotopic division rings with an isotopism (P, Q, R) from D_1 onto D_2. We use the given isotopism to define a mapping σ from $\mathscr{P}(D_1)$ to $\mathscr{P}(D_2)$ as follows:

$$(\infty)^\sigma = (\infty)' \qquad [\infty]^\sigma = [\infty]'$$
$$(m)^\sigma = (mP)' \qquad [m, k]^\sigma = [mP, kR]'$$
$$(a, b)^\sigma = (aQ, bR)' \qquad [k]^\sigma = [kQ].$$

In order to show that σ is an isomorphism from $\mathscr{P}(D_1)$ to $\mathscr{P}(D_2)$ we must show that $mx + y = k$ implies $(mP)(xQ) + yR = kR$. But $kR = (mx + y) R = (mx) R + yR = (mP)(xQ) + yR$ so that the planes are isomorphic.

(ii) Suppose $\mathscr{P}_1, \mathscr{P}_2$ are two isomorphic planes coordinatized by division rings D_1, D_2 respectively. Clearly if one of $\mathscr{P}_1$ or $\mathscr{P}_2$ is a Moufang plane then the other is as well. But, as we saw in Exercise 4.26, the collineation group of a Moufang plane is transitive on triangles. This implies that any two PTR's coordinatizing a given Moufang plane are isotopic (since we may assume they have the same basic triangle (∞), (0), $(0, 0)$).

Thus we may assume that $\mathscr{P}_1$ and $\mathscr{P}_2$ are not Moufang planes which, by Lemma 8.3, implies that D_1 and D_2 are proper division rings and that $\operatorname{Aut}\mathscr{P}_1$ fixes (∞), $[\infty]$ while $\operatorname{Aut}\mathscr{P}_2$ fixes $(\infty)'$, $[\infty]'$. So if ϕ is any isomorphism from $\mathscr{P}_1$ onto $\mathscr{P}_2$ we must have $(\infty)^\phi = (\infty)'$ and $[\infty]^\phi = [\infty]'$. Clearly if θ is any collineation of $\mathscr{P}_2$ then $\phi\theta$ is still an isomorphism from $\mathscr{P}_1$ onto $\mathscr{P}_2$ and so, by Exercise 8.6, by a suitable choice of θ we can get a new isomorphism α from $\mathscr{P}_1$ onto $\mathscr{P}_2$ with $(\infty)^\alpha = (\infty)'$, $(0)^\alpha = (0)'$ and $(0,0)^\alpha = (0,0)'$. Then defining P, Q, R, by $(m)^\alpha = (mP)'$, $(a,0)^\alpha = (aQ,0)'$, $(0,b)^\alpha = (0,bR)'$ and repeating the argument at the beginning of Section 4 we see that (P, Q, R) is an isotopism from D_1 onto D_2. $\square$

The definition of isotopism appears slightly cumbersome as it involves three mappings but, as in the case of autotopisms, these three mappings are not completely independent. If D_1, D_2 are isotopic division rings with respective multiplications $\cdot$, $\odot$ then D_2 is called a *principal isotope* of D_1 if there exist a, b in D_1 such that $x \odot y = x R_a^{-1} \cdot y L_b^{-1}$. The importance of this concept, which involves two very special mappings, is shown by the following exercise.

Exercise 8.9. If D_1, D_2 are any two isotopic division rings show that there is a principal isotope of D_1 which is isomorphic to D_2.

6. On the Solubility of the Autotopism Group

As we remarked in Section 3 the full collineation group of a proper division ring plane is soluble if and only if the autotopism group is. There are no known examples of finite division rings with non-soluble autotopism groups and it seems reasonable to conjecture that the autotopism group is always soluble. In this section we shall prove the conjecture true if the division ring has non-square order. The proof is slightly unsatisfactory in that it uses the powerful and celebrated Feit-Thompson Theorem. Our main reason for including it here is to illustrate the techniques involved. We begin by proving two lemmas which show that the three mappings used to define an autotopism are semi-linear transformations of the division ring (regarded as a vector space over certain of the semi-nuclei [see Exercise 8.2]).

Lemma 8.12. *Let* $\alpha = \alpha(P, Q, R)$ *be an autotopism of a division ring* D *and let* a, b *be the elements of* D *such that* $P = R_a R$ *and* $Q = L_b R$. *If* α_i, *for* $i = r, m, l$, *are the mappings given by*

  (i) $n^{\alpha_r} = (ban) R$ *for* $n \in N_r$,
  (ii) $n^{\alpha_m} = (bna) R$ *for* $n \in N_m$,
  (iii) $n^{\alpha_l} = (nba) R$ *for* $n \in N_l$,

then α_i *is an automorphism of* N_i *for each* i. *We shall call the* α_i *the companion automorphisms of* α.

Proof. We shall prove the lemma for the case $i = r$ and leave the other two cases as an exercise.

If n_1, n_2 are any two elements of N_r then, by the definition of α_r, $(n_1 + n_2)^{\alpha_r} = [ba(n_1 + n_2)] R$. The left distributive law of D together with the additive property of R now gives $(n_1 + n_2)^{\alpha_r} = (ban_1 + ban_2) R = (ban_1) R + (ban_2) R = n_1^{\alpha_r} + n_2^{\alpha_r}$.

In order to establish that α_r is multiplicative on N_r we shall first show that, for any x in D and any n in N_r,

$$(xn) R = x R (ban) R = x R \cdot n^{\alpha_r}. \tag{1}$$

Clearly $(xn) R = [((x R_a^{-1}) a) n] R$ and thus, using the fact that n associates on the right and the equation $(xy) R = (xP) (yQ)$, we have $(xn) R [((x R_a^{-1}) a) n] R = ((x R_a^{-1}) an) R = (x R_a^{-1}) P (an) Q = ((x R_a^{-1}) a) R (ban)R = x R (ban) R$.

For any n_1, n_2 in N_r, $(n_1 n_2)^{\alpha_r} = (ba(n_1 n_2)) R = ((ban_1) n_2) R$ and thus, putting $x = ban_1$ in (1), we have $(n_1 n_2)^{\alpha_r} = (ban_1) R (ban_2) R = n_1^{\alpha_r} n_2^{\alpha_r}$ as required.

If n is any element of N_r then, by definition, $(xy) n = x(yn)$ for all x, y in D. Using Eq. (1) plus the equation $(xy) R = (xa) R (by) R$, we have $((xy) n) R = (xy) R (ban) R = ((xa) R (by) R) (ban) R$. Similarly $(x(yn)) R = (xa) R (b(yn)) R = (xa) R ((by) n) R = (xa) R ((by) R (ban) R)$. Thus, for all x, y in D, we have $((xa) R (by) R) (ban) R = (xa) R ((by) R (ban) R)$. But D is a multiplicative loop which means that as x varies over D so does xa. Thus, since R is a permutation of the elements of D, as x ranges over all values of D, $(xa) R$ also ranges over all values. Similarly for y and $(by) R$, and so we have established that $(xy) (ban) R = x(y(ban) R)$ for all x, y in D or, in other words, that $(ban) R$ is also in N_r.

So far we have shown that α_r is a multiplicative and additive mapping of N_r into itself. In order to prove the lemma we have only to show that α_r is one-to-one and onto. But, since R is already known to be a one-to-one and onto mapping, and since left multiplication by a or b is also one-to-one and onto, the lemma is proved. □

Lemma 8.13. *Let $\alpha = \alpha(P, Q, R)$ be an autotopism of a division ring D. Then R is a semi-linear transformation of the vector space D over N_r and N_l; P is a semi-linear transformation of the vector space D over N_l and N_m; Q is a semi-linear transformation of the vector space D over N_m and N_r.*

Proof. By Eq. (1) of Lemma 8.12, $(xn) R = x R (ban) R = x R n^{\alpha_r}$. Thus since by Lemma 8.12 α_r is an automorphism of N_r, R is a semi-linear transformation of the (right) vector space D over N_r.

There are five other parts of the lemma. We shall prove the two claims for P and leave the remainder as an exercise.

For any x in D and n in N_l, $(nx) P = ((nx) a) R = (nxa) R$. Using a technique similar to that in Lemma 8.12 we have $(nxa) R = (n(b(xa) L_b^{-1})) R$

$= \left(nb(xa)\,L_b^{-1}\right) R = (nba)\,R\,\left(b(xa)\,L_b^{-1}\right) R = n^{\alpha_l}(xa)\,R = n^{\alpha_l}xP$. Thus P is a semi-linear transformation of the (left) vector space D over N_l. Similarly if x is in D and n is in N_m then $(xn)\,P = (xna)\,R = (xa)\,R\,(bna)\,R = xPn^{\alpha_m}$, and hence P is also a semi-linear transformation of the (right) vector space D over N_m. $\Box$

It is worth noting that the triple $\alpha(P, P, P)$ is an autotopism of D if and only if P is an automorphism of D so that an automorphism of D is a semi-linear transformation of D as a vector space over any one of its semi-nuclei.

Since semi-linear transformations of vector spaces are familiar objects, Lemma 8.13 is a very useful result. We now use it in our discussion of the solubility of the autotopism group of a division ring.

Lemma 8.14. *Let D be finite. If Δ_1 is the normal subgroup of Δ consisting of all those autotopisms for which the three mappings α_i of Lemma 8.12 are all the identity, then Δ/Δ_1 is soluble.*

Proof. For any fixed i $(i = r, l, m)$, the mapping which sends each element of Δ onto its companion automorphism of N_i is easily seen to be a homomorphism of Δ into the automorphism group of N_i. Since the automorphism group of any finite field is cyclic this homomorphism has a cyclic image. The group Δ_1 is the intersection of the kernels of the three homomorphisms obtained in this way so that Δ/Δ_1, which is the product of three normal cyclic subgroups, is abelian. $\Box$

Corollary. *Δ is soluble if and only if Δ_1 is.*

Since each of the α_i is the identity for all elements of Δ_1, if $\alpha = \alpha(P, Q, R)$ is in Δ_1 then P, Q, R are each linear transformations of D regarded as a vector space as in Lemma 8.13. Thus for α in Δ_1 we may form the following six determinants: $\det P$ where P is a linear transformation of D over either N_l or N_m; $\det Q$ where Q is a linear transformation of D over either N_m or N_r; $\det R$ where R is a linear transformation of D over either N_r or N_l.

Lemma 8.15. *Δ_1 has a normal subgroup Δ_2 consisting of those autotopisms for which all six of the determinants above are equal to one, and Δ_1/Δ_2 is soluble.*

Proof. We leave the proof as an exercise. (Hint: Note that for any of the six determinants, e.g. $\det P$ where P is a linear transformation of D over N_l, the mapping which sends the autotopism (A, B, C) onto $\det A$ is a homomorphism whose image is a subgroup of the multiplicative group of the finite field N_l and, hence, is cyclic.) $\Box$

Corollary. *Δ is soluble if and only if Δ_2 is.*

If Δ_2 has odd order then, by the Feit-Thompson Theorem, Δ_2 is soluble, so we now consider the possibility of elements of order 2 occurring in Δ_2.

This means that the projective plane $\mathscr{P}(D)$ has an involution fixing the autotopism triangle. Since an elation cannot fix a triangle this involution must, by Theorem 4.3 and Exercise 4.4, be either a homology (in which case the plane has odd order), or fix a Baer subplane pointwise. But in Theorem 8.2 we showed that any homology fixing (∞), (0), $(0, 0)$ is induced by appropriate multiplications from the various semi-nuclei. Thus we have proved

Lemma 8.16. *An element of order 2 in Δ_2 either induces a Baer involution of $\mathscr{P}(D)$ or is one of the mappings (A, A, I), (A, I, A), (I, A, A) where $A = R_{-1} = L_{-1}$.*

This lemma gives very precise information about the involutory homologies in the autotopism group of a finite division ring plane. Such precise information is, unfortunately, not available for Baer involutions. However, the following lemma shows that certain division ring planes cannot admit Baer involutions.

Lemma 8.17. *If Δ_1 contains a Baer involution then the dimension of D over each semi-nuclei is even.*

Proof. Let $\beta = \beta(P, Q, R)$ be a Baer involution in Δ_1. The points $(x, 0)$ which are fixed by β are, on the one hand, equal in number to the square-root of the order of D, and on the other hand are exactly those points for which x is in the subspace V of D belonging to the eigenvalue 1 of the linear transformation Q (over N_m or N_r). Thus $\dim D = 2 \dim V$. A similar argument for fixed points $(0, y)$ proves the claim for N_l. $\square$

Theorem 8.18. *Let D be a finite division ring. If $\mathscr{P}(D)$ has even order and no Baer subplanes, or has odd order and the dimension of D over at least one of its semi-nuclei is odd, then the autotopism group of D is soluble.*

Proof. By the corollary to Lemma 8.15 we need only show that Δ_2 is soluble. In each case we shall show that Δ_2 has odd order and then apply the Feit-Thompson Theorem.

If D has even order then any element of order 2 in Δ_2 must be a Baer involution and this is impossible since $\mathscr{P}(D)$ is assumed to have no Baer subplanes. Thus, in this case, Δ_2 has odd order.

Suppose D has odd order but that its dimension over at least one of its semi-nuclei is odd. Then, by Lemma 8.17, Δ_2 cannot contain a Baer involution. This means that the only possible elements of order 2 are the three listed in Lemma 8.16. But since these elements are all in Δ_2 we must have $\det A = 1$, where the determinant is computed over any one of the semi-nuclei, and A is either R_{-1} or L_{-1}. But this implies $(-1)^{k_i} = 1$, where k_i is the dimension of D over N_i (that is, the number of -1's on the main diagonal of A) and, since at least one k_i is odd, this is a contradiction. Thus Δ_2 has odd order and the theorem is proved. $\square$

*Exercise 8.10. Show that a division ring D is isotopic to a commutative division ring if and only if there is a non-zero element a in D such that $(ax)y = (ay)x$ for all x, y in D.

Exercise 8.11. Generalize the results on isotopy in this chapter to prove: if $\mathscr{P}_1$ and $\mathscr{P}_2$ are projective planes coordinatized by PTR's (R_1, T_1) and (R_2, T_2) respectively, then there is an isomorphism from $\mathscr{P}_1$ to $\mathscr{P}_2$ which sends the points $(\infty)_1$, $(0)_1$ and $(0,0)_1$ onto $(\infty)_2$, $(0)_2$ and $(0,0)_2$ respectively if and only if there is a triple (X, Y, Z) of one-to-one mappings of R_1^* onto R_2^* such that

$$(T_1(a, b, c))Z = T_2(aX, bY, cZ)$$

for all a, b, c in R_1^*.

IX. Examples of Non-desarguesian Projective Planes

1. Introduction

In this chapter we exhibit some non-desarguesian projective planes. Most of these examples are translation planes and we shall show that they exist by exhibiting quasifields which are not skewfields. Each of the quasifields given has finite dimension over its kernel which means (see Theorem 7.3), that it is only necessary to prove that they are weak quasifields. Although we are mainly interested in the finite case, many of our examples are valid in the infinite case too; frequently we give a much less detailed treatment for the infinite case, and the reader may find it interesting to fill in the gaps.

To end the chapter we exhibit a class of planes (the "Hughes planes") which cannot be coordinatized by linear planar ternary rings. The existence of these planes is established by actually constructing the plane itself rather than a coordinatizing planar ternary ring.

In this chapter considerable use is made of the material in the earlier Chapters (III to VIII), but we have avoided using any material which has not been introduced. Consequently, particularly in the case of the Hughes planes, our representation of the planes and proofs of their existence are not always those found in the modern research literature.

2. The Hall Quasifields

Let F be a field and $f(s) = s^2 - as - b$ an irreducible quadratic over F. Let H be a two-dimensional right vector space over F, with basis elements 1 and λ so that H consists of all elements of the form $x + \lambda y$ as x and y vary over F. We wish to define a multiplication on H in such a way that, using the vector space addition, H will be a quasifield. The remarkable way in which this is done is to assume

(i) every element α of H not in F (i.e., not of the form $x + \lambda 0$) satisfies the quadratic equation $f(\alpha) = 0$;

(ii) F is in the kernel of H;

(iii) every element of F commutes with all the elements of H.

Note that we know the addition in H is abelian so that any statement like (iii) is obviously intended to refer to multiplication.

In order to construct the quasifield H we shall assume that we have a quasifield Q which is a two dimensional vector space over F such that its multiplication satisfies (i), (ii), (iii) and determine its rule for multiplication. We then use this rule to define multiplication in H and show that this gives a quasifield with the required properties.

Let $1, \lambda$ be a basis for Q as a vector space over F. If $y \neq 0$ then, since F is contained in the kernel of Q, the elements of F distribute on the right and associate on the right so that $(z + \lambda t) = (x + \lambda y) y^{-1} t + z - xy^{-1} t$, and so $(x + \lambda y)(z + \lambda t) = (x + \lambda y) [(x + \lambda y) y^{-1} t + z - xy^{-1} t] = (x + \lambda y)^2 y^{-1} t + (x + \lambda y)(z - xy^{-1} t)$. But, by assumption, $f(x + \lambda y) = 0$, so that $(x + \lambda y)^2 = a(x + \lambda y) + b$. Thus $(x + \lambda y)(z + \lambda t) = [b + (x + \lambda y) a] y^{-1} t + (x + \lambda y) \cdot (z - xy^{-1} t)$ or

$$(x + \lambda y)(z + \lambda t) = xz - y^{-1} t f(x) + \lambda(yz - xt + at). \qquad (1)$$

We now use (1) to introduce a multiplication in H as follows:

If $\quad y \neq 0, \quad (x + \lambda y)(z + \lambda t) = xz - y^{-1} t f(x) + \lambda(yz - xt + at). \qquad (2)$

If $\quad y = 0, \quad x(z + \lambda t) = xz + \lambda(xt). \qquad (3)$

Clearly (2), (3) completely define a multiplication for H, so we must now show that the axioms for a quasifield are satisfied. From (2) and (3) it is evident that left multiplication is a linear transformation (that is, the right sides of (2) and (3) are linear in z and t), so the left distributive law is satisfied in H. Addition in H is commutative and, writing 0 for $0 + \lambda 0$, it is clear that $0(x + \lambda y) = 0$. Thus, in order to show that H is a weak quasifield, we have only to show that H^* is a multiplicative loop. Clearly $1 = 1 + \lambda 0$ is the identity.

To solve $(x + \lambda y)(z + \lambda t) = p + \lambda q, (x + \lambda y) \neq 0$, for $z + \lambda t$ we must solve the linear equations

$$\begin{aligned} xz - y^{-1} t f(x) &= p, \\ yz - xt + at &= q \end{aligned} \qquad (4)$$

if $y \neq 0$, and

$$xz = p, \quad xt = q, \quad \text{if} \quad y = 0. \qquad (5)$$

If $y = 0$ then, since $x + \lambda y \neq 0$, $x \neq 0$ so that $z = x^{-1} p, t = x^{-1} q$ are unique solutions. If $y \neq 0$ then (4) has a unique solution for z and t if, and only if, the determinant $x(a - x) + y \cdot y^{-1} f(x)$ is non-zero; this determinant simplifies to $-b$ which cannot be zero since, if it were, $f(s) = s^2 - as$ would be reducible.

It only remains to solve $(x + \lambda y)(z + \lambda t) = p + \lambda q$ for x and y given that $z + \lambda t \neq 0$. If $t = 0$ then, since $z + \lambda t \neq 0$, $z \neq 0$ and there is a solution with $y = 0$ if and only if $q = 0$, in which case the solution is $y = 0 (= z^{-1} q)$, $x = z^{-1} p$. Similarly if $z = 0$ there is a solution with $y = 0$ if and only if $p = 0$ when the solution is $y = 0 = (t^{-1} p)$, $x = t^{-1} q$. If $t \neq 0 \neq z$ then there is a solution $y = 0$, $x = pz^{-1}$ if and only if $pz^{-1} = qt^{-1}$. Thus we must show that if either $z=0$ and $p=0$, or $t=0$ and $q=0$, or $pz^{-1}=qt^{-1}$ then there is no solution with $y \neq 0$ but that in any other case there is a unique solution with $y \neq 0$.

Multiplying the first equation of (4) by y, the second by x and subtracting gives

$$bt = py - qx. \qquad (6)$$

Now combining (6) with the second equation of (4) we have the following two linear equations which have the same solutions for x and y as (4):

$$\begin{aligned} py - qx &= bt, \\ zy - tx &= q - at. \end{aligned} \qquad (7)$$

If $t=0$ then (7) has a unique solution $y = qz^{-1}$, $x = pz^{-1}$ which has $y=0$ if and only if $q=0$. If $t \neq 0$, then multiplying the second equation of (7) by qt^{-1} and subtracting from the first gives $y(p - qt^{-1}z) = bt - q^2 t^{-1} + aq$ or

$$y(pt - qz) = -t^2 f(qt^{-1}). \qquad (8)$$

Since $t \neq 0$ and $f(s)$ is irreducible over F the right hand side of (8) is always non-zero. Thus if $pt - qz = 0$ (i.e., if either $p=0$ and $z=0$ or $z \neq 0$ and $pz^{-1} = qt^{-1}$), then the Eqs. (7) are inconsistent and, hence, (4) has no solution with $y \neq 0$. Finally if $pt - qz \neq 0$ then $y = -t^2 f(qt)^{-1})/(pt - qz) \neq 0$, as required.

So far we have shown that H is a weak quasifield. However it is clear from the definition of multiplication that F is contained in the kernel of H and thus, since H is a two-dimensional vector space over F, Theorem 7.3 proves the claim that H is a quasifield. $\Box$

It is evident that the *Hall quasifields*, which we shall call the systems just defined, cannot in general be fields. For if H were a field, then all its elements outside the subfield F would satisfy the same quadratic equation and so, since a quadratic equation has at most two zeros in any field, H would possess at most two elements outside of F, and F would be $GF(2)$. Conversely if $F = GF(2)$ then H has order 4, but, by Exercise 5.9, all projective planes of order 4 are desarguesian. Thus all PTR's of order 4 are fields and, in particular, H is a field. This discussion shows that H is a field if and only if F is $GF(2)$. In fact a slight stronger result is true.

Exercise 9.1. Show that H satisfies the right distributive law if and only if F is the field $GF(2)$.

Lemma 9.1. *A Hall quasifield H is associative (i.e., is a nearfield) if and only if F is $GF(2)$ or F is $GF(3)$ and $f(s) = s^2 + 1$.*

Proof. Suppose H is associative. If x is any non-zero element of F then we compute $\lambda(x\lambda)$ and $(\lambda x)\lambda$ and find conditions for the two expressions to be equal. We have $\lambda(x\lambda) = \lambda(\lambda x) = \lambda^2 x = (b + \lambda a) x = bx + \lambda ax$, and $(\lambda x) \lambda = [(\lambda x)(\lambda x)] x^{-1} = (b + \lambda xa) x^{-1} = bx^{-1} + \lambda a$. Since H is associative

$\lambda(x\lambda) = (\lambda x)\,\lambda$ for all $x \in F$ and thus

$$bx = bx^{-1} \quad \text{for all} \quad x \in F, \quad x \neq 0, \tag{9}$$

$$ax = a \quad \text{for all} \quad x \in F, \quad x \neq 0. \tag{10}$$

If $x \neq 1$ then, from (10), $a = 0$ and, since b cannot be zero (as this would make $f(s)$ reducible), $x^2 = 1$ must hold for all $x \in F, x \neq 0$. In $GF(2)$ there is no element $x \neq 0, 1$ so no conclusion can be drawn from (9) and (10) excepting that $F = GF(2)$ is a possibility (and we have seen above that it actually works). But if there is an element $x \neq 0, 1$ in F then $a = 0$ and all elements x in $F, x \neq 0$, must satisfy $x^2 = 1$. The only field with these properties is $GF(3)$, and the only irreducible quadratic over $GF(3)$ with $a = 0$ is $s^2 + 1$.

The converse problem is already partially solved: We have seen that if F is $GF(2)$ then H is a field. If F is $GF(3)$ and $f(s) = s^2 + 1$ then (9) can be simplified to:

$$(x + \lambda y)(z + \lambda t) = xz - yt(x^2 + 1) + \lambda(yz - xt) \tag{11}$$

since $x^{-1} = x$ for all $x \neq 0$. The demonstration that multiplication in H is associative is left as an exercise. ☐

Exercise 9.2. Show that the Hall quasifield defined over $F = GF(3)$ by $f(s) = s^2 + 1$ is associative.

The following very important exercise gives an interesting characterization of the finite Hall quasifields.

*Exercise 9.3. (a) Let Q be a Hall quasifield over the field K; show that if r, s are in $K, s \neq 0$, then

$$x + \lambda y \rightarrow (x + ry) + \lambda s y$$

is an automorphism of Q, fixing every element of K. Hence show that $\mathrm{Aut}_K Q$, the group of automorphisms of Q fixing every element of K, is transitive (and even *regular*: see Chapter I) on the elements of Q not in K.

(b) Conversely, if Q is a quasifield of dimension two over its kernel K, if every element of K commutes with every element of Q, then show that Q is a Hall quasifield if $\mathrm{Aut}_K Q$ is transitive on the elements of Q not in K.

When we defined a Hall quasifield H over a field F we made H a two-dimensional vector space over F in such a way that F was contained in the kernel K. However H is also a vector space over K and so, clearly, either $H = K$ or $K = F$. Since $H = K$ if and only if H is a field (i.e., if H has order 4) a Hall quasifield of order greater than 4 is of dimension two over its kernel. When we defined a Hall quasifield we also insisted that each element of F commuted with all the elements of H. That this condition is superfluous is shown by the following theorem:

Theorem 9.2. *Let Q be a quasifield of dimension two over its kernel K. If the group of automorphisms of Q which fix every element of K is transitive* *on the elements not in K, then each element of K commutes with all the elements of Q.*

Proof. Let λ be an arbitrary element of Q, not in K. Then $1, \lambda$ is a basis for Q over K, so every element has the form $a + \lambda b$, for some $a, b \in K$. Let $k \in K$; then $k\lambda = a_1 + \lambda b_1$ for some $a_1, b_1 \in K$. But if $c, d \in K$, $d \neq 0$, then there is an element $\alpha \in \mathrm{Aut}_K Q$ such that $\lambda^\alpha = c + \lambda d$, and so, since $(k\lambda)^\alpha = k^\alpha \lambda^\alpha = k\lambda^\alpha$, we have:

$$(a_1 + \lambda b_1)^\alpha = k(c + \lambda d),$$
$$a_1 + (c + \lambda d) b_1 = kc + k \cdot \lambda d,$$
$$a_1 + cb_1 + \lambda \cdot db_1 = kc + k\lambda \cdot d$$
$$= kc + a_1 d + \lambda \cdot b_1 d.$$

Hence $a_1 + cb_1 = a_1 d + kc$ and $db_1 = b_1 d$, for all $c, d \in K$, $d \neq 0$. If $c = 0$, then $a_1 = a_1 d$, and so $a_1 = 0$ or $d = 1$; but if $d = 1$ is the only element in K^*, then $K = GF(2)$ and Q must have four elements, and so is $GF(4)$, which is a contradiction (for then Q is its own kernel). Hence we may conclude that $a_1 = 0$. Then we have $cb_1 = kc$, for all $c \in K$, and $db_1 = b_1 d$, for all $d \in K^*$. If $c = 1$, then putting $b_1 = k$, shows that k commutes with all of K, hence K is a field. Finally, since $a_1 = 0$ and $b_1 = k$, $k\lambda = a_1 + \lambda b_1 = \lambda k$. $\square$

Corollary. *Let Q be a two-dimensional quasifield over its kernel K. Then Q is a Hall quasifield if and only if $\mathrm{Aut}_K Q$ is transitive on the elements of Q not in K.*

The projective planes coordinatized by Hall quasifields will be called *Hall planes* and, as we have just seen, are non-desarguesian provided the order is greater than four. The earliest example of such a plane was given by Veblen and Wedderburn, this being the plane coordinatized by the nearfield of order 9. We shall discuss Hall planes again in Chapter X and, amongst other things, determine some properties of their collineation groups.

3. The André Quasifields

In this section we shall construct a very wide class of quasifields by altering the multiplication of a field in a simple way. First we describe the process in a somewhat abstract form, and then specialize to finite fields where everything is much easier. Let F be a field, Γ a finite group of automorphisms of F and K the subfield of F consisting of all elements fixed by Γ. Then (see for example Result 1.6), the dimension of F over K is equal to the order of Γ, and so is finite. Let v be the "norm" mapping defined by the group Γ; that is

$$x^v = \prod_{\alpha \in \Gamma} x^\alpha.$$

Clearly $x^\nu \in K$ for all $x \in K$ and, in fact, ν is a homomorphism of F^* into K^*. Let $N = (F^*)^\nu$; then the following simple example shows that, in general, $N \neq K^*$ so that ν does not map F^* onto K^*. Take F to be the field of complex numbers and $\Gamma = \langle \gamma \rangle$ where $(a + ib)^\gamma = a - ib$. Then, clearly, K is the field of reals and $(a + ib)^\nu = (a + ib)(a - ib) = a^2 + b^2$. Thus every element of N is positive so that $N \neq K^*$.

Now let ϕ be any mapping from N into the group Γ with the single requirement that ϕ maps the identity of F^* (for the identity must lie in N) onto the identity of Γ. Combining ν and ϕ we have a mapping $\alpha = \nu\phi$ from F^* into Γ. For any $x \in F^*$ we shall write α_x for x^α to emphasize the fact that x^α is an automorphism in Γ.

We now define a system F_ϕ as follows: the elements of F_ϕ are the elements of F, addition in F_ϕ is the same as addition in F while the multiplication $\odot$ is defined by

$$x \odot y = xy^{\alpha_x}, \quad 0 \odot y = 0 \quad \text{for all} \quad x, y \in F, \quad x \neq 0. \tag{1}$$

Since $x \odot (y + z) = x(y + z)^{\alpha_x} = x(y^{\alpha_x} + z^{\alpha_x}) = xy^{\alpha_x} + xz^{\alpha_x} = x \odot y + x \odot z$, F_ϕ satisfies the left distributive law.

Consider the equation $x \odot y = z$; that is $xy^{\alpha_x} = z$. If x and z are given then, clearly, $y = (x^{-1}z)^{\alpha_x^{-1}}$ is the unique solution. On the other hand, if y and z are given then we determine α_x by taking norms and using the fact that $(x^\beta)^\nu = x^\nu$ for all $\beta \in \Gamma$ and all $x \in F$: if $x \odot y = z$ then $(x \odot y)^\nu = z^\nu$. But $(x \odot y)^\nu = (xy^{\alpha_x})^\nu = x^\nu(y^{\alpha_x})^\nu = x^\nu y^\nu$, so that $x^\nu = (zy^{-1})^\nu$ and $\alpha_x = \alpha_w$ where $w = zy^{-1}$. Hence any solution x must satisfy $x = zy^{-\alpha_w}$ where $w = zy^{-1}$. Conversely, it is easy to see that such an x is a solution. Finally $x \odot 1 = x1^{\alpha_x} = x$ and $1 \odot x = 1x^{\alpha_1} = x$, since $\alpha_1 = 1^\alpha = \varepsilon$ (where ε is the identity of Γ) so $(F_\phi^*, \odot)$ is a loop. If k is any element of K then $x \odot k = xk$ for all $x \in F$ and so it is easy to see that K is contained in the kernel of F_ϕ; hence F_ϕ is a weak quasifield which is finite dimensional over its kernel and thus, by Theorem 7.3, is a quasifield, called an *André quasifield*.

Suppose now that F is finite. Then the group Γ is cyclic, and if its order is n then $K = GF(q)$, for some prime power q, while $F = GF(q^n)$. In this case the mapping ν is given by $x^\nu = x^{1 + q + q^2 + \cdots + q^{n-1}}$. The mapping ϕ can be chosen in n^{q-2} ways, since any of the $q - 2$ elements of K not equal to 0 or 1 can have any of the n elements of Γ as its image. This illustrates the size of the class of André quasifields which may be constructed from a given field F. As is to be expected, the structure of F_ϕ depends on the choice of ϕ.

Lemma 9.3. *The André quasifield F_ϕ is associative if and only if ϕ is an anti-homomorphism from the (multiplicative) group N into Γ.*

Proof. The proof is by straight forward computation:

$$x \odot (y \odot z) = x \odot (yz^{\alpha_y}) = x(yz^{\alpha_y})^{\alpha_x} = xy^{\alpha_x}z^{\alpha_y\alpha_x}$$
$$(x \odot y) \odot z = (xy^{\alpha_x}) \odot z = xy^{\alpha_x}z^{\alpha_{xy}}$$

since $(xy^{\alpha x})^\alpha = (x \odot y)^\alpha = (xy)^\alpha$. Hence associativity in F_ϕ requires that $\alpha_{xy} = \alpha_y \alpha_x$, so that α is an anti-homomorphism; but v is a homomorphism and so ϕ must be an anti-homomorphism. The converse is obvious. $\square$

Corollary. *A finite André quasifield is associative if and only if ϕ is a homomorphism from $N = K^*$ into the group Γ.*

Proof. Since F is finite Γ is cyclic so that any anti-homomorphism is a homomorphism. That $N = K^*$ follows by considering the orders (see the next exercise). $\square$

Exercise 9.4. If F is finite show that $N = K^*$.

Lemma 9.4. *The following conditions on the André quasifield F_ϕ are equivalent*

  (i) F_ϕ *is a division ring,*
  (ii) F_ϕ *is a field,*
  (iii) F_ϕ *is isomorphic to F,*
  (iv) $k^\phi = \varepsilon$ *for all k in N where ε is the identity of Γ.*

Proof. Suppose that F_ϕ is a division ring. Then F_ϕ satisfies the right distributive law so that for all $a, b, x \in F^*$ we have

$$ax^\alpha + bx^\beta = (a+b)x^\lambda \tag{1}$$

where $\alpha = \alpha_a$, $\beta = \alpha_b$, $\lambda = \alpha_{a+b}$. Regarding a and b as fixed, (1) is an identity in x and so, replacing x by xy, we have

$$a(x^\alpha y^\alpha) + b(x^\beta y^\beta) = (a+b)(x^\lambda y^\lambda) = [(a+b)x^\lambda]y^\lambda. \tag{2}$$

Using (1), (2) becomes

$$ax^\alpha y^\alpha + bx^\beta y^\beta = (ax^\alpha + bx^\beta)y^\lambda. \tag{3}$$

If we multiply both sides of (3) by $(a+b)$ and use (1) with x replaced by y, we have

$$(a+b)(ax^\alpha y^\alpha + bx^\beta y^\beta) = (ax^\alpha + bx^\beta)(ay^\alpha + by^\beta). \tag{4}$$

Multiplying out both sides of (4) and using the fact that $ab \neq 0$ we have

$$x^\alpha y^\alpha + x^\beta y^\beta = x^\alpha y^\beta + x^\beta y^\alpha, \tag{5}$$

and this simplifies to $(x^\alpha - x^\beta)(y^\alpha - y^\beta) = 0$ for all $x, y \in F$. Hence $x^\alpha = x^\beta$ for all $x \in F$ which implies $\alpha_a = \alpha_b$ for all $a, b \in F^*$. But $\alpha_1 = \varepsilon$ and so $\alpha_a = \varepsilon$ for all a in F^*. Hence (i) implies (ii), (iii) and (iv). But obviously any one of (ii), (iii) or (iv) implies (i). $\square$

In the finite case, since Γ is cyclic, we can replace ϕ by a mapping from K^* into Z_n, the integers modulo n, so that α_x is the automorphism $z \to z^{q^{x\phi}}$. Lemma 9.3 is then replaced by the demand that ϕ be a homomorphism into the additive group Z_n.

Theorem 9.5. *Let* $F = GF(p^n)$ *where* p *is a prime. Then there exists an André quasifield of order* p^n *which is not a field if and only if* (a) p *is odd and* $n \geq 2$ *or* (b) $p = 2$ *and* $n \geq 2$ *is not a prime.*

Proof. In view of Lemma 9.4, it is only necessary to show that we can find a mapping ϕ from some subfield K of F, $K \neq F$, into the integers modulo m, where m is the dimension of F over K, such that some element of K other than 0 or 1 is mapped onto something other than 0. If p is odd and $n \geq 2$ then we choose $K = GF(p)$ and send -1 onto 1 in Z_n. If $p = 2$ and n is a prime, then $GF(2)$ is the only subfield possessed by F, and ϕ is completely determined on $GF(2)$ by $1^\phi = 0$ and so F_ϕ must be a field. If, however, n is not a prime then there is a field $GF(2^s)$ properly contained between $GF(2)$ and $GF(2^n)$; let $K = GF(2^s)$, then K^* contains a non-identity element which we can map onto 1 in $Z_{n/s}$. $\square$

Since an associative quasifield is a nearfield we shall call an associative André quasifield an *André nearfield*.

Exercise 9.5. Find the finite fields $F = GF(q)$ for which there exist André nearfields F_ϕ which are not fields.

One case in Exercise 9.5 is so important that we include it here as a lemma:

Lemma 9.6. *If* q *is an odd prime power and* $F = GF(q^2)$, *then there always exists an André nearfield* F_ϕ *whose kernel is* $GF(q)$ *and such that the elements of the kernel commute with every element of* F_ϕ.

Proof. Taking Γ as the group of order two generated by the automorphism $\alpha : x \to x^q$, we have $K = GF(q)$. The multiplicative group of K^* is cyclic of order $q - 1$ and exactly half its elements are squares. For any $k \in K^*$ let k^ϕ be defined by $k^\phi = 0$ if k is a square, $k^\phi = 1$ otherwise. Then it is easily seen that ϕ is a homomorphism from K^* onto the additive group Z_2 and so F_ϕ must be associative and not a field. Since the kernel of F_ϕ cannot be all of F_ϕ but must contain $GF(q)$, it must be exactly $GF(q)$. If b is any element of K^* then, for any x in F, $x \odot b = xb^{\alpha_x} = xb$; since $b^\nu = b^{1+q}$, which is a square in K^*, $(b^\nu)^\phi = 0$ so that $\alpha_b = 1$ and thus $b \odot x = bx = xb = x \odot b$ as required. $\square$

As an example, let us consider $F = GF(9)$. Then Γ must be the group of order two generated by $\alpha : x \to x^3$ and $K = GF(3)$. F_ϕ will be determined when we define ϕ on $GF(3)^*$. But $1^\phi = 0$ is obligatory so the only choice we have is for $(-1)^\phi$. But, by Lemma 9.4, if we put $(-1)^\phi = 0$ then F_ϕ is $GF(9)$ again and so if we wish to avoid F_ϕ being a field we must have $(-1)^\phi = 1$. But this is exactly the situation of Lemma 9.6 which means that the only André quasifields of order 9 are associative; one is a field and one is not. (Also, compare Exercise 7.10 and 7.11.)

Exercise 9.6. Show that the André quasifield of order 9 which is not a field is isomorphic to the Hall quasifield of order 9 defined by $f(s) = s^2 + 1$.

As a further example, we could see that there are eight ways to construct F_ϕ if $F = GF(25)$, one of which is a field, one of which is an associative system which is not a field and the other six of which are not associative. But note that the somewhat deeper problem of deciding whether the seven non-desarguesian planes coordinatized by these quasifields are all perhaps isomorphic is not settled by this discussion: this suggests that quasifields are not totally satisfactory objects when we want to ask questions about isomorphism between translation planes.

We shall see in Chapter X that all finite Hall planes are contained in the class of André planes (that is, the planes coordinatized by André quasifields). It is also true, although we shall not prove it, that the only finite André planes with non-soluble collineation groups are the desarguesian planes and the Hall planes.

Exercise 9.7. Determine all elements x of multiplicative order two in an André quasifield F_ϕ.

4. A Class of Division Rings

Among the many classes of division rings known there is one which we shall exhibit since it includes many interesting special cases. We shall just pull it out of the hat, as it were, instead of building it up as we did in the two preceding sections.

Let F be a skewfield with centre Z (so that Z is a field), and let θ be an anti-automorphism of F of finite order m and a, b elements of F, all such that

$$F \text{ has finite dimension over } Z \text{ and } xx^\theta \text{ is in } Z \text{ for all } x \text{ in } F; \quad (1)$$

$$a = x^{1+\theta} + xb \text{ has no solution for } x \text{ in } F. \quad (2)$$

(If $xx^\theta = y$ then, since y is in Z, $x^\theta = x^{-1}y = yx^{-1}$ so that $xx^\theta = y = x^\theta x$; thus we may write $x^{1+\theta}$ or $x^{\theta+1}$ for xx^θ.)

Now let D be a two-dimensional vector space over F with basis elements 1 and λ; we shall define a multiplication on D so that, using the vector space addition, it becomes a division ring. The rule for multiplication is

$$(x + \lambda y)(z + \lambda t) = (xz + aty^\theta) + \lambda(zy + x^\theta t + y^\theta bt). \quad (3)$$

Theorem 9.7. *D is a division ring.*

Proof. Clearly each of left and right multiplication is additive (i.e., $(x + y)T = xT + yT$ if T is either left or right multiplication), and, consequently, D satisfies both distributive laws.

Now since θ has order m as an anti-automorphism of F, and since θ must fix Z, it has order dividing m as an automorphism of Z. If K is the subfield of Z consisting of elements fixed by θ, then Z has finite dimension over K, so F has finite dimension over K as well; let k be the dimension of F over K. Then D has dimension $2k$ over K, and K is easily seen to be contained in the centre of D. Thus, since D is finite dimensional over its centre, in order to show that $(x+\lambda y)(z+\lambda t)=(p+\lambda q)$ has a unique solution for one factor on the left, given the other and $p+\lambda q$, it is only necessary to show that the only solutions of $(x+\lambda y)(z+\lambda t)=0(=0+\lambda 0)$ are $x+\lambda y=0$ or $z+\lambda t=0$ (see Exercise 8.3). Putting the right hand side of (3) equal to 0 gives:

$$xz + aty^\theta = 0,$$
$$zy + x^\theta t + y^\theta bt = 0. \tag{4}$$

Multiplying the first equation of (4) on the right by y, the second on the left by x and subtracting yields

$$aty^{1+\theta} - x^{1+\theta}t - xy^\theta bt = 0. \tag{5}$$

Now if $t=0$ the equations of (4) give $xz = zy = 0$ so that either $x = y = 0$ or $z = 0$. So we may assume that $t \neq 0$ and then, since $y^{1+\theta}$ is in Z, (5) becomes

$$ay^{1+\theta} - x^{1+\theta} - xy^\theta b = 0. \tag{6}$$

If $y = 0$ then (6) implies that $x = 0$; so assume $y \neq 0$. Let $x = x_1 y$ so that $x^\theta = y^\theta x_1^\theta$; then $x^{1+\theta} = x_1 y^{1+\theta} x_1^\theta = x_1^{1+\theta} y^{1+\theta}$, since $y^{1+\theta}$ is in Z. Then (6) becomes $ay^{1+\theta} - x_1^{1+\theta} y^{1+\theta} - x_1 y^{1+\theta} b = 0$ or

$$(a - x_1^{1+\theta} - x_1 b) y^{1+\theta} = 0. \tag{7}$$

The first factor of (7) cannot be zero by condition (2) and so $y^{1+\theta} = 0$ is the only solution. But this implies $y = 0$ and then, from (6), $x = 0$ as well. This completes the proof except for the observation that $1 = 1 + \lambda 0$ is the identity for D. ▯

There are many very interesting special cases, and we now discuss a few of them.

Case (a). Let F be the field of reals and choose $\theta = 1$, $a = -1$, $b = 0$; then D is (isomorphic to) the complex numbers.

Case (b). Let F be the field of complex numbers and choose $a = -1$, $b = 0$ and θ such that $(x + iy)^\theta = x - iy$: then D is the quaternion skewfield, which is not a field.

Exercise 9.8. Prove the assertions of (a) and (b) above.

Exercise 9.9. Show that if F is the field of real numbers, then any allowable choice of a, b, θ leads to the complex numbers (here you may use the fact that F has only the identity automorphism).

Exercise 9.10. If F is the field of complex numbers decide whether or not there exist choices of a, b and θ such that D is not associative.

*Exercise 9.11. Show that the quaternions, as defined in (b) above, could also have been defined as a four-dimensional vector space Q over the reals, with the basis elements $1, i, j, k$ and multiplication defined by the rules $1x = x1 = x$ for all x in Q; $ij = -ji = k$; $jk = -kj = i$; $ki = -ik = j$; $i^2 = j^2 = k^2 = -1$; if x is in the reals then x commutes with each of i, j, k.

Exercise 9.12. Show that the mapping θ defined on the quaternions by $(x + iy + jz + kt)^\theta = x - ij - jz - kt$ is an anti-automorphism, and that $(x + iy + jz + kt)^{1+\theta}$ is a real number and, hence, is in the centre of Q.

Taking the quaternions as our skewfield F we give a further example.

Case (c). Let F be the quaternions and choose $a = -1$, $b = 0$, θ as in Exercise 9.12; then D is a non-associative division ring, called the *Cayley-Dickson algebra*.

*Exercise 9.13. Show that the Cayley-Dickson algebra as defined in (c) satisfies both inverse properties, but is not associative.

(Properly interpreted, the Cayley-Dickson algebras are the only IP division rings which are not associative: this is the Bruck-Kleinfeld Theorem, see Chapter VI.)

We now come to the examples which are most interesting for the study of finite projective planes. If F is finite then it is a field and consequently θ must be an automorphism. This forces the division ring D to have some additional properties which we now study.

In any quasifield Q the *associator* $[r, s, t]$ of three elements r, s, t of Q was defined by $[r, s, t] = (rs)t - r(st)$ (see Section 7 of Chapter VI). If F is a field and we compute the associator in D then we find

$$[x + \lambda y, z + \lambda t, h + \lambda k] = at^\theta k((x^{\theta^2} - x) + (b^\theta y^{\theta^2} - by^\theta)) \\ + \lambda t^\theta k((a^\theta y^{\theta^2} - ay) + b(x^{\theta^2} - x^\theta) + b(b^\theta y^{\theta^2} - by^\theta)). \tag{8}$$

We now observe that, since F is a field, F is its own centre and the demand that θ satisfies $x^{1+\theta} \in Z$ for all x in F is empty. The three semi-nuclei of D can be computed from (8) and two of them are particularly simple.

Lemma 9.8. *If F is finite then, provided $\theta \ne 1$, F is the right and middle nucleus of D.*

Proof. It is easy to see that putting either t or k equal to zero in (8) gives 0 for the right hand side for all x, y, z, h. Thus F is contained in each of the right and middle nuclei. However each semi-nucleus is itself a skew-field and so, since D is finite, a field (see Chapter VIII); thus if the right or middle nucleus were larger than F, then D would be a field and the right hand side of (8) would be zero for all values of all the variables. We shall assume that D is a field and show that this leads to a contradiction. We first note that a is non-zero since otherwise $k = 0$ would be a solution to the equation in (2). Since D is associative, every associator is zero and so, putting $t = k = 1$ and $y = 0$, (8) tells us that $x^{\theta^2} = x$ and $b(x^{\theta^2} - x^\theta) = 0$ for all x. Since $\theta \neq 1$ this implies $\theta^2 = 1$ and $b = 0$. Putting $b = 0$ in (8) we have $t^\theta k(a^\theta - a) y = 0$ for all t, k, y. But this is true if and only if $a^\theta = a$; i.e., if and only if a is in the subfield K of elements fixed by θ. Since $\theta^2 = 1$, K must be $GF(q)$ where F is $GF(q^2)$ and θ is given by $x^\theta = x^q$ for all x in F. But as x varies over F, $x^{1+\theta}$, which is x^{1+q}, varies over all of K so that there is a value of x in F such that $x^{1+\theta} = a$; this violates (2) (recall that $b = 0$), and proves the lemma. □

Note that we have proved more:

Lemma 9.9. *If F is finite, then D is associative if and only if $\theta = 1$.*

To construct D we have assumed the existence of elements a, b of F and an anti-automorphism θ satisfying certain conditions. It is now natural to see how restrictive our conditions are. Since $\theta \neq 1$, if F is finite then F must be $GF(p^n)$ where p is a prime and $n \neq 1$. We now show that this condition is sufficient for the existence of a non-trivial example.

Theorem 9.10. *If $F = GF(p^n)$ where p is a prime and $n > 1$, then there is a choice of a, b and θ such that D is not associative.*

Proof. If p is odd, we choose $x^\theta = x^p$, $b = 0$ and a to be a non-square in F. Then, since x^{1+p} is always a square, condition (2) is satisfied. If $p = 2$, we choose $x^\theta = x^2$ and choose any element b whatsoever except $b = 0$. Suppose that no choice of a is possible, then $x^3 + bx = a$ has a solution for all a in F. But then the mapping $x \rightarrow x^3 + bx$ is one-to-one and onto which is impossible since $x^3 + bx = 0$ has two solutions; namely $x = 0$ and $x = d$ where d is the (unique) element in F satisfying $d^2 = b$ (here we are using the fact that every element of a finite field of characteristic two is a square). Thus there is a choice for a and the theorem is proved. □

*Exercise 9.14. Find the conditions on F, a, b, θ that F should also be the left nucleus of D. If F is also the left nucleus of D, what is the centre of D?

Extra significance is given to the finite members of this class by the following:

Theorem 9.11. *Let D be a finite division ring whose right and middle nuclei are equal and such that the dimension of D over its right nucleus is two. Then D is isomorphic to a division ring given by the multiplication rule (3).*

The proof of this theorem is difficult and we shall not include it here. However it might not be beyond the capabilities of the interested student and can be found in [5]. Finally we remark that the planes coordinatized by the finite division rings of this class have been studied in considerable detail, and it has been shown that their full collineation groups are soluble.

5. The Dickson Commutative Division Rings

Our last example of a division ring will be commutative but not associative. The construction is similar to that of the previous section, in the sense that we let F be a finite field, D a two-dimensional vector space over F with basis elements 1, λ and we define a multiplication on D.

Let $F = GF(p^n)$ where p is an odd prime and $n > 1$ and let a be any element which is not a square in F. If θ is an automorphism of F given by $x^\theta = x^{p^r}$, $1 \leq r < n$ then we define multiplication in D by:

$$(x + \lambda y)(z + \lambda t) = (xz + ay^\theta t^\theta) + \lambda(yz + xt). \tag{1}$$

Theorem 9.12. *D is a commutative division ring and is never associative.*

Proof. Clearly, from the definition, multiplication in D is commutative. As in the proof of Theorem 9.7, to prove that D is a division ring we need only verify that it has no non-zero divisors.

If we put the right hand side of (1) equal to zero we have

$$\begin{aligned} xz + ay^\theta t^\theta &= 0, \\ yz + xt &= 0. \end{aligned} \tag{2}$$

If $z = 0$ then either $t = 0$ or $x = y = 0$, so we may assume $z \neq 0$. Solving the first equation of (2) for x and substituting in the second we get $yz - ay^\theta t^{1+\theta} z^{-1} = 0$, which yields

$$z^2 = ay^{\theta-1}t^{\theta+1} = ay^{p^r-1}t^{p^r+1}, \quad \text{if} \quad y \neq 0. \tag{3}$$

But every factor of (3) is a square except a and so z must be zero. This is a contradiction and, hence, $y = 0$; it is now easy to see that x must also be zero. Thus D is a commutative division ring. To show that D is non-associative, i.e., not a field, we prove a lemma.

Lemma 9.13. *F is the middle nucleus of D and, if K is the subfield of F of elements fixed by θ, K is the right and left nucleus; hence K is also the nucleus and centre.*

(Here we identify the element x in F with $x + \lambda \cdot 0$ in D, for instance.)

Proof. If we compute the associator we find $[x + \lambda y, z + \lambda t, h + \lambda k]$ $= at^\theta(y^\theta(h - h^\theta) + k^\theta(x^\theta - x)) + \lambda at^\theta(y^\theta k - yk^\theta)$.

Putting $t = 0$ clearly makes the right hand side zero. We leave the rest of the proof as an exercise. (Note that the lemma will be proved once the right nucleus is shown to be K. The commutativity of D will imply that the left nucleus is also K and, since $K \neq D$, D cannot be a field.) $\square$

Exercise 9.15. Show that the set of all elements $h + \lambda 0$, where $h^\theta = h$, is the right nucleus of D.

The planes coordinatized by these division rings have also been studied and their full collineation groups are soluble.

6. The Hughes Planes

In the preceeding sections we have exhibited non-desarguesian projective planes by finding quasifields which are not skewfields. In this section we exhibit a class of finite projective planes, called the *Hughes planes*, which cannot be coordinatized by any linear PTR. For any odd prime power q, there exists a Hughes planes of order q^2 and the smallest, i.e., of order 9, was first discovered by Veblen and Wedderburn.

A second property which distinguishes the Hughes plane from all the other examples given is that, for any given Hughes plane $\mathcal{H}$, the full collineation group of $\mathcal{H}$ does not fix a point or line of $\mathcal{H}$. The Hughes planes are, in fact, the only known finite non-desarguesian projective planes with this property.

If q is any odd prime power then, by Lemma 9.6, there is always at least one nearfield N of order q^2 whose kernel is $F = GF(q)$ and such that every element of F commutes with each element of N. We shall use the elements of N to construct a Hughes plane $\mathcal{H}$ of order q^2.

Let $V = \{(x_1, x_2 x_3) | x_i \in N, i = 1, 2, 3\}$. Define addition of V by $(x_1, x_2, x_3) + (y_1, y_2, y_3) = (x_1 + y_1, x_2 + y_2, x_3 + y_3)$ (note that we are really using the symbol $+$ for two distinct operations, namely addition in V and addition in N; however this should not cause any confusion), and define a left scalar multiplication by the elements of N by $k(x_1, x_2, x_3) = (kx_1, kx_2, kx_3)$. In other words we make V into a sort of "left vector space over N". The points of $\mathcal{H}$ will now be the elements of V, other than $(0, 0, 0)$, with the identification $(x_1, x_2, x_3) = (kx_1, kx_2, kx_3)$ for all non-zero k in N. If $x = (x_1, x_2, x_3)$ then we denote its point in $\mathcal{H}$ by $\langle x \rangle$. If $A = (a_{ij})$ is any element of $GL_3(q)$, i.e., any 3×3 matrix with entries from $F = GF(q)$, then the mapping $\theta : (x_1, x_2, x_3) \to (a_{11}x_1 + a_{12}x_2 + a_{13}x_3, a_{21}x_1 + a_{22}x_2 + a_{23}x_3, a_{31}x_1 + a_{32}x_2 + a_{33}x_3)$ is an automorphism of the additive group of V. But since N is associative and every element of F commutes with the elements of N, $(kx_1, kx_2, kx_3) \to (a_{11}(kx_1) + a_{12}(kx_2) + a_{13}(kx_3), a_{21}(kx_1) + a_{22}(kx_2) + a_{23}(kx_3), a_{31}(kx_1 + a_{32}(kx_2) + a_{33}(kx_3)) = (k(a_{11}x_1 + a_{12}x_2 + a_{13}x_3), k(a_{21}x_1 + a_{22}x_2 + a_{23}x_3), k(a_{31}x_1 + a_{32}x_2 + a_{33}x_3))$ so that θ induces a permutation on the points of $\mathcal{H}$. We shall use vector space notation and represent θ by $x' \to Ax'$ or, if P is the point $\langle x \rangle$ we shall write PA for the image of P under θ.

Exercise 9.16. Show that the group induced on the points of $\mathcal{H}$ by $GL_3(q)$ has exactly two orbits. (Hint: Show that one orbit is the set $\mathcal{H}_0$ of all points of the form $\langle x \rangle$ where $x = (x_1, x_2, x_3)$ and $x_i \in F$ ($i = 1, 2, 3$).)

If t is an arbitrary element of N then $x_1 + x_2 t + x_3 = 0$ if and only if $kx_1 + (kx_2) t + kx_3 = 0$ for all k in N. Thus the set of all elements (x_1, x_2, x_3) of V such that $x_1 + x_2 t + x_3 = 0$ is made up of $(0, 0, 0)$ together with a set of points of $\mathcal{H}$. We denote this set of points by $L(t)$. As we saw in Chapter II (see Corollary to Theorem 2.63) $GL_3(q)$ contains a matrix A such that

$A^{q^2+q+1} = kI$, for some k in $GF(q)$, and no smaller power of A has this property. Thus A has order $q^2 + q + 1$ as a permutation on the points of $\mathscr{H}$. We now define lines of $\mathscr{H}$ to be the point sets

$$L(t) A^m = \{\langle A^m x' \rangle \mid \langle x \rangle \in L(t)\}$$

for $0 \leq m \leq q^2 + q$, where either $t = 1$, or t is an arbitrary element of $N \setminus F$.

Theorem 9.14. *If $\mathscr{H}$ is the set of points and lines defined above with incidence given by set theoretic inclusion then $\mathscr{H}$ is a finite projective plane of order q^2.*

Proof. The number of points of $\mathscr{H}$ is equal to the number of elements of V other than $(0, 0, 0)$ divided by the number which give the same point of $\mathscr{H}$. Since this latter is the same as the number of non-zero elements of N, the number of points of $\mathscr{H}$ is $q^6 - 1/q^2 - 1 = q^4 + q^2 + 1$.

Clearly for any t the number of points on $L(t)$ is $q^2 + 1$ and thus, since every other line is the image of one of these point sets under an element of $GL_3(q)$, every line contains exactly $q^2 + 1$ points.

The number of lines is merely the number of distinct $L(t)$ multiplied by the number of possible powers of A, i.e., $(q^2 - q + 1)(q^2 + q + 1) = q^4 + q^2 + 1$. Thus, by Exercise 3.11, the theorem will be proved if we show that any two distinct lines intersect in a unique point.

Whether $\mathscr{H}$ is a projective plane or not it is clear that each mapping A^i preserves incidence in $\mathscr{H}$. Thus, in order to show that any pair of distinct lines intersect in a unique point, it is sufficient to show that any pair of distinct lines $L(t) A^m$ and $L(s)$ do. Let A^{-m} be the matrix (a_{ij}). A point $P = \langle (x, y, z) \rangle$ is on the lines $L(t) A^m$ and $L(s)$ if and only if P is on $L(s)$ and PA^{-m} is on $L(t)$. Thus P is on both lines if and only if

$$\begin{aligned}(a_{11}x + a_{12}y + a_{13}z) + (a_{21}x + a_{22}y + a_{23}z)t \\ + (a_{31}x + a_{32}y + a_{33}z) = 0,\end{aligned} \tag{1}$$

$$x + ys + z = 0. \tag{2}$$

So we must show (1) and (2) have a unique solution (x, y, z) up to a left multiple by N for x, y, z in N (recall that the a_{ij} are all in F).

Solving (2) for x and substituting in (1) gives

$$yu + za + (yv + zb)t = 0 \tag{3}$$

where

$$\begin{aligned}u = a_{12} + a_{32} - (a_{11} + a_{31})s, \quad v = a_{22} - a_{21}s, \\ a = a_{13} + a_{33} - (a_{11} + a_{31}), \quad b = a_{23} - a_{21}.\end{aligned} \tag{4}$$

Since the a_{ij} are all in F, a and b both belong to F and, consequently, commute with every element of N. We now have several cases to consider.

Case (a): $b \neq 0$. In this case (3) can be rewritten as:

$$(yv + zb) b^{-1} a + y(u - vb^{-1} a) + (yv + zb)t = 0$$

and this simplifies to

$$(yv + zb)(b^{-1}a + t) + y(u - vb^{-1}a) = 0. \tag{5}$$

If $t = 1$ then it is easy to see that (2) and (3) have a unique common solution for the point $\langle(x, y, z)\rangle$. But if $t \neq 1$ then t is not in F and so, in this case, $w = b^{-1}a + t$ is non-zero. Thus (5) becomes $(yv + zb)w = -y(u - vb^{-1}a)$ or, multiplying through by w^{-1} and collecting terms

$$y(v + (u - vb^{-1}a)w^{-1}) + zb = 0. \tag{6}$$

Since $b \neq 0$, (6) and (2) now have a unique common solution for the point $\langle(x, y, z)\rangle$.

Case (b): $b = 0$, $a \neq 0$. In this case (3) becomes

$$y(u + vt) + za = 0 \tag{7}$$

and now, since $a \neq 0$, (7) and (2) have a unique common point.

Case (c): $a = b = 0$. These conditions imply

$$a_{13} + a_{33} = a_{11} + a_{31} \quad \text{and} \quad a_{23} = a_{21}. \tag{8}$$

But now consider the point $P = \langle(1, 0, -1)\rangle$. From (8) it is immediate that $PA^{-m} = (c, 0, -c)$, where $c = a_{11} - a_{13}$, and, since A^{-m} is non-singular, $c \neq 0$ so that $PA^{-m} = P$. Thus, by the original choice of A, $m \equiv 0 \pmod{q^2 + q + 1}$ and $L(t)A^m = L(t)$. But now it is easy to see that $L(t)$ and $L(s)$ have only the point $\langle(1, 0, -1)\rangle$ in common. $\square$

Lemma 9.15. *The set $\mathscr{H}_0$ of all points of the form $\langle x \rangle$, where $x = (x_1, x_2, x_3)$ with $x_i \in F$, $i = 1, 2, 3$, together with the lines joining them, forms a desarguesian subplane of $\mathscr{H}$.*

Proof. From Eq. (1) in the proof of Theorem 9.14 we note that the line $L(t)A^m$ of $\mathscr{H}$ can be represented by an equation of the form $xa + yb + zc + (xa' + yb' + zc')t = 0$, where a, b, c, a', b', c', are in F. Thus any line of the form $L(1)A^m$ has an equation of the form $xa + yb + zc = 0$ where a, b, c are in F. Furthermore since there are exactly $q^2 + q + 1$ distinct lines $L(1)A^m$, any equation of the form $xa + yb + zc = 0$ where a, b, c are in F represents one of the lines $L(1)A^m$.

It now follows from the discussion following Exercise 2.6 that the points of $\mathscr{H}_0$ and the lines $L(1)A^m$, $m = 0, 1, ..., q^2 + q$ form a desarguesian plane of order q. $\square$

From the construction of the plane it is clear that each of the matrices A^i induces a collineation of $\mathscr{H}$ and that the cyclic group which they generate does not fix a point or line of $\mathscr{H}$. Thus if $\mathscr{H}$ contains a translation line it must contain more than one. But in Chapter VI we proved that if a finite plane has two translation lines then it is desarguesian. Hence either $\mathscr{H}$ is desarguesian or it cannot be coordinatized by a quasifield.

In order to show that $\mathscr{H}$ is non-desarguesian we will now coordinatize
it in such a way that the resulting PTR is not even linear. Using ordered
triples to have the same meaning as in the earlier parts of this section, we
now coordinatize $\mathscr{H}$ by the method introduced in Chapter V. Let (∞) be
$(0, 0, 1)$; (0) be $(1, 0, 0)$; $(0, 0)$ be $(0, 1, 0)$ and (1) be $(1, 0, -1)$. The line l_∞
is, in the earlier representation, the line $y = 0$ and the new x-axis and y-axis
are the earlier lines $z = 0$ and $x = 0$ respectively. We can now label the
points of the new y-axis by letting $(0, 1, v)$ be $(0, v)$ in our new system.
Every line through (1), or $(1, 0, -1)$, is a line of the form $x + yt + z = 0$.
The point $(v, 0)$ on the x-axis will be the point $(u, 1, 0)$ which is collinear
with $(1, 0, -1)$ and $(0, 1, v)$. But $(1, 0, -1)$ and $(0, 1, v)$ are on the line $L(-v)$,
if $v \notin F$, from which it follows that $u + 1(-v) + 0 = 0$ or $u = v$. If $v \in F$, then
it is obvious that $u = v$ and thus $(v, 0)$ is the point $(v, 1, 0)$.

Since the point (m) is on l_∞, it will be a point of the form $(1, 0, v)$.
Furthermore (m) is the point on l_∞ which is collinear with $(1, 1, 0)$ and
$(0, 1, m)$. Let $xa + yb + zc + (xa' + yb' + zc')t = 0$ be the line joining $(1, 1, 0)$
and $(0, 1, m)$. Then:

$$a + b + (a' + b')t = 0, \tag{1}$$

$$b + mc + (b' + mc')t = 0. \tag{2}$$

Since $a, a', b, b' \in F$, (1) implies that either $t = 1$ and $a + a' = -(b + b')$, or
$t \neq 1$ and $a + b = a' + b' = 0$. If $t = 1$, then $a + a' + v(c + c') = 0$, and (2)
becomes

$$a + a' + (-m)(c + c') = 0 ;$$

thus $v = -m$. If $t \neq 1$, then $a + vc + (a' + vc')t = 0$, and (2) becomes
$-a + mc + (-a' + mc')t = 0$, so again $v = -m$. Thus (m) is the point
$(1, 0, -m)$.

Consider the point (u, v) which lies on the line m_1 joining $(0, 0, 1)$ and
$(u, 1, 0)$ and on the line m_2 joining $(1, 0, 0)$ and $(0, 1, v)$. Then:

$$m_1 : xa + yb + zc + (xa' + yb' + zc')t = 0,$$
$$m_2 : xd + ye + zf + (xd' + ye' + zf')s = 0, \tag{3}$$

where

$$c + c't = d + d's = ua + b + (ua' + b')t = e + vf + (e' + vf')s = 0. \tag{4}$$

Thus as before, $t = 1$ and $c = -c'$, or $t \neq 1$ and $c = c' = 0$; similarly, $s = 1$
and $d = -d'$, or $s \neq 1$ and $d = d' = 0$.

There are four cases to check, but all of them are easy. If $t \neq 1$, $s \neq 1$,
then it is simple verification, using (4), that $(u, 1, v)$ is on both of the lines
m_1 and m_2. If $t = 1$, $s \neq 1$, then m_1 is

$$x(a + a') + y(b + b') = 0,$$

where $u(a + a') + b + b' = 0$, so $(u, 1, v)$ is on m_1. The line m_2 becomes

$$ye + zf + (ye' + zf')s = 0,$$

where $e + vf + (e' + vf')s = 0$, so $(u, 1, v)$ is on m_2. In all cases, we find that $(u, 1, v)$ is on both m_1 and m_2, so (u, v) is $(u, 1, v)$. So we have:

Lemma 9.16. *If $\mathscr{H}$ is coordinatized as above, then (u, v) is $(u, 1, v)$, (m) is $(1, 0, -m)$ and (∞) is $(0, 0, 1)$.*

Now we shall investigate the ternary ring for $\mathscr{H}$, where we use $T(a, b, c)$ for the ternary operation, and let $a \oplus b = T(1, a, b)$, $a \otimes b = T(a, b, 0)$. In order to find the value of $T(m, u, v)$, we consider the line l which contains $(1, 0, -m)$ and $(u, 1, v)$, and let $(0, 1, k)$ be the intersection of l with the y-axis; then $k = T(m, u, v)$. Let l be the line

$$xa + yb + zc + (xa' + yb' + zc')t = 0.$$

Then:

$$a - mc + (a' - mc')t = 0, \tag{5}$$

$$ua + b + vc + (ua' + b' + vc')t = 0. \tag{6}$$

Lemma 9.17. *For all a and b, $a \oplus b = a + b$.*

Proof. Let $m = 1$ in (5). Then $a - c + (a' - c')t = 0$. So if $t = 1$, we have $a + a' = c + c'$, and (6) becomes $u(a + a') + (b + b') + v(c + c') = 0$, or $(u + v) \cdot (a + a') + (b + b') = 0$. But then the point $(0, 1, u + v)$ is on l so $k = u \oplus v = u + v$. If $t \neq 1$, then $a = c$ and $a' = c'$, and (6) becomes

$$(u + v)a + b + [(u + v)a' + b']t = 0,$$

whence again $(0, 1, u + v)$ is on l, so $u \oplus v = u + v$.   ☐

Theorem 9.18. *The ternary ring (N, T) is not linear.*

Proof. Refering to (5) and (6), let m and u be arbitrary, $u \neq 0$, and let v be chosen so that $k = T(m, u, v) = 0$. Then $(0, 1, 0)$ is on l, so $b + b't = 0$. Suppose $t = 1$, whence $b = -b'$. Then (5) is $a + a' - m(c + c') = 0$, and (6) is

$$u(a + a') + v(c + c') = 0,$$

and this implies $u^{-1}v = -m$, or $um + v = 0$. Suppose $t \neq 1$, whence $b = b' = 0$. Then (5) is $a - mc + (a' - mc')t = 0$, and (6) can be written

$$a + u^{-1}vc + (a' + u^{-1}vc') = 0,$$

and so again $u^{-1}v = -m$, or $um + v = 0$. Now we assume that (N, T) is linear. For arbitrary m and u, $u \neq 0$, let $p = m \otimes u$. Then $m \otimes u \oplus (-p) = T(m, u, -p) = 0$, so by the above $um + (-p) = 0$, or $um = p = m \otimes u$. Thus (N, T) is multiplicatively anti-isomorphic to the nearfield N, and so (N, T) is itself a right nearfield. As pointed out earlier, any finite projective plane which may be coordinatized by a right quasifield and whose full collineation group does not fix a point must be desarguesian. Thus $\mathscr{H}$ is desarguesian and N must be a field. Hence, since N was chosen not to be a field we have a contradiction and the PTR (N, T) cannot be linear.   ☐

Corollary 1. *The Hughes planes are non-desarguesian.*

The Hughes planes have many other interesting properties, for example the desarguesian subplane $\mathscr{H}_0$ of the plane $\mathscr{H}$ is left invariant by Aut $\mathscr{H}$, but we have not yet established sufficient machinery to prove them all. For our purposes the main interest at this stage is that we have shown the existence of a finite projective plane which is not a translation plane. In fact, by Exercise 4.28, we also have

Corollary. 2. *A Hughes plane is not (V, l)-transitive for any choice of V or l and hence is not a translation plane or a dual translation plane.*

References

Each class of planes introduced in this chapter has been the object of much research and, not surprisingly, much more is known about them than the basic properties which we have established. In particular their collineation groups have been studied in detail. The list of references given here is in no sense complete, and the interested student should consult Dembowski [1] before reading the specialized papers.

Theorem 9.2 is due to Jha [6], while Theorem 9.11 is proved in [5]. Finally we remark that the construction of the André planes has been generalised by Foulser [2] to give a larger class of planes.

1. Dembowski, H. P.: Finite geometries. Berlin-Heidelberg-New York: Springer 1968.
2. Foulser, D. A.: A generalization of André's systems. Math. Z. **100**, 380–395 (1967).
3. Hughes, D. R.: Collineation groups of non-desarguesian planes. I. The Hall Veblen-Wedderburn systems. Am. J. Math. **81**, 921–938 (1959).
4. — Collineation groups of non-desarguesian planes. II. Some seminuclear division algebras. Am. J. Math. **82**, 113–119 (1960).
5. — Kleinfeld, E.: Seminuclear extensions of Galois fields. Am. J. Math. **82**, 389–392 (1960).
6. Jha, V.: On the automorphism groups of quasifields. PhD Thesis (University of London) 1971.
7. Rosati, L. A.: I gruppi di collineazioni dei piani di Hughes. Boll. Un. Mat. Ital. **13**, 505–513 (1958).
8. Zappa, G.: Sui gruppi di collineazioni dei piani di Hughes. Boll. Un. Mat. Ital. **12**, 507–516 (1957).

X. Derivation

1. Introduction

In this chapter we shall investigate, in a simple special form, a method of construction of finite projective planes called "derivation". We shall show that Hall planes and some André planes can be constructed from desarguesian planes in this way and, as a corollary, we obtain the result that every Hall plane is an André plane. We shall also prove a number of results relating the collineation groups of a plane and its "derived plane": these results are useful in establishing the structure of the "derived plane". In particular, we use these results to study the collineation group of the Hall plane in considerable detail.

2. The Basic Results and Definitions

Let $\mathscr{A}$ be a finite affine plane of order n^2, and let l_∞ be its special line (so that $\mathscr{A} = \mathscr{P}^{l_\infty}$, for the appropriate projective plane $\mathscr{P}$). Let $\mathscr{S}$ be a subset of $n+1$ special points such that, for every pair of distinct affine points X, Y of $\mathscr{A}$ for which XY meets l_∞ in $\mathscr{S}$, there is a Baer subplane of $\mathscr{P}$ containing X, Y and $\mathscr{S}$. Then we will say that $\mathscr{S}$ is a *derivation set* for $\mathscr{A}$ (or for $\mathscr{P}$). If $\mathscr{B}$ is a Baer subplane of $\mathscr{P}$ containing $\mathscr{S}$ then we say that $\mathscr{B}$, or $\mathscr{B}^{l_\infty}$, *belongs* to $\mathscr{S}$.

Lemma 10.1. *Let $\mathscr{S}$ be a derivation set of an affine plane $\mathscr{A} = \mathscr{P}^{l_\infty}$, and let $\{\mathscr{B}_1, \mathscr{B}_2, \ldots\}$ be the Baer subplanes of $\mathscr{P}$ which belong to $\mathscr{S}$. If $\mathscr{A}_i = \mathscr{B}_i^{l_\infty}$ for each i, then $\mathscr{A}_i \cap \mathscr{A}_j$ consists of 0 or 1 points, if $i \neq j$.*

Proof. Suppose there are two distinct points X and Y in $\mathscr{A}_i \cap \mathscr{A}_j$. Then $\mathscr{C} = \mathscr{B}_i \cap \mathscr{B}_j$ contains X, Y and the $n+1$ points of $\mathscr{S}$. Since $n+1 > 3$, $\mathscr{C}$ contains a quadrangle and, hence, is a subplane of $\mathscr{P}$. Furthermore, since $\mathscr{C}$ contains the $n+1$ collinear points of $\mathscr{S}$, the order of $\mathscr{C}$ is at least n. Thus $\mathscr{C} = \mathscr{B}_i = \mathscr{B}_j$, or $\mathscr{A}_i = \mathscr{A}_j$. $\square$

Note, by the way, that the $\mathscr{A}_i$ are uniquely determined by $\mathscr{S}$, and that any pair of affine points of $\mathscr{A}$ are in at most one Baer subplane belonging to $\mathscr{S}$. Now let $\mathscr{S}'$ be the complement of $\mathscr{S}$ on l_∞ and define $\mathscr{D}_{\mathscr{S}'}(\mathscr{A})$ to be the collection of "points" and "lines" as follows:

a point of $\mathscr{D}_{\mathscr{S}'}(\mathscr{A})$ is a point of $\mathscr{A}$;

a line of $\mathscr{D}_{\mathscr{S}'}(\mathscr{A})$ is either a line of $\mathscr{A}$ which meets l_∞ in $\mathscr{S}'$ or is a Baer subplane $\mathscr{A} = \mathscr{B}^{l_\infty}$ such that $\mathscr{B}$ contains $\mathscr{S}$;

incidence in $\mathscr{D}_{\mathscr{S}'}(\mathscr{A})$ is the natural containment relation.

Theorem 10.2. $\mathscr{D}_{\mathscr{S}'}(\mathscr{A})$ *is an affine plane of order n^2.*

Proof. Certainly $\mathscr{D}_{\mathscr{S}'}(\mathscr{A})$ has n^4 points, since its points are those of $\mathscr{A}$. Each of its lines is either a line of $\mathscr{A}$, in which case it has exactly n^2 points, or is an affine Baer subplane of $\mathscr{A}$, which also has n^2 points.

Let X and Y be any two distinct points of $\mathscr{A}$ (or of $\mathscr{D}_{\mathscr{S}'}(\mathscr{A})$). If XY meets l_∞ in $\mathscr{S}'$, then X and Y are on the line XY of $\mathscr{D}_{\mathscr{S}'}(\mathscr{A})$ and are in no Baer subplane containing $\mathscr{S}$; if XY meets l_∞ in $\mathscr{S}$ then the line XY of $\mathscr{A}$ is not a line of $\mathscr{D}_{\mathscr{S}'}(\mathscr{A})$, while the unique Baer subplane containing X, Y and $\mathscr{S}$ is. So X and Y are on a unique line of $\mathscr{D}_{\mathscr{S}'}(\mathscr{A})$.

Thus we have shown that $\mathscr{D}_{\mathscr{S}'}(\mathscr{A})$ is a set of n^4 points such that each line contains n^2 points and every pair of distinct points is on a unique line, and so, by Exercise 3.14, $\mathscr{D}_{\mathscr{S}'}(\mathscr{A})$ is an affine plane of order n^2. []

Since Exercise 3.14 is so important to this proof we will now sketch a solution to it.

Exercise 3.14. A set $\mathscr{C}$ of n^2 points with a collection of subsets called lines such that each line has n points and each pair of points is on exactly one common line is an affine plane of order n, if $n > 1$.

Proof. Let X be a point of $\mathscr{C}$ and suppose there are t lines of $\mathscr{C}$ containing X. Each of these lines contains $n - 1$ points other than X and thus, since every point is on exactly one line through X, we have $t(n - 1) = n^2 - 1$, or $t = n + 1$.

Let m be a line of $\mathscr{C}$ and let A be any point not on m. Then A is joined to each of the n points of m by n necessarily distinct lines, and so there is a unique line through A which fails to meet m. This, plus the condition $n > 1$, is enough to guarantee that $\mathscr{C}$ is an affine plane. []

$\mathscr{D}_{\mathscr{S}'}(\mathscr{A})$ is called the *derived plane* of $\mathscr{A}$ (with respect to $\mathscr{S}$); the reason for using $\mathscr{S}'$, as opposed to $\mathscr{S}$, will become clear below. First there is a special line l'_∞ of $\mathscr{D}_{\mathscr{S}'}(\mathscr{A})$, but its points are not exactly the points of l_∞: however the points of $\mathscr{S}'$ have a natural correspondence to certain points of l'_∞, since the lines of $\mathscr{A}$ through a point of $\mathscr{S}'$ are all lines of $\mathscr{D}_{\mathscr{S}'}(\mathscr{A})$ and pass through a common point of l'_∞. We shall denote this subset of l'_∞ by the same symbol $\mathscr{S}'$. The remaining points of l'_∞ do not correspond in any natural way to the points of $\mathscr{S}$; we shall denote this subset of l'_∞ by $\mathscr{D}_{\mathscr{S}'}(\mathscr{S})$, and shall often write $\mathscr{S}^* = \mathscr{D}_{\mathscr{S}'}(\mathscr{S})$.

Lemma 10.3. *If k is a line of $\mathscr{A}$ meeting l_∞ in $\mathscr{S}$, then the points of k are the points of a Baer subplane $\mathscr{D}_{\mathscr{S}'}(k)$ in $\mathscr{D}_{\mathscr{S}'}(\mathscr{A})$ and the lines of $\mathscr{D}_{\mathscr{S}'}(k)$ all meet l'_∞ in $\mathscr{D}_{\mathscr{S}'}(\mathscr{S})$.*

Proof. There are n^2 points of $\mathscr{D}_{\mathscr{S}'}(k)$, and its "lines" are the intersections of k with the Baer subplanes of $\mathscr{A}$ belonging to $\mathscr{S}$. But two distinct points of k are in a unique Baer subplane $\mathscr{A}_i$ belonging to $\mathscr{S}$, and since $k \cap \mathscr{A}_i$ contains more than one point, it is a line of $\mathscr{A}_i$. Thus $k \cap \mathscr{A}_i$ consists of exactly n points. But now by Exercise 3.14 $\mathscr{D}_{\mathscr{S}'}(k)$ is an affine plane of order n.

A "line" of $\mathscr{D}_{\mathscr{S}'}(k)$ is the intersection of k with a Baer subplane of $\mathscr{A}$ belonging to $\mathscr{S}$, and so does not meet l'_∞ in $\mathscr{S}'$. ☐

Theorem 10.4. $\mathscr{D}_{\mathscr{S}'}(\mathscr{S})$ is a derivation set for $\mathscr{D}_{\mathscr{S}'}(\mathscr{A})$, and $\mathscr{D}_{\mathscr{S}'}\mathscr{D}_{\mathscr{S}'}(\mathscr{A})$ is naturally isomorphic to $\mathscr{A}$.

Proof. An immediate corollary of Lemma 10.3. (Remember that we use $\mathscr{S}'$ for the complement of $\mathscr{D}_{\mathscr{S}'}(\mathscr{S})$ on l'_∞.) ☐

Clearly a collineation of $\mathscr{A}$ must send a derivation set onto a derivation set. Furthermore we have

Theorem 10.5. If $\alpha \in \operatorname{Aut}\mathscr{A}$, then $\mathscr{D}_{\mathscr{S}'}(\mathscr{A}) \cong \mathscr{D}_{\mathscr{S}'^\alpha}(\mathscr{A})$.

Proof. Almost obvious. We leave it as a simple exercise. ☐

Theorem 10.6. If $\alpha \in \operatorname{Aut}\mathscr{A}$ and $\mathscr{S}^\alpha = \mathscr{S}$ then α induces an automorphism of $\mathscr{D}_{\mathscr{S}'}(\mathscr{A})$.

Proof. Obvious, since α maps the Baer subplanes belonging to $\mathscr{S}$ onto themselves. ☐

Corollary 1. If α is a perspectivity of $\mathscr{A}$ with axis l_∞, then α induces a collineation of $\mathscr{D}_{\mathscr{S}'}(\mathscr{A})$.

Corollary 2. If α is a perspectivity of $\mathscr{A}$ with axis $m \neq l_\infty$, and if $\mathscr{S}^\alpha = \mathscr{S}$ then α induces a perspectivity of $\mathscr{D}_{\mathscr{S}'}(\mathscr{A})$ with axis m if $m \cap l_\infty \in \mathscr{S}'$, and induces a Baer collineation of $\mathscr{D}_{\mathscr{S}'}(\mathscr{A})$ if $m \cap l_\infty \in \mathscr{S}$.

Corollary 3. If α is a Baer collineation of $\mathscr{A}$, whose Baer subplane belongs to $\mathscr{S}$, then α induces a perspectivity of $\mathscr{D}_{\mathscr{S}'}(\mathscr{A})$ whose centre is in $\mathscr{D}_{\mathscr{S}'}(\mathscr{S})$ and whose axis passes through $\mathscr{D}_{\mathscr{S}'}(\mathscr{S})$.

Theorem 10.7. If $\mathscr{A}$ is a translation plane with a derivation set $\mathscr{S}$ then $\mathscr{D}_{\mathscr{S}'}(\mathscr{A})$ is also a translation plane.

Proof. Let X be a point in $\mathscr{S}'$, then $\mathscr{A}$ is (X, l_∞)-transitive. But, by Corollary 1 to Theorem 10.6, each (X, l_∞)-elation α induces a collineation α^* of $\mathscr{D}_{\mathscr{S}'}(\mathscr{A})$. Since α fixes every line of $\mathscr{A}$ through X and since each line of $\mathscr{A}$ through X is also a line of $\mathscr{D}_{\mathscr{S}'}(\mathscr{A})$, α^* is an (X, l'_∞)-translation of $\mathscr{D}_{\mathscr{S}'}(\mathscr{A})$. Hence $\mathscr{D}_{\mathscr{S}'}(\mathscr{A})$ is (X, l'_∞)-transitive for all X in $\mathscr{S}'$, and so, by Theorem 4.19, $\mathscr{D}_{\mathscr{S}'}(\mathscr{A})$ is a translation plane. ☐

We now set an exercise which improves Corollary 1 of Theorem 10.6 and gives an alternative proof to Theorem 10.7.

Exercise 10.1. Show that if α is a translation of $\mathscr{A}$ then α induces a translation of $\mathscr{D}_{\mathscr{S}'}(\mathscr{A})$. (Note that in the proof of the last theorem we showed this if the centre of α is in $\mathscr{S}'$.)

3. Desarguesian Planes

Of course there is no reason why an arbitrary plane should have a derivation set, but the desarguesian planes do.

Theorem 10.8. $\mathscr{A}_2(q^2)$ *always has a derivation set.*

Proof. Let $\mathscr{S}$ be the subset of all points (b) on l_∞, $b \in GF(q)$, plus (∞). We must show that if X and Y are distinct points of $\mathscr{A} = \mathscr{A}_2(q^2)$ and if $XY \cap l_\infty \in \mathscr{S}$, then X, Y and $\mathscr{S}$ are in a Baer subplane of $\mathscr{A}$. Since the translations of $\mathscr{A}$ fix $\mathscr{S}$ and are transitive on the points of $\mathscr{A}$, we may assume that $X = (0, 0)$. The line XY passes through $\mathscr{S}$ and the group of (X, l_∞)-homologies fixes $\mathscr{S}$ and is transitive on the points of XY, other than X. Thus we may assume that Y is any point on a line joining $(0, 0)$ to a point of $\mathscr{S}$ and, in particular, that Y lies in the subplane of $\mathscr{A}$ coordinatized by $GF(q)$ and thus, since $\mathscr{A}_2(q)$ is a Baer subplane of $\mathscr{A}_2(q^2)$, the theorem is proved. □

Now that we know that desarguesian planes are derivable we are immediately faced with the problem of determining their derived planes. As a first step in this direction we prove

Theorem 10.9. *If $\mathscr{A}$ is desarguesian then, unless $\mathscr{A}$ has order 4, $\mathscr{D}_{\mathscr{S}'}(\mathscr{A})$ is not.*

Proof. Suppose $\mathscr{A}$ is desarguesian of order q^2, with $\mathscr{S}$ as a derivation set. Every Baer subplane of $\mathscr{A}$ is isomorphic to $\mathscr{A}_2(q)$, and so we may coordinatize $\mathscr{A}$ so that (∞) is in $\mathscr{S}$ and (b) is in $\mathscr{S}$ if, and only if, b is in $GF(q)$.

But the shear σ_a:

$$(x, y) \to (x, -ax + y)$$
$$(m) \to (m + a)$$
$$[m, k] \to [m + a, k]$$
$$[k] \to [k]$$

with a in $GF(q)$, fixes $\mathscr{S}$ and, consequently, induces a collineation of $\mathscr{D}_{\mathscr{S}'}(\mathscr{A})$. Since the axis of σ_a is $[0]$ it induces a Baer collineation of $\mathscr{D}_{\mathscr{S}'}(\mathscr{A})$ (see Corollary 2 to Theorem 10.6). Furthermore all the shears σ_a, $a \in GF(q)$, will induce Baer collineations with the same Baer subplane $\mathscr{D}_{\mathscr{S}'}([0])$. If $\mathscr{D}_{\mathscr{S}'}(\mathscr{A})$ is desarguesian then the group of collineations fixing a given Baer subplane pointwise has order 2 (see Theorem 2.12). Thus, since the group of all shears σ_a is isomorphic to the additive group of $GF(q)$, $q = 2$ and $\mathscr{A}$ has order 4. ☐

In view of the last two theorems we now have a way of constructing non-desarguesian planes. Our next problem is to identify them. But first we set two exercises.

Exercise 10.2. Let K be a right quasifield of order n^2, and let $\mathscr{A}$ be a dual translation plane coordinatized by K. If $\mathscr{A}$ has a derivation set $\mathscr{S}$ containing (∞), then show that $\mathscr{D}_{\mathscr{S}'}(\mathscr{A})$ has a Baer subplane fixed pointwise by a group of collineations of order n. (Hint: try to imitate the proof of Theorem 10.9, first finding a subgroup of $(K, +)$ of order n.)

Exercise 10.3. Ket $\mathscr{S}$ and $\mathscr{T}$ be derivation sets for an affine plane $\mathscr{A}$, with $\mathscr{S} \cap \mathscr{T} = \emptyset$. Show that $\mathscr{T}$ is a derivation set for $\mathscr{D}_{\mathscr{S}'}(\mathscr{A})$, and that $\mathscr{D}_{\mathscr{S}'}\mathscr{D}_{\mathscr{T}'}(\mathscr{A}) = \mathscr{D}_{\mathscr{T}'}\mathscr{D}_{\mathscr{S}'}(\mathscr{A})$.

As a result of Exercise 10.3 (whose proof is easy, by the way), we can speak of multiple derivation leading to $\mathscr{D}_{\mathscr{S}_1', \mathscr{S}_2', \ldots, \mathscr{S}_r'}(\mathscr{A})$, where $\mathscr{S}_1, \mathscr{S}_2, \ldots, \mathscr{S}_r$ are mutually disjoint derivation sets. Note that the order in which the $\mathscr{S}_i$ are written is irrelevant and so, by Theorem 10.4, if two were the same they could be deleted.

Before determining the plane $\mathscr{D}_{\mathscr{S}'}(\mathscr{A})$ of Theorems 10.8 and 10.9, we first look for other derivation sets in $\mathscr{A} = \mathscr{A}_2(q^2)$. If we wish to show that a given set $\mathscr{S}$ of special points is a derivation set, then, as in the proof of Theorem 10.8, it is sufficient to show that there is just one Baer subplane of $\mathscr{A}$ which belongs to $\mathscr{S}$. This leads to a simple restriction on $\mathscr{S}$.

Lemma 10.10. *In $\mathscr{A}_2(q^2)$, $\mathscr{S}$ is a derivation set if, and only if, in some coordinatizing system $\mathscr{S}$ consists of (∞) plus the points (b) where $b \in GF(q)$.*

Proof. Refer back to the proof of Theorem 10.9. ☐

Corollary. *If $\mathscr{S}$ and $\mathscr{T}$ are derivation sets for $\mathscr{A} = \mathscr{A}_2(q^2)$, then $\mathscr{D}_{\mathscr{S}'}(\mathscr{A}) \cong \mathscr{D}_{\mathscr{T}'}(\mathscr{A})$.*

Now let $K = GF(q^2)$, $F = GF(q)$ and let σ be the automorphism of K given by $x^\sigma = x^q$ for all x in K. Then we know that $x^{1+\sigma} \in F$, and in fact that every element of F has the form $x^{1+\sigma}$ for some x in K (see Result 1.7). So for each t in F^*, we can define in $\mathscr{A} = \mathscr{A}_2(q^2) = \mathscr{A}_2(K)$ the set $\mathscr{S}_t$:

$$\mathscr{S}_t = \{(x) \text{ on } l_\infty \mid x^{1+\sigma} = t\} .$$

Theorem 10.11. *$\mathscr{S}_t$ is a derivation set.*

Proof. Using homogeneous coordinates with l_∞ as $z = 0$, $\mathscr{S}_t$ consists of the points $(x, y, 0)$ such that $(y/x)^{1+\sigma} = t$, or $y^{1+\sigma} = tx^{1+\sigma}$ (see Chapter II). The set $\mathscr{S}$ of the proof of Theorem 10.8 consists of the points $(a, b, 0)$, where a, b are in F. Choose an element k in K such that $k^\sigma \neq k$ and let α

be the collineation of $\mathscr{A}$ defined by the matrix:

$$\begin{vmatrix} k^\sigma & sk & 0 \\ 1 & s & 0 \\ 0 & 0 & 1 \end{vmatrix}$$

where $s^{1+\sigma} = t$. (The determinant of this matrix is not zero since $k^\sigma \neq k$.) Then it is straightforward to verify that $\cdot\alpha$ maps $\mathscr{S}$ onto $\mathscr{S}_t$, and this proves the theorem. $\square$

It is obvious that $\mathscr{S}_t$ and $\mathscr{S}_u$ are disjoint if $t \neq u$, and we are now in a position to perform multiple derivation on the desarguesian plane.

Theorem 10.12. Let $K = GF(q^2)$, $F = GF(q) \subset K$ and $t_1, t_2, \ldots, t_r$ be distinct elements of F^*. If $\mathscr{A} = \mathscr{A}_2(K)$ then $\mathscr{D}_{\mathscr{S}_{t_1}, \ldots, \mathscr{S}_{t_r}}(\mathscr{A})$ is an André plane.

Proof. We prove some auxilliary results first.

(a) If $a \neq 0$ in K, then $x^{q-1} = a$ has 0 or $q - 1$ solutions.

Proof. If both g and h in K are solutions, that is $g^{q-1} = h^{q-1} = a$, then clearly $(gh^{-1})^{q-1} = 1$. But since K^* is a cyclic group of order $q^2 - 1$, it has a unique cyclic subgroup of order $q - 1$, and so $x^{q-1} = 1$ has $q - 1$ solutions; it is now easy to see that each of the elements gx satisfies $(gx)^{q-1} = a$, and there are no other solutions.

(b) If $a \neq 0$ in K, if b, c are elements of K, then

$$ax^q + bx + c = 0$$

has 0, 1, or q solutions.

Proof. If the equation has more than one solution, say g and h are solutions; then $ag^q + bg + c = 0 = ah^q + bh + c$, and so

$$a(g-h)^q + b(g-h) = 0.$$

But $ax^q + bx = 0$ has the solution $x = 0$, and all other solutions are solutions of $x^{q-1} = -ba^{-1}$. So $g - h$ is a solution of this last equation, hence it has $q - 1$ solutions; if y is one of these, then $g + y$ is a solution of $ax^q + bx + c = 0$, and conversely.

Suppose first that no t_i is equal to 1; we consider the André plane defined as follows: $K = GF(q^2)$, $F = GF(q) < K$, and the mapping μ (see Chapter IX, Section 3) from F into Z_2 is given by: $\mu(t_1) = \mu(t_2) = \cdots = \mu(t_r) = 1$, $\mu(k) = 0$ otherwise. (We do the other case at the end.) The lines of the André plane are:

(i) points (x, y) such that $mx + y = k$, if $m^{q+1} \neq t_i$;

(ii) points (x, y) such that $mx^q + y = k$, if $m^{q+1} = t_i$.

We shall show that the lines of the second type are, in the original plane $\mathscr{A}$, Baer subplanes whose sets of points on $[\infty]$ consist of exactly $\mathscr{S}_{t_i}$. This will finish the proof in this case.

So we are given m in $\mathscr{S}_{t_i}$, and k, and wish to consider the set of points (x, y) in $\mathscr{A}$ satisfying $mx^q + y = k$; i.e., the set of points $(x, k - mx^q)$, as x ranges over F. If a line of $\mathscr{A}$ contains two such points, let the line be $[a, b]$ and the points $(x, k - mx^q)$ and $(y, k - my^q)$. So $ax + k - mx^q = b = ay + k - my^q$; but we now have an equation of the type of (b) above, with at least two solutions, and so it has exactly q solutions. Thus our line contains exactly q points in the distinguished set. So the set consists of q^2 points, every two points are on a common line and each of these common lines contains exactly q points of the set. So the set is an affine plane (see Exercise 3.14). Its points at infinity are the points (a) satisfying $ax + k - mx^q = ay + k - my^q$, or $a(x - y) = m(x - y)^q$, or $a = m(x - y)^{q-1}$. But then $a^{q+1} = m^{q+1}(x - y)^{q^2-1} = t_i$, since $(x - y)^{q^2-1} = 1$ for all x, y if $x \neq y$. So (a) is in $\mathscr{S}_{t_i}$. Since at least $q + 1$ points on $[\infty]$ must be on lines joining two points of our special set, all $q + 1$ of the points of $\mathscr{S}_{t_i}$ will occur in this way.

Now if one of the t_i is 1, but some element of F^* is not a t_j, then the above method does not work, since we are required in the André construction to define $\mu(1) = 0$. Here we shall find a collineation which maps the collection of the sets $\mathscr{S}_t$ onto itself (that is, permutes them) but which sends the particular sets $\mathscr{S}_{t_1}, \mathscr{S}_{t_2}, ..., \mathscr{S}_{t_r}$ onto another collection of r such $\mathscr{S}_t$, none of which have $t = 1$. To do this, choose a w in F^* such that $w \neq t_i$ for any i, and choose an element d in K such that $d^{q+1} = w$. Then let α be the collineation of $\mathscr{A}$ given by:

$$\alpha: (x, y) \to (d_x^{-1}, y),$$

$$(m) \to (md),$$

$$[m, k] \to [md, k]$$

$$[k] \to [d^{-1}k].$$

Then it is easy to see that α permutes the derivation sets $\mathscr{S}_t$ among themselves and sends $\mathscr{S}_w$ onto $\mathscr{S}_1$. So the images $\mathscr{S}_{s_1}, \mathscr{S}_{s_2}, ..., \mathscr{S}_{s_r}$ of the original derivation sets $\mathscr{S}_{t_i}$ do not include $\mathscr{S}_1$. Now the first part of the proof assures us that we can multiply derive and get an André plane, which is isomorphic to that one derived on the original $\mathscr{S}_{t_i}$.

Finally we must consider the case that we are multiply deriving on the set of all the $\mathscr{S}_t$. None of the above techniques works, but in fact we demonstrate that in this case the derived plane is isomorphic to the desarguesian plane $\mathscr{A}$. To do this we find a mapping of the desarguesian plane $\mathscr{A}$ onto itself which is a permutation of the points and maps the lines not through (0) or (∞) onto Baer subplanes; we will then leave it as a simple exercise for the student to demonstrate that these Baer subplanes are exactly all the subplanes belonging to all the derivation sets $\mathscr{S}_t$. The desired mapping is $(x, y) \to (x, y^q)$. A line $[m, k]$ of $\mathscr{A}$ consists of the points $(x, k - mx)$, as x ranges over K, and under the mapping above, this becomes the point set $\{(x, k^q - m^q x^q)\}$. But (see above in this proof) this is exactly a Baer subplane, and belongs to $\mathscr{S}_t$, where $t = m^{q+1}$. $\square$

We now analyse the plane $\mathscr{D}_{\mathscr{S}'}(\mathscr{A})$, where $\mathscr{A} = \mathscr{A}_2(q^2)$, and $\mathscr{S}$ is the derivation set of Theorem 10.8. We already know that this plane is an André plane (by Theorem 10.12), but we shall show that it is also a Hall plane. Anticipating this result, we denote it by $\mathscr{A}_H$. Let m be a line of $\mathscr{A}$ which meets l_∞ in a point of $\mathscr{S}$, then the subgroup of Aut$\mathscr{A}$ which fixes $\mathscr{S}$ and consists of perspectivities with axis m is merely the subgroup of $P\Gamma L_2(q)$ consisting of perspectivities with axis m and centres on l_∞. This group, which we denote by Λ, has order $q(q-1)$. The group induced on $\mathscr{A}_H$ by Λ fixes a Baer subplane pointwise and, since any Baer subplane is a maximal subplane, only the identity can fix a point not in the Baer subplane. Thus, as $\mathscr{S}'$ has exactly $q(q-1)$ points, the group induced by Λ is transitive (and even regular) on $\mathscr{S}'$.

Next let X be an affine point of $\mathscr{A}$, or of $\mathscr{A}_H$ since the point sets are the same. If Σ denotes the group of (X, l_∞)-homologies of $\mathscr{A}$ then Σ is cyclic of order $q^2 - 1$ and, since it is transitive on the affine points on any line through X (other than X itself of course), Σ induces a transitive permutation group on the Baer subplanes which belong to $\mathscr{S}$ and contain X. Thus Σ induces on $\mathscr{A}_H$ a collineation group which is (i) transitive on lines through X which meet $\mathscr{S}^*$, and (ii) fixes every line through X which meets $\mathscr{S}'$. There are $q+1$ lines of $\mathscr{A}_H$ which pass through X and meet $\mathscr{S}^*$ and so, if Σ' is the subgroup of Σ fixing one of them, $|\Sigma'| = q-1$ (by Result 1.13). But Σ is abelian and so, by Result 1.17, it induces a regular permutation group on $\mathscr{S}^*$, which implies that Σ' fixes every point of $\mathscr{S}^*$. Hence Σ' induces a group of (X, l'_∞)-homologies of $\mathscr{A}_H$ of order $q-1$. But by Lemma 7.9 this must be a subgroup of the multiplicative group of the kernel of $\mathscr{A}_H$, and so the kernel of $\mathscr{A}_H$ is at least $GF(q)$. Thus, since we already know that $\mathscr{A}_H$ is non-desarguesian for $q \neq 2$, the kernel of $\mathscr{A}_H$ is $GF(q)$, if $q \neq 2$. This brief discussion proves most of the following theorem.

Theorem 10.13. *Let Q be a quasifield which coordinatizes $\mathscr{A}_H$ with (∞) and (0) in $\mathscr{S}^*$, and (-1) in $\mathscr{S}^*$ and let $|Q| > 4$. Then Q has dimension two over its kernel K and the automorphism group of Q fixing every element of K is transitive on the elements of Q not in K.*

Proof. We have proved that Q has dimension two over K, since we have shown $|K| = q$ while $|Q| = q^2$. The group of homologies, of order $q-1$, coming from K^* are inherited from homologies of $\mathscr{A}$, and hence fix every line through the origin in $\mathscr{A}$; thus they fix every Baer subplane belonging to $\mathscr{S}^*$ in $\mathscr{A}_H$. Now we coordinatize so that (-1) is in $\mathscr{S}^*$, which means that the line joining $(0,0)$ to $(1,1)$ passes through $\mathscr{S}^*$. The $q-1$ images of $(1,1)$ under the homology group are all in the Baer subplane $\mathscr{B}$ determined by $(0,0)$ and $(1,1)$ and belonging to $\mathscr{S}^*$.

Now there is also a "kernel subplane", that is the subplane $\mathscr{K}$ of all points in $\mathscr{A}_H$ both of whose coordinates are in the kernel K; by considerations of order, $\mathscr{K}$ is a Baer subplane, and contains the point $(1,1)$

as well as $(0, 0)$, (0), and (∞). It also must contain the $q-1$ points on $[-1, 0]$ which are images of $(1, 1)$ under the group of homologies above (since these are really induced by multiplication by elements of the kernel K). So $\mathscr{K}$ and $\mathscr{B}$ intersect in (0), (∞), $(0, 0)$ and q points on the line $[-1, 0]$. This is enough to guarantee that $\mathscr{K} = \mathscr{B}$.

Now the subgroup Λ of Baer collineations fixing every point of $\mathscr{B}$ has order $q(q-1)$. No non-identity element of Λ can fix any point on the special line outside of $\mathscr{S}^*$, so Λ must be transitive on the special points not in S^*. But it is also clear that Λ is induced by a group of automorphisms of Q (since it fixes $(1, 1)$) and this group fixes every element of K (since $\mathscr{K} = \mathscr{B}$). It is immediate that Λ is transitive (and regular) on the elements of Q not in K. ⬜

Corollary. $\mathscr{A}_H$ is a Hall plane.

Proof. See the Corollary to Theorem 9.2. ⬜

Exercise 10.4. Generalize the proof of Theorem 10.12 to show: if the elements of F^* are partitioned into two sets $t_1, \ldots, t_r$ and $t_{r+1}, \ldots, t_{q-1}$, then the multiply derived plane on $\mathscr{S}_{t_1}, \ldots, \mathscr{S}_{t_r}$ is isomorphic to the multiply derived plane on $\mathscr{S}_{t_{r+1}}, \ldots, \mathscr{S}_{t_{q-1}}$.

4. General Derivation

Before studying the derived (or multiply derived) desarguesian planes in more detail, we prove in this section that the class of derivable planes is quite large.

Theorem 10.14. *Let $\mathscr{A}$ be an affine plane of order q^2 coordinatized by a PTR (R, T) satisfying:*

(a) *(R, T) has a sub-PTR (F, T) which is isomorphic to $GF(q)$;*

(b) *if $\alpha \in F$ and $x, y \in R$, then $T(\alpha, x, y) = \alpha x + y$;*

(c) *$(R, +)$ is an abelian group;*

(d) *if $a \in R^*$, then in either of the equations $\alpha a = \beta(\delta a)$ or $\alpha a = \beta a + \delta a$ if two of α, β, δ are in F then so is the third;*

(e) *if $a, b \in R$, $a - b \neq 0$, then in the equation $\alpha(a - b) = \beta a - \beta b$ if one of α, β is in F then so is the other.*

Then if $\mathscr{S}$ is the set of points (α), with α in F, plus the point (∞), $\mathscr{S}$ is a derivation set. If a, b, c are in R, $a \neq 0$, then the Baer subplanes belonging to $\mathscr{S}$ are (as point sets) the sets $\mathscr{B}(a, b, c)$ where $\mathscr{B}(a, b, c)$ is the set of all points $(\alpha a + b, \beta a + c)$ as α and β vary over F.

Proof. First we show that each $\mathscr{B}(a, b, c)$ is the set of points of a Baer subplane. Clearly $\mathscr{B}(a, b, c)$ contains exactly q^2 points.

We shall prove the theorem by showing that each line contains exactly q points and appealing to Exercise 3.14 (since the lines of $\mathscr{B}(a, b, c)$ are the various subsets which are collinear in $\mathscr{A}$, any two points must be on a unique common line). Let $X = (\alpha_1 a + b, \beta_1 a + c)$ and $Y = (\alpha_2 a + b, \beta_2 a + c)$ be any two points of $\mathscr{B}(a, b, c)$. If $\alpha_1 = \alpha_2$ then X and Y are on the line $[\alpha_1 a + b]$ which clearly contains the q points $(\alpha_1 a + b, \beta a + c)$ of $\mathscr{B}(a, b, c)$. If $\alpha_1 \neq \alpha_2$, then X and Y are certainly on a line $[m, k]$ of $\mathscr{A}$; we want to show that m is in F so that the line XY will have a "linear" equation.

By (d), the elements α, β, δ in R which satisfy $\alpha a + \alpha_2 a = \alpha_1 a$, $\beta a + \beta_1 a = \beta_2 a$ and $\delta(\alpha a) = \beta a$ are all in F. Furthermore, by (e), there is an element λ in F such that

$$\lambda(\alpha_1 a + b) - \lambda(\alpha_2 a + b) = \delta[(\alpha_1 a + b) - (\alpha_2 a + b)] \,.$$

Now it is easy to show that

$$\lambda(\alpha_1 a + b) + (\beta_1 a + c) = \lambda(\alpha_2 a + b) + (\beta_2 a + c) = k$$

say. Thus X and Y are on the line $[\lambda, k]$ which meets l_∞ in $\mathscr{S}$. This line contains the point $(\alpha a + b, \beta a + c)$ of $\mathscr{B}(a, b, c)$ if and only if $\lambda(\alpha a + b) + \beta a + c = k$. But for each α in F, there will be a unique element β in R such that

$$\begin{aligned}\beta a &= k - \lambda(\alpha a + b) - c \\ &= \lambda(\alpha_1 a + b) + \beta_1 a + c - \lambda(\alpha a + b) - c \\ &= \lambda(\alpha_1 a + b) - \lambda(\alpha a + b) + \beta_1 a \,.\end{aligned}$$

Now $\lambda(\alpha_1 a + b) - \lambda(\alpha a + b) = \delta((\alpha_1 a + b) - (\alpha a + b))$ has a solution for δ in F; thus $\lambda(\alpha_1 a + b) - \lambda(\alpha a + b) = \delta_1(\alpha_1 a - \alpha a)$ for some δ_1 in F. By (e) there is an ε in F such that $\delta_1(\alpha_1 a - \alpha a) = \varepsilon(\alpha_1 a) - \varepsilon(\alpha a)$ and finally, by (d), there are elements η, η_1 in F such that $\varepsilon(\alpha_1 a) = \eta_1 a$ and $\varepsilon(\alpha a) = \eta a$.

The above equation for β now becomes

$$\beta a = \eta_1 a - \eta a + \beta_1 a \,.$$

But $\eta_1 a - \eta a = \mu a$ for some μ in F, by (d), and, again using (d), $\mu a + \beta_1 a = \beta a$ implies that β is itself in F. So for each α in F the element β of R such that $\lambda(\alpha a + b) + \beta a + c = k$ is also in F, and hence, if the line $[\lambda, k]$ contains a point of $\mathscr{B}(a, b, c)$ it contains exactly q such points. Thus, by Exercise 3.14 $\mathscr{B}(a, b, c)$ is the set of points of a Baer subplane.

It only remains to show that any two points P, Q of $\mathscr{A}$ are in a unique $\mathscr{B}(a, b, c)$ if the line PQ meets l_∞ in $\mathscr{S}$. Since they cannot lie in two such subplanes, it will suffice to show that they lie in at least one. The condition that PQ meets l_∞ in $\mathscr{S}$ is exactly that the line which joins them is a line $[\lambda, k]$ with λ in F, or a line $[k]$. Let $P = (x_1, y_1)$ and $Q = (x_2, y_2)$; then the line PQ is $[k]$ if and only if $x_1 = x_2 = k$, and we leave it as an exercise to show that there is a $\mathscr{B}(a, b, c)$ containing P and Q in this case.

Now suppose that $x_1 \neq x_2$, but that both points lie on $[\lambda, k]$. Then from (b) $\lambda x_1 + y_1 = \lambda x_2 + y_2 = k$. By (e) there is a μ in F such that $\mu(x_1 - x_2) = \lambda x_1 - \lambda x_2$ and now it is easy to see that both P and Q are in $\mathscr{B}(x_1 - x_2,$

x_2, y_1) since:

$$(x_1, y_1) = (1 \cdot (x_1 - x_2) + x_2, 0(x_1 - x_2) + y_1),$$
$$(x_2, y_2) = (0(x_1 - x_2) + x_2, \mu(x_1 - x_2) + y_1).$$

The latter equation follows from

$$\mu(x_1 - x_2) + y_1 = \lambda x_1 - \lambda x_2 + y_1 = k - \lambda x_2 = y_2. \quad \square$$

Corollary 1. *If Q is a right quasifield of dimension two over its kernel and $\mathscr{A}$ is its affine plane, then $\mathscr{A}$ is derivable.*

Corollary 2. *If D is a division ring of dimension two over its left or middle nucleus and $\mathscr{A}$ is its affine plane, then $\mathscr{A}$ is derivable.*

Exercise 10.5. Using the notation of Theorem 10.14, show that $\mathscr{B}(a, b, c)$ and $\mathscr{B}(d, e, f)$ are equal if and only if $F^*a = F^*d$, $F^*a + b = F^*d + e$ and $F^*a + c = F^*d + f$; and that $\mathscr{B}(a, b, c)$ and $\mathscr{B}(d, e, f)$ are "parallel" if and only if $F^*a = F^*d$.

The derived plane of the second corollary will, of course, be a translation plane, but it is easy to see that those obtained from the first corollary are not translation planes in general; in fact they are not dual translations planes either. This gives us further examples (see Chapter IX) of planes which have no translation lines or points. It is even possible to show that the Hughes planes are derivable, and their derived planes form yet another new class of planes (with the exception of order 9).

Theorem 10.14 gives conditions for a given plane to be derivable, and if we add still further conditions then it is possible to compute the PTR for the derived plane.

Theorem 10.15. *Let $\mathscr{A}$ be an affine plane coordinatized by a PTR satisfying:*

(a) (R, T) *has a sub-PTR (F, T) which is isomorphic to $GF(q)$;*
(b) *if α is in F, and x, y are in R then $T(\alpha, x, y) = \alpha x + y$;*
(c) $(R, +)$ *is an abelian group;*
(d) *if α, β are in F and a is in R then $(\alpha + \beta) a = \alpha a + \beta a$;*
(e) *if α is in F and a, b are in R then $\alpha(a + b) = \alpha a + \alpha b$;*
(f) *if α, β are in F and a is in R then $\alpha(\beta a) = (\alpha\beta) a$.*

If t is any element of R not in F, then every element of R has a unique representation as $\alpha t + \beta$ for some α, β in F. Then if $\mathscr{S}$ is the set of points (α), α in F, plus (∞), $\mathscr{S}$ is a derivation set for $\mathscr{A}$ (as in Theorem 10.14) and $\mathscr{A}_1 = \mathscr{D}_{\mathscr{S}}(\mathscr{A})$ can be coordinatized so that a point $(\alpha_1 + \alpha_2 t, \beta_1 + \beta_2 t)$ in $\mathscr{A}$ (with the α_i, β_i in F), has coordinates $(\alpha_1 + \beta_1 t, \alpha_2 + \beta_2 t)'$ in $\mathscr{A}_1$. Also the new ternary T_1, the new addition $\oplus$ and the new multiplication $*$ satisfy:

(1) $x \oplus y = x + y$ *for all x, y in R;*
(2) (F, T_1) *is a sub-PTR of (R, T_1) and is isomorphic to (F, T);*

(3) *if λ is in F and x is in R then $\lambda * x = \lambda x$;*
(4) *$\mathscr{S}^* = \mathscr{D}_{\mathscr{S}'}(\mathscr{S})$ consists of the points $(\lambda)'$, λ in F, and (∞);*
(5) *if λ is in F and x, y are in R then $T_1(\lambda, x, y) = \lambda x + y = \lambda * x \oplus y$.*

(Here we are writing $(x, y)'$, $(m)'$ etc. for the points of $\mathscr{A}_1$ (or the points of its special line) in the new coordinate system.)

Proof. We coordinatize $\mathscr{A}_1$ so that $[0, 0]'$ is the Baer subplane $\mathscr{B}(1, 0, 0)$, and $[0]'$ is $\mathscr{B}(t, 0, 0)$. Then we label the points of $[0, 0]'$ so that $(\alpha + \beta t, 0)'$ is (α, β) and we put $(0, 1)' = (t, 0)$. (Note that this is all the choice we have in setting up the coordinate system for $\mathscr{A}_1$.) Since $(1, 0)' = (1, 0) = (0(t-1) + 1, 0(t-1) + 0)$ and $(0, 1)' = (t, 0) = (1(t-1) + 1, 0(t-1) + 0)$, the line $[1, 1]'$ is $\mathscr{B}(t - 1, 1, 0)$. Thus, by Exercise 10.5, the lines $[1, k]'$ have the form $\mathscr{B}(t - 1, b, c)$. A point $(\alpha + \beta t, 0)'$ is on $\mathscr{B}(t - 1, b, c)$ if (α, β) is in $\mathscr{B}(t - 1, b, c)$, and thus $(\alpha + \beta t, 0)'$ is in $\mathscr{B}(t - 1, \alpha, \beta)$. The point $(0, \alpha + \beta t)'$ will be that point on $\mathscr{B}(t, 0, 0)$ which is also on the line of the form $[1, k]'$ which passes through $(\alpha + \beta t, 0)'$, i.e., the line $\mathscr{B}(t - 1, \alpha, \beta)$. Thus $(0, \alpha + \beta t)'$ is a point $(\gamma t, \delta t)$ such that γt is in $F(t - 1) + \alpha$ and δt is in $F(t - 1) + \beta$. But $\alpha t = \alpha(t - 1) + \alpha$ and $\beta t = \beta(t - 1) + \beta$, and so $(0, \alpha + \beta t)'$ is $(\alpha t, \beta t)$.

A line in $\mathscr{A}$ parallel to $[0]'$ will have the form $\mathscr{B}(t, x, y)$ and one parallel to $[0, 0]'$ will be of the form $\mathscr{B}(1, u, v)$. Clearly $(\alpha + \beta t, \gamma + \delta t)$ is on both $\mathscr{B}(t, \alpha, \gamma)$ and $\mathscr{B}(1, \beta t, \delta t)$. But $\mathscr{B}(t, \alpha, \gamma)$ meets $[0, 0]'$ in $(\alpha, \gamma) = (\alpha + \gamma t, 0)'$ while $\mathscr{B}(1, \beta t, \delta t)$ meets $[0]'$ in $(\beta t, \delta t) = (0, \beta + \delta t)'$. Thus $(\alpha + \beta t, \gamma + \delta t)' = (\alpha + \gamma t, \beta + \delta t)$ proving the first part of the theorem.

To determine the operation $\oplus$ we use the fact that if $k = (\alpha + \beta t) \oplus (\gamma + \delta t)$ then $[1, k]'$ passes through $(\alpha + \beta t, \gamma + \delta t)'$. A line $[1, k]'$ is of the form $\mathscr{B}(t - 1, b, c)$ and contains $(\alpha + \beta t, \gamma + \delta t)' = (\alpha + \gamma t, \beta + \delta t)$ if and only if $\alpha + \gamma t$ is in $F(t - 1) + b$ and $\beta + \delta t$ is in $F(t - 1) + c$. Since $\alpha + \gamma t = \alpha(t - 1) + (\alpha + \gamma)$ and $\beta + \delta t = \beta(t - 1) + (\beta + \delta)$, we have $b = \alpha + \gamma$ and $c = \beta + \delta$. But now $\mathscr{B}(t - 1, \alpha + \gamma, \beta + \delta)$ meets $[0]' = \mathscr{B}(t, 0, 0)$ in $(\eta t, \mu t)$, where ηt is in $F(t - 1) + \alpha + \gamma$ and μt is in $F(t - 1) + \beta + \delta$. Solving for η and μ gives $\eta = \alpha + \gamma$ and $\mu = \beta + \delta$, and so $(0, k)' = (0, (\alpha + \gamma) + (\beta + \delta) t)'$. Hence $(\alpha + \beta t) \oplus (\gamma + \delta t) = (\alpha + \gamma) + (\beta + \delta) t$ and (1) is proved.

If α is any element of F then the line $[\alpha, 1]'$ passes through $(0, \alpha)' = (\alpha t, 0)$ and $(1, 0)' = (1, 0)$. Since $(\alpha t, 0) = (1(\alpha t - 1) + 1, 0(\alpha t - 1) + 0)$ and $(1, 0) = (0(\alpha t - 1) + 1, 0(\alpha t - 1) + 0)$, $\mathscr{B}(\alpha t - 1, 1, 0)$ contains both $(0, \alpha)'$ and $(1, 0)'$ and, consequently, is the line $[\alpha, 1]'$. Hence by Exercise 10.5 the affine lines through $(\alpha)'$ are of the form $\mathscr{B}(\alpha t - 1, b, c)$. To evaluate $k = \alpha * (\beta + \gamma t)$ we must find $[\alpha, k]'$ passing through $(\beta + \gamma t, 0)' = (\beta, \gamma)$. But (β, γ) is in $B(\alpha t - 1, \beta, \gamma)$ and hence we want to find a point of the form $(0, \eta + \mu t)' = (\eta t, \mu t)$ on $B(\alpha t - 1, \beta, \gamma)$. Straightforward verification shows that $((\alpha \beta) t, (\alpha \gamma) t) = (\beta(\alpha t - 1) + \beta, \gamma(t - 1) + \gamma)$ so that $\eta = \alpha \beta$, $\mu = \gamma \delta$ is the solution and $\alpha * (\beta + \gamma t) = \alpha \beta + (\alpha \gamma) t$ which proves (3).

The proof of (4) is now a matter of routine computation and we leave it to the reader; instead we pass on to (5), which will include (2). To prove (5) we choose points $(\lambda)'$ and $(\alpha + \beta t, \gamma + \delta t)'$ and find the point $(0, \eta + \mu t)'$

which is collinear with them. Imitating the proof of (3), it is an easy exercise to show that $\eta + \mu t = \lambda(\alpha + \beta t) + \gamma + \delta t$, and this proves (5). ⬚

Exercise 10.6. Prove (4) and complete the proof of (5) in Theorem 10.15.

We already know that the derived plane of a derivable translation plane is again a translation plane. In this case, if we knew the rule for general multiplication, the preceding theorem would completely specify the new quasifield.

Theorem 10.16. *If the PTR (R, T) of Theorem 10.15 is a quasifield, the multiplication in (R, T_1) is given by:*

$$(\alpha + \beta t)*(\gamma + \delta t) = -\eta - \mu t, \quad \text{if } \beta \neq 0$$

where

(1) $(\varrho + \sigma t)(-1 + \alpha t) = -\beta t$,

(2) $(\varrho + \sigma t)(\gamma + \eta t) = -\delta - \mu t$,

and by

$$a*(\gamma + \delta t) = \alpha\gamma + (\alpha\delta) t, \quad \text{if } \beta = 0.$$

Proof. The line $[\alpha + \beta t, 0]'$ is on $(0, 0)' = (0, 0)$ and on $(-1, \alpha + \beta t)' = (-1 + \alpha t, \beta t)$. If $\beta \neq 0$, then $[\alpha + \beta t, 0]'$ is not a line of the form $\mathscr{B}(x, y, z)$ (since $(0, 0)$ is on $\mathscr{B}(x, y, z)$ if and only if $\mathscr{B}(x, y, z) = \mathscr{B}(w, 0, 0)$ for some w, and $(-1 + \alpha t, \beta t)$ is on no line of the form $\mathscr{B}(w, 0, 0)$ if $\beta \neq 0$). So $[\alpha + \beta t, 0]'$ is a line $(\varrho + \sigma t, 0]$, and $\varrho + \sigma t$ must satisfy $(\varrho + \sigma t)(-1 + \alpha t) + \beta t = 0$. Then $(\gamma + \delta t, \eta + \mu t)' = (\gamma + \eta t, \delta + \mu t)$ is on $[\varrho + \sigma t, 0]$ if and only if $(\varrho + \sigma t) \cdot (\gamma + \eta t) + \delta + \mu t = 0$. ⬚

We now apply the last theorem to the desarguesian plane $\mathscr{A} = \mathscr{A}_2(q^2)$. We already know that the result of deriving $\mathscr{A}$ is a Hall plane, no matter which derivation set we use. However we shall now show that, by choosing our derivation set correctly, we can get any Hall plane of the correct order.

Theorem 10.17. *Let $f(x) \equiv t + wx + x^2$ be any irreducible quadratic over $GF(q)$. Then it is possible to derive $\mathscr{A} = \mathscr{A}_2(q^2)$ in such a way that a quasifield for the derived plane is the Hall quasifield with multiplication*

$$(a + \lambda b)*(c + \lambda d) = ac + b^{-1} df(a) + \lambda(bc - ad - wd) \quad \text{if} \quad b \neq 0$$

and

$$a*(c + \lambda d) = ac + \lambda ad.$$

Proof. If λ is in $GF(q^2)$, but not in $GF(q)$, then λ satisfies an (irreducible) quadratic $\lambda^2 - v\lambda - u = 0$, with u, v in $GF(q)$. So $GF(q^2) = \{a + \lambda b \mid a, b \text{ are in } GF(q)\}$, and using Theorem 10.16, we have (if $b \neq 0$):

$$(a + \lambda b)*(c + \lambda d) = h + \lambda m$$

where

(1) $(r + \lambda s)(-1 + \lambda a) = -\lambda b$,

(2) $(t + \lambda s)(c - \lambda h) = -d + \lambda m$.

Now Eq. (1) yields;

$$-r + asu = 0,$$
$$ar - s + vas = -b,$$

and these can be solved for r and s to give

$$r = -abu(a^2 u + av - 1)^{-1},$$
$$s = -b(a^2 u + av - 1)^{-1}.$$

Substituting for r and s in (2) gives;

$$b(acu - hu)(a^2 u + av - 1)^{-1} = d,$$
$$b(c - ahu - hv)(a^2 u + av - 1)^{-1} = -m,$$

and solving these two equations for r and m gives;

$$h = ac - b^{-1} dg(a),$$
$$m = bc - ad - dv/u,$$

where $g(a) = a^2 + (v/u)a - 1/u$. Every irreducible quadratic over $GF(q)$ is satisfied by some element in $GF(q^2)$, and so $\lambda^2 - v\lambda - u$ is an arbitrary irreducible quadratic over $GF(q)$. Hence (see Exercise 10.7), g is also arbitrary.

It is now immediate that we can choose λ to give us any Hall quasi-field for $\mathscr{A}_H$, since $a*(c + \lambda d) = ac + \lambda ad$ follows from Theorem 10.16. □

Exercise 10.7. Show that any irreducible monic quadratic $g(x)$ over $GF(q)$ has the form $g(x) = x^2 + (v/u)x - 1/u$, where $x^2 - vx - u$ is also irreducible.

Corollary. *Any two Hall planes of the same order are isomorphic.*

In fact it is true that two infinite Hall planes are isomorphic if the quadratics which define them are equivalent. (Irreducible quadratics over a field F are equivalent if they have zeros in the same quadratic extension field.) Quite probably the converse is true, but it has not been resolved yet.

We end this section with some exercises.

Exercise 10.8. If $\mathscr{S}$ and $\mathscr{T}$ are distinct derivation sets for $\mathscr{A} = \mathscr{A}_2(q^2)$, show that $|\mathscr{S} \cap \mathscr{T}| \leq 2$.

Exercise 10.9. Show that any three special points of $\mathscr{A}_2(q^2)$ are contained in a unique derivation set.

Exercise 10.10. Show that the special points of $\mathscr{A}_2(q^2)$ form the points of an "inversive plane" with the derivation sets as circles. (An "inversive plane" is a collection of $n^2 + 1$ points, with certain subsets called *circles* satisfying: every circle has $n + 1$ points, and any three points are on a unique circle.)

5. Inherited Collineation Groups

In Theorem 10.6 we observed that if $\mathscr{S}$ is a derivation set for an affine plane $\mathscr{A}$ and if α is a collineation of $\mathscr{A}$ with $\mathscr{S}^{\alpha} = \mathscr{S}$ then α induces a collineation of $\mathscr{D}_{\mathscr{S}}(\mathscr{A}) = \mathscr{A}_1$. We call α an *inherited collineation* of $\mathscr{A}_1$ and define the *inherited collineation group* of $\mathscr{A}_1$ as the collineation group induced by all collineations of $\mathscr{A}$ leaving $\mathscr{S}$ invariant. Often a determination of the inherited group gives useful information about the derived plane. As an example we shall investigate the inherited group of the Hall plane $\mathscr{A}_H$ regarded as the derived desarguesian plane. As we shall see the inherited group is especially important in this case because (a) it is (with two exceptions) the entire collineation group of the Hall plane, and (b) it is non-soluble, with the same exceptions.

We have already seen that the inherited group of $\mathscr{A}_H$ has the following properties: (i) it has a subgroup fixing every point of $\mathscr{S}'$ and transitive on $\mathscr{S}^*$; (ii) it has a subgroup fixing every point of $\mathscr{S}^*$ and transitive on $\mathscr{S}'$.

In general the inherited group will only be a subgroup of the full collineation group of the derived plane. However, in the case of a finite Hall plane of order greater than 9 it is the full collineation group which means that the full collineation group of the Hall plane is the subgroup of $\mathrm{Aut}\,\mathscr{A}_2(q^2)$ which leaves $\mathscr{S} = \{(m)\,|\,m = \infty \text{ or } m \in GF(q)\}$ invariant.

Theorem 10.18. *The full collineation group of a Hall plane of order $q^2 > 9$ has two orbits of special points: one of length $q + 1$ the other of length $q^2 - q$.*

Proof. Let $\Pi = \mathrm{Aut}\,\mathscr{A}_H$ where $\mathscr{A}_H$ is a given Hall plane. If we consider $\mathscr{A}_H$ as $\mathscr{D}_{\mathscr{S}}(\mathscr{A}_2(q^2))$ then Π has at most two orbits of special points, namely $\mathscr{S}'$ and its complement. Thus either $\mathscr{S}'^{\Pi} = \mathscr{S}'$ or Π is transitive on the special points of $\mathscr{A}_H$. Suppose that $\mathscr{S}'^{\Pi} \neq \mathscr{S}'$; i.e., that there is an α in Π with $\mathscr{S}'^{\alpha} \neq \mathscr{S}'$. Then, $\mathscr{T} = (\mathscr{S}^*)^{\alpha}$ is a derivation set of $\mathscr{A}_H$ and $\mathscr{D}_{\mathscr{T}}(\mathscr{A}_H) \cong \mathscr{A}_2(q^2)$. This means that $\mathscr{D}_{\mathscr{T}}(\mathscr{A}_H)$ and $\mathscr{A}_2(q^2)$ are two desarguesian affine planes of order q^2 which are defined on the same affine points, and we can represent the operations of their respective coordinate systems by $\oplus$ and $\odot$ (for $\mathscr{D}_{\mathscr{T}}(\mathscr{A}_{\mathscr{H}})$) and "ordinary" addition and multiplication for the second. But note that both coordinate systems are fields. Since $|\mathscr{S} \cup \mathscr{T}| \leq 2q + 2$, the two affine planes have at least $q^2 + 1 - (2q+2) = q^2 - 2q - 1$ parallel classes of lines in common. If $q > 3$ then $q^2 - 2q - 1 > 3$ so that we may coordinatize $\mathscr{D}_{\mathscr{T}}(\mathscr{A})$ and $\mathscr{A}_2(q^2)$ with the same lines for $[0]$, $[0,0]$ and $[1,0]$. Furthermore we may assign the elements of $GF(q^2)$ to the points of $[0,0]$ in the same way in each plane. Since $[1,0]$ is the same line in each plane, (1) is the same special point and, consequently, the elements of $GF(q^2)$ are also assigned to the points of $[0]$ in an identical way in each plane. Now it is immediate (draw a picture!) that for any b, c in $GF(q^2)$, $b \oplus c = b + c$. The coordinate of the special point Q is (m) if Q is on the line joining $(1, 0)$ to $(0, m)$ and since the two planes have at least $q^2 - 2q - 1$ parallel classes in common, there are at least $q^2 - 2q - 1$ common special points which have the same coordinates in each plane. Let $\mathcal{M}$ be the set of elements of $GF(q^2)$ assigned to the common special points. Then, since in any linear PTR the product $a*b$ of any two elements is given by $(0, a*b)$ which is the intersection of $[0]$ with the line joining (a) to $(b, 0)$, $s \odot x = sx$ for all $s \in \mathcal{M}$ and all $x \in GF(q^2)$. Furthermore, since by the definition of $\mathcal{M}$, there are no other common parallel classes, $\mathcal{M} = \{s \in GF(q^2) \mid s \odot x = sx \text{ for all } x \in GF(q^2)\}$. But this implies that $\mathcal{M}$ is closed under both addition and multiplication, i.e., that $\mathcal{M}$ is a subfield of $GF(q^2)$. However no proper subfield of $GF(q^2)$ can have more than q elements and so, as $|\mathcal{M}| \geq q^2 - 2q - 1 > q$, $\mathcal{M}$ must be $GF(q^2)$. This is a contradiction thus proving the theorem. $\square$

Corollary. *The full collineation group of a finite Hall plane of order greater than 9 is the group inherited under derivation from the desarguesian plane.*

Exercise 10.11. Show that for $q > 3$ the full collineation group of a Hall plane of order q^2 is non-soluble. (For $q \leq 3$ the full collineation group of a Hall plane behaves in a very different way, although the group is still non-soluble (for details see [1]).)

We have discussed derivation in considerable detail. This is mainly because it has given many "new" translation planes and the reader should feel that he has enough machinery at his disposal to perform the routine operations necessary to construct these "new planes". We have given the derivation of the desarguesian planes special attention for two reasons; firstly because it is the simplest example and secondly because derivation gives an elegant proof that a Hall plane is an André plane and an equally elegant determination of the full collineation group of the Hall plane. All this had been known before derivation was introduced but the proofs were considerably more complex.

6. Partial Spreads

We now discuss briefly a more general construction technique, which includes derivation of translation planes but does not yet seem to have been fully exploited. It would appear that derivation should be regarded as a special case of this technique and the more general method of net replacement. (We shall not even define net replacement here. For the definitions and a detailed survey of the theory of net replacement see [2].)

If V is a vector space of dimension $2d$ over $GF(q)$ then a *partial spread* of V is a set of d-dimensional subspaces $W_1, W_2, \ldots, W_t$ such that $W_i \cap W_j = \{0\}$ for $i \neq j$. Thus a spread, as defined in Chapter VII, is a partial spread such that each vector of V is in one of the subspaces W_i. The subspaces W_i are called the *components* of the partial spread.

In the exercises at the end of Chapter VII we showed that the elements of V may be regarded as the points of a translation plane $\mathscr{A}$ whose lines are the additive cosets of the components of a spread; each component of the spread defining a parallel class of $\mathscr{A}$. If we construct a new translation plane $\mathscr{A}'$ from $\mathscr{A}$ by removing the lines of some parallel classes in $\mathscr{A}$ and replacing them by "new lines", then the spread corresponding to $\mathscr{A}'$ will have certain components in common with the spread of $\mathscr{A}$. This observation motivates the following definition; two partial spreads $W_1, W_2, \ldots, W_t$, and $U_1, U_2, \ldots, U_t$ are called *replacements* for each other if $W_1 \cup W_2 \cup \cdots \cup W_t = U_1 \cup U_2 \cup \quad \cup U_t$ and $W_i \neq U_j$ for any i or j. Either one of the partial spreads is said to be *replaceable*.

Exercise 10.12. If $W_1, W_2, \ldots, W_s$ is a partial spread and if $U_1, U_2, \ldots, U_t$ is a replacement for $W_1, W_2, \ldots, W_t$ $(t \leqq s)$ show that $U_1, U_2, \ldots, U_t, W_{t+1}, \ldots, W_s$ is a partial spread.

Let $W_1, \ldots, W_{q^d+1}$ be a spread and let $\mathscr{A}$ be the translation plane obtained from the W_i. If any partial spread of the given spread is replaceable, then replacing that partial spread by any one of its replacements will give a new spread and, hence, a new translation plane defined on the same points and having the parallel classes of the non-replaced components in common with $\mathscr{A}$. The reader should convince himself that derivation is a particular instance of this construction with $d = 2$. Examples with $d \neq 2$ are provided by the André planes of non-square order, but no other value has received the same detailed attention that has been given to $d = 2$.

References

As we said before, the object of this chapter was to give a specific example of a general construction. We hope that the reader will be sufficiently interested to consult Ostrom's survey article [2].

Apart from the discussion of the collineation group of the Hall plane of order 9, the chapter is completely self-contained. The determination of this particular collineation group is in [1].

1. Hughes, D. R.: Collineation groups of non-desarguesian planes 1. The Hall Veblen-Wedderburn systems. Am. J. Math. **81**, 921–938 (1959).
2. Ostrom, T. G.: Finite translation planes. Lecture notes in mathematics, 158. Berlin-Heidelberg-New York: Springer 1971.

XI. Free Closures

1. Introduction

We want to consider here certain "free" methods by which projective planes can be generated. This leads us to consider problems about embedding configurations in planes, and about collineations. The basic ideas are very simple, and were first introduced in 1943 by Hall: let any configuration generate successively bigger and bigger configurations by introducing new elements each to be the "meet" of exactly two old elements which did not yet have an element in common. The union of the process will (usually) be a projective plane, called the *free closure* of the given configuration. This enables us to embed arbitrary configurations in projective planes and it enables us to construct projective planes with, for instance, only trivial collineations. We can characterize, under a finiteness condition, projective planes which are free closures, and we can decide, again under the same finiteness condition, when two configurations have isomorphic free closures; the collineation groups of these free closures are very difficult to identify, but in some cases it is easy to prove some surprising "Sylow-like" theorems about their finite subgroups (i.e. every finite subgroup is contained in a maximal finite subgroup, and all these maximal finite subgroups are conjugate). We say a little about homomorphisms and observe that polarities are very unlike the desarguesian case.

2. Configurations

A set $\mathscr{C}$ of points and lines, with an incidence relation, is a *configuration* if every pair of distinct points of $\mathscr{C}$ are incident with at most one line of $\mathscr{C}$. By Exercise 3.2 this condition implies its own dual so that any two lines of $\mathscr{C}$ intersect in at most one point of $\mathscr{C}$. We shall say that a configuration $\mathscr{C}$ is .*closed* if it satisfies the definition of Chapter III, that is, every pair of distinct elements of the same type are incident with a common element. A closed configuration is a projective plane if and only if it contains a quadrangle, and it is known (see Exercise 3.1) which closed configurations are not projective planes. Such closed configurations are called *degenerate*. A configuration $\mathscr{A}$ is called a *sub-configuration* of a configuration $\mathscr{B}$ if the points and lines of $\mathscr{A}$ are all points and lines of $\mathscr{B}$ and if, for any point P and line m of $\mathscr{A}$, P is on m in $\mathscr{B}$ if and only if P is on m in $\mathscr{A}$. (This prevents the over-configuration $\mathscr{B}$ from changing the incidences between elements already in $\mathscr{A}$.) If $\mathscr{A}$ is a sub-configuration of $\mathscr{B}$ we write $\mathscr{A} < \mathscr{B}$. A configuration $\mathscr{C}$ is called *confined* if every one of its elements is incident with at least three elements, and $\mathscr{C}$ is *open* if none of its sub-configurations is confined.

Exercise 11.1. Show that a confined configuration has at least seven points and seven lines, and find a confined configuration with exactly seven points and seven lines.

The following is an example of a confined configuration with eight points and eight lines: the points are the symbols $0, 1, \ldots, 7$, and the lines are the point sets $\{0, 1, 2\}$, $\{5, 6, 7\}$, $\{0, 6, 3\}$, $\{5, 1, 4\}$, $\{1, 7, 3\}$, $\{6, 2, 4\}$, $\{0, 7, 4\}$, $\{5, 2, 3\}$.

Exercise 11.2. Show that the example above is a sub-configuration of $\mathscr{A}_2(3)$.

Almost as an immediate consequence of the definitions we have:

Lemma 11.1. *Let* $\{\mathscr{A}_\alpha\}$ *be a collection of sub-configurations of a configuration* $\mathscr{C}$*. Then*

(i) *if the* $\mathscr{A}_\alpha$ *are all open then* $\bigcap_\alpha \mathscr{A}_\alpha$ *is open,*

(ii) *if the* $\mathscr{A}_\alpha$ *are all confined then* $\bigcup_\alpha \mathscr{A}_\alpha$ *is confined.*

Proof. Both claims are easy; for instance, if all the $\mathscr{A}_\alpha$ are confined then each element of the union has at least three incidences in some particular $\mathscr{A}_\alpha$ and, therefore, certainly has three incidences in $\bigcup_\alpha \mathscr{A}_\alpha$. $\square$

In view of Lemma 11.1. we define the *confined core*, or the *core* $\mathscr{K} = \mathscr{K}(\mathscr{C})$ of a configuration $\mathscr{C}$ to be the union of all the finite confined sub-configurations of $\mathscr{C}$. Clearly $\mathscr{K}(\mathscr{C})$ need not be finite but it is uniquely defined for any given $\mathscr{C}$. A configuration $\mathscr{C}$ is said to be *openly finite* (which we abbreviate to o.f.), if the number of elements of $\mathscr{C}$ not in $\mathscr{K}(\mathscr{C})$ is finite. A very important concept in the study of o.f. configurations is that of open rank: If $\mathscr{C}$ is an o.f. configuration with confined core $\mathscr{K}$ then the *open rank* $r_0(\mathscr{C})$ is defined as follows:

$$r_0(\mathscr{C}) = 2(j_1 + j_2) - i,$$

where

$j_1 =$ number of points of $\mathscr{C}$ not in $\mathscr{K}$,

$j_2 =$ number of lines of $\mathscr{C}$ not in $\mathscr{K}$,

$i =$ number of flags (X, m) where at least one of X or m is not in $\mathscr{K}$. (A *flag* is an incident point-line pair.)

In order to be sure that this definition makes sense it is necessary to show that i is, in fact, a finite integer: if i is infinite then some element of $\mathscr{C}$ not in $\mathscr{K}$ must have infinitely many incidences with elements of $\mathscr{K}$; suppose the point P is such an element. If l, m, n are any three lines of $\mathscr{K}$ which contain P then, by the definition of $\mathscr{K}$, each of these lines is contained in a finite confined configuration and, by Lemma 11.1., the union of these three finite confined configuration is, again, finite and confined. Adding P to this union gives a new configuration which is clearly finite and, since the only new element has at least three incidences, confined. But this implies that this new configuration is in $\mathscr{K}$ and therefore, in particular, that P is in $\mathscr{K}$. This contradiction shows that i is finite. Note that in fact we have shown that in an o.f. configuration each element outside of $\mathscr{K}$ has at most two incidences with elements of $\mathscr{K}$.

3. Generation

If $\mathscr{A} < \mathscr{P}$ where $\mathscr{A}$ is a configuration and $\mathscr{P}$ is a closed configuration then the set $\{\mathscr{C}_\alpha\}$ of closed configurations in $\mathscr{P}$ which contain $\mathscr{A}$ is non empty. Thus we define the *configuration generated by* $\mathscr{A}$ *in* $\mathscr{P}$, written $\langle \mathscr{A} \rangle_{\mathscr{P}}$ to be the intersection of all the closed sub-configurations of $\mathscr{P}$ which contain $\mathscr{A}$. We shall also call $\langle \mathscr{A} \rangle_{\mathscr{P}}$ the *closure* of $\mathscr{A}$ in $\mathscr{P}$; if there is no danger of ambiguity we shall omit the "in $\mathscr{P}$" and the subscript and write $\langle \mathscr{A} \rangle$. The terminology suggests that $\langle \mathscr{A} \rangle_{\mathscr{P}}$ is closed; we shall prove that this is so and then examine the way in which $\mathscr{A}$ actually generates $\langle \mathscr{A} \rangle_{\mathscr{P}}$. But first, notice that if $\mathscr{A}$ is a quadrangle and $\mathscr{P}$ is $\mathscr{P}_2(2)$ then $\langle \mathscr{A} \rangle_{\mathscr{P}} = \mathscr{P}_2(2)$, while if $\mathscr{A}$ is a quadrangle and $\mathscr{P}'$ is $\mathscr{P}_2(3)$ then $\langle \mathscr{A} \rangle_{\mathscr{P}'} = \mathscr{P}_2(3)$; so that $\langle \mathscr{A} \rangle_{\mathscr{P}}$ obviously depends on $\mathscr{P}$. We shall make statements like $\langle \mathscr{A} \rangle$ is closed to mean $\langle \mathscr{A} \rangle_{\mathscr{P}}$ is closed for all $\mathscr{P}$ and write $\langle \mathscr{A} \rangle$ when the property of $\langle \mathscr{A} \rangle_{\mathscr{P}}$ under discussion is independent of $\mathscr{P}$.

Lemma 11.2. $\langle \mathscr{A} \rangle$ *is closed.*

Proof. Suppose X and Y are distinct points of $\langle \mathscr{A} \rangle$. Then every closed configuration which contains $\mathscr{A}$ contains a line joining X and Y. Thus the line XY is in $\langle \mathscr{A} \rangle$. Similarly any distinct lines of $\langle \mathscr{A} \rangle$ intersect in a point of $\langle \mathscr{A} \rangle$ so that $\langle \mathscr{A} \rangle$ is a closed configuration. $\square$

Now how does $\mathscr{A}$ generate $\langle \mathscr{A} \rangle_{\mathscr{P}}$? Intuitively, we intersect (in $\mathscr{P}$) any pair of like elements in $\mathscr{A}$ which do not meet in $\mathscr{A}$ and we adjoin all these new elements to $\mathscr{A}$; we then repeat this process over and over again and finally take the union of all the configurations obtained. This, it seems reasonable to hope, should lead to a closed sub-configuration of $\mathscr{P}$ containing $\mathscr{A}$. Certainly any closed sub-configuration of $\mathscr{P}$ which contains $\mathscr{A}$ must include all the elements obtained in this manner. We now formalize this intuitive process.

If $\mathscr{C}$ is a sub-configuration of a configuration $\mathscr{C}_1$ such that every element of $\mathscr{C}_1$ which is not in $\mathscr{C}$ is incident with at least two elements of $\mathscr{C}$, than we say that $\mathscr{C}_1$ is a *one-step extension* of $\mathscr{C}$. (If $\mathscr{C}$ and $\mathscr{C}_1$ are both sub-configurations of a configuration $\mathscr{P}$ then we call $\mathscr{C}_1$ a one-step extension of $\mathscr{C}$ in $\mathscr{P}$. However, as will be apparent from the definitions, it is not always necessary to assume the existence of an over-configuration $\mathscr{P}$.) If $\mathscr{C} = \mathscr{C}_0 < \mathscr{C}_1 < \mathscr{C}_2 < \cdots < \mathscr{C}_n$ is a sequence of configurations such that each $\mathscr{C}_{i+1}$ is a one-step extension of $\mathscr{C}_i$, then we say that $\mathscr{C}_n$ is an *n-step* extension of $\mathscr{C}$. Often we shall merely say that $\mathscr{C}_n$ is an *extension* of $\mathscr{C}$. Clearly $\mathscr{C}$ is an n-step extension of itself for any n; to avoid having to worry about this we shall usually be interested in some special types of extensions which we now define. If $\mathscr{C}_1$ is a one-step extension of $\mathscr{C}$ such that every pair of distinct like elements in $\mathscr{C}$ always meet in $\mathscr{C}_1$ then we call $\mathscr{C}_1$ a *full* one-step extension of $\mathscr{C}$. *Full n-step extensions* are defined in the obvious way. The point of this is that in a full extension we have made sure that everything we could add on has been added; thus, while every configuration is an extension of itself, a configuration is a full extension of itself if and only if it is closed. The next special type of extension is: if $\mathscr{C}_1$ is a one-step extension of $\mathscr{C}$ such that every element of $\mathscr{C}_1$ not in $\mathscr{C}$ is incident with exactly two elements of $\mathscr{C}_1$ then $\mathscr{C}_1$ is a *free one-step extension* of $\mathscr{C}$. Again *free n-step extensions* (or *free extensions*) are defined accordingly, and a *free full* extension is the appropriate combination of the two. It is important to note that an arbitrary configuration $\mathscr{C}$ in a closed configuration $\mathscr{P}$ may have no free extensions in $\mathscr{P}$ at all excepting, trivially, $\mathscr{C}$ itself. We are now in a position to formalize the intuitive construction for $\langle \mathscr{A} \rangle_{\mathscr{P}}$.

Theorem 11.3. *If $\mathscr{C}$ is a sub-configuration of a closed configuration $\mathscr{P}$ and if $\mathscr{C} = \mathscr{C}_0 < \mathscr{C}_1 < \mathscr{C}_2 < \cdots < \mathscr{C}_n < \cdots$ is a sequence of full extensions of $\mathscr{C}$ in $\mathscr{P}$, then $\langle \mathscr{C} \rangle_{\mathscr{P}} = \bigcup_i \mathscr{C}_i$.*

Proof. Since the elements of $\mathscr{C}_{i+1}$ are either in $\mathscr{C}_i$ or are joins of elements of $\mathscr{C}_i$ they must lie in every closed configuration containing $\mathscr{C}_i$. Thus $\bigcup_i \mathscr{C}_i < \langle \mathscr{C} \rangle_{\mathscr{P}}$. But clearly $\bigcup_i \mathscr{C}_i$ is closed and, therefore, must contain $\langle \mathscr{C} \rangle_{\mathscr{P}}$. $\square$

Exercise 11.3. Let F be a field and let K be the prime field contained in F. If $\mathscr{P} = \mathscr{P}_2(F)$ and if $\mathscr{C}$ is an arbitrary quadrilateral in $\mathscr{P}$, show that $\langle \mathscr{C} \rangle_{\mathscr{P}}$ is isomorphic to $\mathscr{P}_2(K)$.

If $\mathscr{C} < \mathscr{P}$ and if each full n-step extension $\mathscr{C}_n$ of $\mathscr{C}$ in $\mathscr{P}$ is also free, then we say that $\langle \mathscr{C} \rangle_{\mathscr{P}}$ is a *free closure* of $\mathscr{C}$, and write $\mathscr{F}(\mathscr{C}) = \langle \mathscr{C} \rangle_{\mathscr{P}}$. The existence of free closures is guaranteed by:

Theorem 11.4. *If $\mathscr{C}$ is a configuration, then there exists a closed configuration $\mathscr{P}$ containing $\mathscr{C}$ such that the closure of $\mathscr{C}$ in $\mathscr{P}$ is a free closure. If $\mathscr{P}_1$ and $\mathscr{P}_2$ are closed configurations, both containing $\mathscr{C}$, in which the closure of $\mathscr{C}$ is free, then the two closures are isomorphic.*

Proof. We shall construct $\mathscr{P}$ so that it will, itself, be $\mathscr{F}(\mathscr{C})$. Let $\mathscr{C} = \mathscr{C}_0$, and define a sequence $\mathscr{C}_i$ of configurations as follows: if $\mathscr{C}_i$ is given, then $\mathscr{C}_{i+1}$ will consist of all the elements of $\mathscr{C}_i$ (with their incidences) and in addition for each pair X and Y of two like elements of $\mathscr{C}_i$ which are not incident with a common element in $\mathscr{C}_i$, we define an element XY, which is unlike X and Y and is incident with X and Y only, and is to be in $\mathscr{C}_{i+1}$. It is then easy to see that each $\mathscr{C}_i$ is an i-step full free extension of $\mathscr{C}$, and if we let $\mathscr{P} = \bigcup_i \mathscr{C}_i$, then $\mathscr{P}$ is closed, and $\langle \mathscr{C} \rangle_{\mathscr{P}} = \mathscr{F}(\mathscr{C}) = \mathscr{P}$. The uniqueness of $\mathscr{F}(\mathscr{C})$, up to isomorphism, is easy; we omit the proof. []

Note that, as a result of Theorem 11.4., when we are discussing free extensions we may forget the over-configuration in which the extensions are taken.

Suppose $\mathscr{C}$ is contained in the closed configuration $\mathscr{P}$ and let $\mathscr{B} = \langle \mathscr{C} \rangle_{\mathscr{P}}$; let X be an element of $\mathscr{B}$. Then since $\mathscr{B}$ is the union of full extensions of $\mathscr{C}$, there is an integer n such that X is in the full n-step extension $\mathscr{C}_n$ of $\mathscr{C}$, but not in $\mathscr{C}_{n-1}$ (if X is in $\mathscr{C}$, then $n = 0$). We define n to be the $\mathscr{C}$-*stage* of X, written $s_{\mathscr{C}}(X) = n$. In addition, we define the $\mathscr{C}$-*socle* $\mathscr{S}_{\mathscr{C}}(X)$ as follows: if an element is in $\mathscr{C}$, then its $\mathscr{C}$-*socle* consists of itself alone, while if $s_{\mathscr{C}}(X) = n > 0$, then X is incident with a certain set $\{y_\alpha\}$ of elements of $\mathscr{C}_{n-1}$, at least two in number. Then $\mathscr{S}_{\mathscr{C}}(X) = X \cup \left\{ \bigcup_\alpha \mathscr{S}_{\mathscr{C}}(y_\alpha) \right\}$. I.e., the $\mathscr{C}$-socle of X is itself plus its set of ancestors, and might depend on $\mathscr{C}$, just as stage does. If the configuration $\mathscr{C}$ is clear, we sometimes speak merely of the socle.

Exercise 11.4. Show that in $\mathscr{F}(\mathscr{C})$ every element not in $\mathscr{C}$ has exactly two incidences inside its own socle. Conversely if every element of $\langle \mathscr{C} \rangle$ not in $\mathscr{C}$ has exactly two incidences inside its own socle, show that $\langle \mathscr{C} \rangle = \mathscr{F}(\mathscr{C})$. (Hence in $\mathscr{F}(\mathscr{C})$ every element not in $\mathscr{C}$ has exactly two incidences with elements of lower stage; all its other incidences are with elements of higher stage.)

Earlier we had defined degenerate and non-degenerate closed configurations. As a consequence of this last exercise we now extend this definition to arbitrary configurations. A configuration $\mathscr{C}$ is *degenerate* or *non-degenerate* according as its free closure $\mathscr{F}(\mathscr{C})$ is a degenerate or non-degenerate projective plane.

Exercise 11.5. If $\mathscr{C}$ is degenerate then show that $\mathscr{F}(\mathscr{C})$ is itself an n-step extension of $\mathscr{C}$, and $n < 3$. Conversely, if $\mathscr{F}(\mathscr{C})$ is an n-step extension of $\mathscr{C}$ for some n, show that either $\mathscr{C}$ is degenerate or $\mathscr{C}$ is a projective plane.

Lemma 11.5. *If $\mathscr{A}$ is a finite confined configuration contained in $\mathscr{F}(\mathscr{B})$, then $\mathscr{A} < \mathscr{B}$.*

Proof. Suppose that $\mathscr{A}$ is a finite confined configuration in $\mathscr{F}(\mathscr{B})$, but $\mathscr{A}$ is not contained in $\mathscr{B}$. Then there is an element of $\mathscr{A}$ of maximal $\mathscr{B}$-stage $n > 0$ and this element has at most two incidences with elements of lower $\mathscr{B}$-stage; since this element must have at least three incidence in $\mathscr{A}$, it has an incidence with at least one element of $\mathscr{A}$ whose $\mathscr{B}$-stage is greater than n, and this contradicts the maximality of n. ☐

Corollary. *For any configuration $\mathscr{B}$, $\mathscr{K}(\mathscr{F}(\mathscr{B})) = \mathscr{K}(\mathscr{B})$.*

Exercise 11.6. Show that there is only one confined configuration with seven points and seven lines, none with seven points and eight lines, and find one with eight points and eight lines.

*Exercise 11.7. Find all confined configurations with nine points and nine lines, and show that there are none with nine points and fewer than nine lines.

Lemma 11.6. *Suppose $\mathscr{A}$, $\mathscr{B}$, $\mathscr{C}$ are configurations, $\mathscr{A} < \mathscr{B} < \mathscr{C}$. Then if any two of the three statements below are true, so is the third:*
(a) *$\mathscr{C}$ is a free extension of $\mathscr{A}$,*
(b) *$\mathscr{C}$ is a free extension of $\mathscr{B}$,*
(c) *$\mathscr{B}$ is a free extension of $\mathscr{A}$.*

Proof. That (b) and (c) imply (a) is of course trivial.

Suppose (a) and (c) are true. Then if $\mathscr{D}$ is a maximal free extension of $\mathscr{B}$ such that $\mathscr{D} \subset \mathscr{C}$, we will be done if we can show that $\mathscr{D} = \mathscr{C}$. Suppose $\mathscr{D} \neq \mathscr{C}$ so that the addition of any element of $\mathscr{C}$ not in $\mathscr{D}$ must result in a configuration that is not a free extension of $\mathscr{B}$, hence not of $\mathscr{D}$. Choose an element X in $\mathscr{C}$, not in $\mathscr{D}$, of minimal $\mathscr{A}$-stage, and let $\mathscr{D}_1 = \mathscr{D} \cup X$. Then $\mathscr{D}_1$ can fail to be a free extension of $\mathscr{D}$ only because X is not incident with exactly two elements of $\mathscr{D}$. But X is incident with exactly two elements in $\mathscr{C}$ of lower $\mathscr{A}$-stage, and so, since X was chosen to have minimal $\mathscr{A}$-stage of all elements not in $\mathscr{D}$, these two must lie in $\mathscr{D}$. Hence X must be incident with a third element y say of $\mathscr{D}$, and this third element must have higher $\mathscr{A}$-stage than X. Thus X is one of the two elements of lower $\mathscr{A}$-stage incident with y; but since y is in $\mathscr{D}$, and $\mathscr{D}$ is a free extension of $\mathscr{A}$, the entire socle of y must lie in $\mathscr{D}$, and in particular X must. This is a contradiction.

Suppose (a) and (b) are true. Choose $\mathscr{C}_1$ to be a configuration minimal in $\mathscr{C}$ with the property of being a free extension of both $\mathscr{A}$ and $\mathscr{B}$ (by assumption $\mathscr{C}$ has this property, so such a $\mathscr{C}_1$ exists). If $\mathscr{C}_1 \neq \mathscr{B}$ then there is an element X in $\mathscr{C}_1$, not in $\mathscr{B}$, and hence certainly not in $\mathscr{A}$, which has exactly two incidences in $\mathscr{C}_1$ (i.e. an element of maximal $\mathscr{A}$-stage in $\mathscr{C}_1$ but not in $\mathscr{B}$). Now if $\mathscr{C}_2$ is $\mathscr{C}_1$ with this element deleted, then $\mathscr{C}_2$ is still a free extension of $\mathscr{A}$, and even of $\mathscr{B}$, since the deleted element must have had maximal $\mathscr{B}$-stage as well. This is a contradiction. ☐

The student should observe that finiteness conditions are not necessary in the above proof; they are really built into the assumption that $\mathscr{B}$ and $\mathscr{C}$ are free extensions in various ways, for extension always means extension of a finite number of steps. But all three configurations can be themselves infinite.

Exercise 11.8. If $\mathscr{B}$ is an extension of $\mathscr{A}$ show that $\mathscr{A}$ is finite if and only if $\mathscr{B}$ is.

The word "free" is crucial in Lemma 11.6. We now give two exercises to show what happens if freeness is not assumed.

Exercise 11.9. Show that if the word "free" is deleted in (a), (b), (c) of Lemma 11.6 then (a) and (c) imply (b) and (b) and (c) imply (a).

That (a) and (b) do not imply (c) is shown by the following:

Exercise 11.10. Let $\mathscr{A}$ be an affine plane, k and m distinct parallel lines of $\mathscr{A}$ and Y a point of $\mathscr{A}$ not on k or m. Let $\mathscr{C}$ be a configuration consisting of $\mathscr{A}$, the point X and the line n where X is on k, m and n while n passes through X and Y. Let $\mathscr{B}$ be the union of $\mathscr{A}$ and n. Then show:
 (i) $\mathscr{A} < \mathscr{B} < \mathscr{C}$ (this is trivial);
 (ii) $\mathscr{C}$ is an extension (free) of $\mathscr{A}$ and an extension of $\mathscr{B}$;
 (iii) $\mathscr{B}$ is not an extension of $\mathscr{A}$.

Lemma 11.7. *Let $\mathscr{A}$ and $\mathscr{B}$ be configurations in the closed configuration $\mathscr{P}$ such that each is contained in an extension of the other. Then there is a sub-configuration $\mathscr{C}$ of $\mathscr{P}$ which is a common extension of each. If $\mathscr{P} = \mathscr{F}(\mathscr{A}) = \mathscr{F}(\mathscr{B})$, then $\mathscr{C}$ can be chosen to be a free extension (of both $\mathscr{A}$ and $\mathscr{B}$) and to be minimal in the sense that any configuration which is a free extension of both $\mathscr{A}$ and $\mathscr{B}$ also contains $\mathscr{C}$ (and hence, in view of Lemma 11.6, $\mathscr{P}$ is a free closure of $\mathscr{C}$ as well).*

Proof. Let $\mathscr{B}_1$ be an extension of $\mathscr{B}$ which contains $\mathscr{A}$. Then since $\mathscr{A}$ has an extension which contains $\mathscr{B}$, it has one which contains $\mathscr{B}_1$; let $\mathscr{A}_1$ be such a one. Now let $\mathscr{B}_2$ be a maximal extension of $\mathscr{B}$, inside of $\mathscr{A}_1$, and suppose, if possible, that $\mathscr{B}_2 \neq \mathscr{A}_1$. Choose an element X in $\mathscr{A}_1$ which is not in $\mathscr{B}_2$ and has minimal $\mathscr{A}$-stage with this property. Then X is incident with at least two elements k, m of lower $\mathscr{A}$-stage, which must, therefore, lie in $\mathscr{B}_2$. Hence $\mathscr{B}_2 \cup X$ is an extension of $\mathscr{B}_2$, and so of $\mathscr{B}$, contained in $\mathscr{A}_1$. This proves that $\mathscr{B}_2 = \mathscr{A}_1$, i.e. that $\mathscr{A}_1$ is a common extension of $\mathscr{A}$ and $\mathscr{B}$.

Now suppose that $\mathscr{P} = \mathscr{F}(\mathscr{A}) = \mathscr{F}(\mathscr{B})$, then the extensions $\mathscr{B}_1$ and $\mathscr{B}_2$ in the argument above can be chosen to be free. But we must show that $\mathscr{B}_2 \cup X$ is a free extension of $\mathscr{B}_2$, i.e., that X has no other incidences in $\mathscr{B}_2$. However suppose that this time we choose X in $\mathscr{A}_1$, not in $\mathscr{B}_2$, but of minimal $\mathscr{B}$-stage with this property (this is possible since $\mathscr{P} = \mathscr{F}(\mathscr{B})$). Then we will find two elements in $\mathscr{B}_2$ as before, but there could be no third incident with X since any third must have X in its $\mathscr{B}$-socle and this is impossible.

Now in the case $\mathscr{P} = \mathscr{F}(\mathscr{A}) = \mathscr{F}(\mathscr{B})$, suppose we had chosen $\mathscr{B}_1$ to be the union of the $\mathscr{B}$-socles of all the elements of $\mathscr{A}$. Then the free extension $\mathscr{A}_1$ of $\mathscr{A}$ has been shown to be a free extension of $\mathscr{B}_1$, and so, by Lemma 11.6, $\mathscr{B}_1$ is itself a free extension of $\mathscr{A}$. But any free extension of $\mathscr{B}$ that includes $\mathscr{A}$ must certainly include $\mathscr{B}_1$, and clearly we could have chosen $\mathscr{A}_1 = \mathscr{B}_1$, so $\mathscr{B}_1$ is the configuration $\mathscr{C}$ of the last part of the lemma. ☐

Corollary. *If* $\mathscr{A}_1, \mathscr{A}_2, \ldots, \mathscr{A}_n$ *are configurations in the closed configuration* $\mathscr{P}$, *such that* $\mathscr{P} = \mathscr{F}(\mathscr{A}_i)$ *for each* $i = 1, 2, \ldots, n$, *and each* $\mathscr{A}_i$ *is contained in a free extension of any* $\mathscr{A}_j$, *then there is a unique configuration* $\mathscr{B}$ *which is a free extension of all the* $\mathscr{A}_i$ *such that any configuration which is a free extension of all the* $\mathscr{A}_i$ *must contain* $\mathscr{B}$.

We are now in a position to begin to attack the problem of determining when two configurations $\mathscr{A}$ and $\mathscr{B}$ have the same (or isomorphic) free closures. Certainly if there is a configuration $\mathscr{C}$ which is simultaneously a free extension of both $\mathscr{A}$ and $\mathscr{B}$, then $\mathscr{F}(\mathscr{A}) = \mathscr{F}(\mathscr{B})$; in view of Lemma 11.7 we have the converse, which we state as a theorem, in more general form.

Theorem 11.8. *Let* $\mathscr{A}$ *and* $\mathscr{B}$ *be openly finite configurations. Then* $\mathscr{F}(\mathscr{A})$ *and* $\mathscr{F}(\mathscr{B})$ *are isomorphic if and only if there is a sequence* $\mathscr{C}_1, \mathscr{C}_2, \ldots, \mathscr{C}_n$ *of configurations such that* $\mathscr{C}_1 \cong \mathscr{A}$, $\mathscr{C}_n \cong \mathscr{B}$, *and each* $\mathscr{C}_{i+1}$ *is a free extension of* $\mathscr{C}_i$ *or* $\mathscr{C}_i$ *is a free extension of* $\mathscr{C}_{i+1}$.

Proof. If such a sequence exists, then clearly $\mathscr{F}(\mathscr{C}_i) = \mathscr{F}(\mathscr{C}_{i+1})$ for each $i = 1, 2, \ldots, n-1$, and so $\mathscr{F}(\mathscr{A}) \cong \mathscr{F}(\mathscr{B})$. (Note that this half of the theorem does not even require finiteness conditions.) Now suppose that $\mathscr{A}$ and $\mathscr{B}$ are o.f., and let us assume (without loss of generality) that $\mathscr{P} = \mathscr{F}(\mathscr{A}) = \mathscr{F}(\mathscr{B})$. Then $\mathscr{K}(\mathscr{P}) = \mathscr{K}(\mathscr{A}) = \mathscr{K}(\mathscr{B})$ from the corollary to Lemma 11.5.; let us write $\mathscr{K} = \mathscr{K}(\mathscr{P})$. Since each element of $\mathscr{B}$, say, occurs in some free extension of $\mathscr{A}$, and since at most only the finitely many elements of $\mathscr{B}$ not in $\mathscr{K}$ are not already in $\mathscr{A}$, there is a free extension of $\mathscr{A}$ which contains $\mathscr{B}$, and similarly there is a free extension of $\mathscr{B}$ which contains $\mathscr{A}$. Hence by Lemma 11.7. there is a common free extension $\mathscr{C}$ and, with $\mathscr{C}_1 = \mathscr{A}$, $\mathscr{C}_2 = \mathscr{C}$, $\mathscr{C}_3 = \mathscr{B}$, this proves the theorem. ☐

Corollary. *If* $\mathscr{A}$ *and* $\mathscr{B}$ *are o.f., then* $\mathscr{F}(\mathscr{A}) \cong \mathscr{F}(\mathscr{B})$ *implies* $r_0(\mathscr{A}) = r_0(\mathscr{B})$ *and* $\mathscr{K}(\mathscr{A}) \cong \mathscr{K}(\mathscr{B})$.

Proof. That the two confined cores are isomorphic is a consequence of the corollary to Lemma 11.5. That the open ranks are the same follows from considering a single free extension of an o.f. configuration, even a one-step extension: each element added increases the number of elements by one and the number of incidences by two, hence leaves the open rank unchanged. ☐

The corollary gives us some invariants for free closures; one of our goals will be to show that the core and the open rank are a complete set of invariants in the openly finite case.

4. Classification Problems

In this section we want to investigate two problems: the first is to decide when two free closures are isomorphic (i.e., the *isomorphism problem*), and the second is to decide whether a given projective plane is a free closure or not. We cannot give answers to these problems in general, indeed none are known at this time; but with the condition of "openly finite" added on, complete solutions can be found. In view of Theorem 11.8. we define two configurations $\mathscr{A}$ and $\mathscr{B}$ to be *free equivalent* if there is a sequence $\mathscr{C}_0, \mathscr{C}_1, \mathscr{C}_2, \ldots, \mathscr{C}_n$ of configurations, such that:

(i) $\mathscr{A} \cong \mathscr{C}_0, \mathscr{B} \cong \mathscr{C}_n$,

(ii) for each $i = 0, 1, \ldots, n-1$, one of $\mathscr{C}_i, \mathscr{C}_{i+1}$ is a free extension of the other.

It is easy to see that "free equivalence" is an equivalence relation, and Theorem 11.8. can then be rephrased as: if $\mathscr{A}$ and $\mathscr{B}$ are o.f., then $\mathscr{F}(\mathscr{A}) \cong \mathscr{F}(\mathscr{B})$ if and only if $\mathscr{A}$ and $\mathscr{B}$ are free equivalent. We shall attack our first problem therefore, by finding a "canonical" configuration which is free equivalent to a given one, and noting that this canonical configuration depends only on the open rank and the core of the given configuration (if it is o.f.).

Lemma 11.9. *If $\mathscr{A}$ is a non-degenerate finite open configuration, then $\mathscr{A}$ is free equivalent to a configuration $\mathscr{W}_t$, consisting of a line k with t points on k, and two points not on k. Furthermore, $t = r_0(\mathscr{A}) - 6$.*

[Before proving the free equivalence of $\mathscr{A}$ and $\mathscr{W}_t$ we make some definitions useful in this and the next lemma. In a configuration $\mathscr{C}$ we say that an element X is "*i*-incident", $i = 0, 1, 2$, or 3, according as X has no incidences, 1 incidence, 2 incidences, or at least 3 incidences. When free equivalence is under consideration, 2-incident elements can be removed (or added) at will.]

Proof. Since $\mathscr{A}$ is non-degenerate, it has a free extension that contains a set of four points, no three collinear. This motivates us to define a *2-configuration* as a pair $(\mathscr{C}, \mathscr{Q})$, where $\mathscr{C}$ is a configuration and $\mathscr{Q} < \mathscr{C}$ is a configuration consisting of four distinct points P_1, P_2, P_3, P_4 and two distinct lines k_1, k_2; the only incidences in $\mathscr{Q}$ are to be that k_1 contains P_1 and P_2, while k_2 contains P_3 and P_4. Clearly there are 2-configurations such that $\mathscr{C}$ is free equivalent to $\mathscr{A}$, and we shall say then that the 2-configuration is free equivalent to $\mathscr{A}$. Among all the 2-configurations free equivalent to $\mathscr{A}$, choose one, $(\mathscr{C}^*, \mathscr{Q}^*)$ such that:

(i) $\mathscr{C}^*$ has the smallest possible number of lines;

(ii) among those satisfying (i), $\mathscr{Q}^*$ has the maximal number of 3-incident points (in $\mathscr{C}^*$).

Now suppose, if possible, that m is a 0-incident line in $\mathscr{C}^*$. By intersecting m with k_1 and k_2 (that is, adding the points mk_1 and mk_2) and then deleting m (now a 2-incident line), we would have a $\mathscr{Q}$-configuration which is free equivalent to $\mathscr{A}$ but has fewer lines than $\mathscr{C}^*$, violating (i). So $\mathscr{C}^*$ has no 0-incident lines. Next, suppose m is a 1-incident line, incident with the point X say. If X is not on both k_1 and k_2, then suppose X is not on k_2; we adjoin the point mk_2 and delete m; again this gives a $\mathscr{Q}$ configuration free equivalent to $\mathscr{A}$ but with fewer lines than $\mathscr{C}^*$ which violates (i), so m did not exist. Suppose X is the intersection point k_1k_2. If there is already a line on P_1 and P_3, let it be y; otherwise add the line $y = P_1P_3$. In either case adjoin the point $Z = ym$ and delete m; if the line y had already existed this would violate (i) again, so we assume that y was added as the union of P_1 and P_3. If we adjoin the line $w = ZP_2$, and delete Z; then we can delete y. Now the situation is that instead of the 1-incident line m, we have the 1-incident line w, but incident with the point P_2; since P_2 is not on k_2 this case has already been shown to lead to a contradiction, and so we may assert:

(1) $\mathscr{C}^*$ contains no 0- or 1-incident lines, and all lines other than k_1 and k_2 are 3-incident.

The assertion follows from the fact that $\mathscr{C}^*$ could contain no 2-incident lines other than perhaps the lines k_1 and k_2. (Otherwise we could delete them to get a free equivalent $\mathscr{Q}$-configuration with fewer lines.)

If $\mathscr{C}^*$ contains 0-incident points, we can "move them into a corner" for a while, and ignore them; we shall return to them later. Suppose $X \neq P_i$ is a 1-incident point, not on k_1. If X is on k_2, we add the lines $s = P_1P_3$ and $t = P_2P_4$, if necessary, then the point $Z = st$, again if necessary; then we add $ZX = u$, $X' = uk_1$. Now delete X, then delete u, and finally delete all of the elements Z, s and t which were not present and were actually added in the free extensions above. So we have "moved" X to X' on k_1. Suppose on the other hand, that X is not on k_2, but is on the line m. Since m cannot contain both of P_3 and P_4, suppose P_4 is not on m, and add the line $u = P_4X$, delete X, add $X' = uk_1$ and delete u; so again we have moved X to k_1.

Hence:

(2) Any 1-incident points can be assumed to be on k_1 or k_2, as we choose. All points not on k_1 and k_2 are 0- or 3-incident.

Now suppose that there are points not on k_1 or k_2, and not 0-incident; such points must be 3-incident. Suppose furthermore that there is a point, P_4 say, of $\mathscr{Q}^*$ which is not 3-incident. Let X be a 3-incident point not on either k_i; if there is a line $u = XP_3$, then if neither of P_1 and P_2

lie on u, we can replace the quadrilateral $\mathscr{Q}^*$ by $\mathscr{Q}'$, where we write X instead of P_4 and u instead of k_2. This contradicts (ii). So P_1, say, must lie on u; there is another line w on X, such that w does not contain P_4 (since X lies on at least 3 lines), and w is a 3-incident line. Again, we distinguish two cases: P_2 is on u or not. If it is on u, then let R be a third point on w; R is not 1-incident (since 1-incident points are on one of the k_i), so R must be 3-incident. Hence we replace $\mathscr{Q}^*$ by $\mathscr{Q}'$, where the points are P_1, P_3, P_2, R, and the lines of $\mathscr{Q}'$ are u and w. This also violates the assumption (ii). So P_2 is not on w, and we can choose a 3-incident point R on w, $R \neq X$. Then we replace both P_3 and P_4 in $\mathscr{Q}^*$ by X and R, replacing k_2 by w at the same time; this certainly increases the number of 3-incident points, and so violates (ii).

Consequently we now can suppose that there is no line $u = X P_3$. It is easy to see that there must be a line w on X which contains a point Y not on any k_i; so as above, we discard P_3, P_4, k_2, replacing them with X, Y and w. So we have:

(3) If there are any 3-incident points not on the k_i, or if there are any lines other than the k_i, then all the points P_i are 3-incident.

This follows from observing that if there were a line other than one of the k_i, then it would have to contain a 3-incident point not on either k_i.

But now we recall that $\mathscr{A}$ is open; so certainly $\mathscr{C}^*$ is open. If we disregard all the 0- and 1-incident points, then what is left cannot be confined; but if there are points off the k_i in what is left, then everything is 3-incident except perhaps the k_i; so at least one of them is 2-incident; suppose it is k_2. Then we easily see that there is a 3-incident line other than k_1 which contains neither of the points P_3 and P_4, and so we can replace $\mathscr{Q}^*$ by a configuration $\mathscr{Q}'$ which contains neither of P_3 and P_4. Then we could delete the line k_2, and hence violate (i). So there cannot be 3-incident points off of the k_i, nor are there any other lines except the k_i.

Our configuration now looks like a $\mathscr{W}_t$, except for possible 0-incident points. But let X be a 0-incident point. Add the lines $P_3 X = v$, $P_4 X = w$, delete X, add $X_1 = v k_1$, $X_2 = w k_1$, delete v and w: thus X has been "moved" into two 1-incident points on k_1. Hence $\mathscr{C}$ can be assumed to have the form stated in the lemma.

That $t = r_0(\mathscr{A}) - 6$ follows from the fact that $\mathscr{W}_t$ and $\mathscr{A}$ are free equivalent and, hence, have the same open rank. Thus, since it is uniquely determined by t, $\mathscr{W}_t$ is a canonical configuration of the type sought. $\quad\square$

Lemma 11.10. *Let $\mathscr{A}$ be an o.f. configuration whose core $\mathscr{K} = \mathscr{K}(\mathscr{A})$ is not empty. Let $\mathscr{B}$ be the configuration consisting of $\mathscr{K}$ plus $r_0(\mathscr{A})$ points on some (arbitrarily chosen) line k of $\mathscr{K}$. Then $\mathscr{A}$ is free equivalent to $\mathscr{B}$.*

Proof. Among all the o.f. configurations free equivalent to $\mathscr{A}$, choose one, $\mathscr{C}$ say, which has (1) the minimal number of 1-incident elements which are not incident with an element of $\mathscr{K}$, and (2) has no 2-incident elements. That (2) can be satisfied is evident, since any decreasing sequence of free equivalent o.f. configurations must be stationary after a certain point.

Suppose $\mathscr{C}$ has a 1-incident element not incident with $\mathscr{K}$, and let X be such a one, incident with m say. Choose Q in K and not on m, construct the configuration $\mathscr{C}_1$ consisting of $\mathscr{C}$ plus the line QX, and then the configuration $\mathscr{C}_2$ which results from deleting X from $\mathscr{C}_1$. Clearly $\mathscr{C}_2$ is free equivalent to $\mathscr{C}$, and hence to $\mathscr{A}$, and has fewer 1-incident elements not incident with $\mathscr{K}$. So all the 1-incident elements of $\mathscr{C}$ are in fact incident with elements of $\mathscr{K}$.

Suppose that X is a 3-incident element, incident with k, m, n say, and suppose that X is not in $\mathscr{K}$. Then none of k, m, n can be 1-incident elements, since then X would be their only incidence, which is impossible, as X is not in $\mathscr{K}$. So all of k, m, n are 3-incident.

Now the core of $\mathscr{C}$ must be $\mathscr{K}$. But if we let $\mathscr{L}$ be the set of all 3-incident elements of $\mathscr{C}$, then certainly $\mathscr{L}$ includes $\mathscr{K}$ as well as the finitely many 3-incident elements (as above) not in $\mathscr{K}$. But let $\mathscr{S}$ be the set of all 3-incident elements in $\mathscr{L}$ and not in $\mathscr{K}$; $\mathscr{S}$ is finite, so if we adjoin to it any finite set of elements from $\mathscr{K}$, we will still have a finite set. Each element of $\mathscr{S}$ is 3-incident; if it is not 3-incident with elements of $\mathscr{S}$, then there are elements of $\mathscr{K}$ with which it is incident. Thus if necessary we can adjoin a finite set of elements from $\mathscr{K}$ to $\mathscr{S}$, obtaining a configuration $\mathscr{T}$ in which every element of $\mathscr{S}$ is 3-incident; if there are elements of $\mathscr{T}$ which are not 3-incident in $\mathscr{T}$, then they are in $\mathscr{K}$, and are contained in finite confined configurations in $\mathscr{K}$. So adjoining these finitely many finite confined configurations, we obtain finally a finite configuration $\mathscr{V}$ which is confined and which contains $\mathscr{S}$. But this implies that $\mathscr{V}$, hence $\mathscr{S}$, in contained in $\mathscr{K}$, and hence $\mathscr{S}$ must have been empty.

So now our configuration $\mathscr{C}$ looks like this: it may have some 0-incident elements, which we can ignore for a moment, but all its other elements are either in $\mathscr{K}$ or are 1-incident and are incident with $\mathscr{K}$. Now we can "shift" all these 1-incident elements onto one line, where they will appear as points. Choose a line k in $\mathscr{K}$, and suppose that x is a 1-incident line of $\mathscr{C}$, incident with the point P in $\mathscr{K}$. If P is not on k, then we adjoin the point kx and then delete x; if P is on k, then we choose a line y in $\mathscr{K}$, y not on P, and adjoin yx, delete x, resulting in a configuration with a point on y instead of a line on P. We will be done when we can see how to shift points onto k. But to do this, we first shift them to lines through a point not on k, then (as above) to points on k. (The reader will find this all rather easy to follow if he makes a few sketches.) Now all 1-incident elements are on k. Finally the 0-incident elements can be shifted to points on k as follows: suppose X is a 0-incident point, then choose two points P and Q in K and adjoin the two lines PX and QX, then delete X; now shift to points on k, as above. Note that a 0-incident element will contribute two points on k (as it must since open rank stays fixed). Now we have proved our lemma. (The open rank of $\mathscr{B}$ will be the number of points on k but not in $\mathscr{K}$.) $\quad\Box$

Theorem 11.11. *Let $\mathscr{A}$ and $\mathscr{B}$ be o.f. configurations. Then $\mathscr{F}(\mathscr{A}) \cong \mathscr{F}(\mathscr{B})$ if and only if $r_0(\mathscr{A}) = r_0(\mathscr{B})$ and $\mathscr{K}(\mathscr{A}) \cong \mathscr{K}(\mathscr{B})$.*

Proof. The theorem follows immediately from the two preceding lemmas, plus the Corollary to Theorem 11.8 since we have seen that the "canonical form" for an o.f. configuration depends only upon its open rank and its core. $\quad\Box$

Now we are ready to attack the next problem: if $\mathscr{P}$ is openly finitely generated, when is it a free closure? Suppose $\mathscr{P} = \langle \mathscr{A} \rangle_\mathscr{P}$, where $\mathscr{A}$ is o.f.; then let $\mathscr{K} = \mathscr{K}(\mathscr{P})$, and $\mathscr{B} = \mathscr{A} \cup \mathscr{K}$. Since $\mathscr{A}$ is o.f., only finitely many of its elements are not in the core of $\mathscr{A}$, hence only finitely many are not in $\mathscr{K} > \mathscr{K}(\mathscr{A})$. So $\mathscr{B}$ is o.f., $\mathscr{P} = \langle \mathscr{B} \rangle_\mathscr{P}$, and $\mathscr{B} > \mathscr{K}(\mathscr{P})$. Among all the o.f. configurations which contain $\mathscr{K}$ and which generate $\mathscr{P}$, choose one with minimal open rank; let this one be $\mathscr{C}$. We wish to show that $\mathscr{P} = \mathscr{F}(\mathscr{C})$. So let $\mathscr{C} = \mathscr{C}_0 < \mathscr{C}_1 < \mathscr{C}_2 < \cdots$ be a sequence of full extensions in $\mathscr{P}$ of $\mathscr{C}$; $\mathscr{P}$ will be a free closure unless some $\mathscr{C}_i$ is not a free extension of the preceding term. Let n be chosen minimally so that $\mathscr{C}_{n+1}$ is not a free extension of $\mathscr{C}_n$, if that is possible. Then there is an element X in $\mathscr{C}_{n+1}$, not in $\mathscr{C}_n$, such that X has (at least) three incidences in $\mathscr{C}_{n+1}$. Let $l, k,$ and m be three elements incident with X; if there are any others, we ignore them. Each of l, k, m has at least two incidences with elements of lower $\mathscr{C}$-stage, and again, we choose two elements for each of the three. These last elements are definitely in $\mathscr{C}_n$ and not in $\mathscr{C}_{n+1}$, so have exactly two incidences with elements of lower $\mathscr{C}$-stage, and so proceeding in this manner we construct a finite set $\mathscr{D}$ of elements which is similar to (but not exactly the same as, for we may have ignored some elements at the beginning) the socle. The configuration $\mathscr{D}$ "leads back" to $\mathscr{C}$, and so if we adjoin $\mathscr{D}$ to $\mathscr{C}$ we obtain an o.f. configuration $\mathscr{E}$ which also generates $\mathscr{P}$.

Lemma 11.12. *If $\mathscr{E}$ is constructed as above, then $r_0(\mathscr{E}) < r_0(\mathscr{C})$.*

Proof. Let $\mathscr{E}_i = \mathscr{C}_i \cap \mathscr{E}$, for $i = 0, 1, 2, \ldots, n+1$. Then clearly each $\mathscr{E}_{i+1}$ is an extension of $\mathscr{E}_i$, and if $i < n$, $\mathscr{E}_{i+1}$ is even a free extension of $\mathscr{E}_i$ (for the elements of $\mathscr{E}_{i+1}$ not in $\mathscr{E}_i$ are the elements of $\mathscr{D}$ of $\mathscr{C}$-stage $i+1$, and so have exactly two incidences with elements of lower $\mathscr{C}$-stage if $i < n$). So $r_0(\mathscr{E}_n) = r_0(\mathscr{C})$. But $\mathscr{E}_{n+1}$ has a new element, namely X, with three incidences, so in passing from $\mathscr{E}_n$ to $\mathscr{E}_{n+1}$ the open rank must go down by at least one and the lemma is proved. $\quad\Box$

This lemma violates the minimality of the open rank of $\mathscr{C}$. So each $\mathscr{C}_i$ is a free extension, and $\mathscr{P} = \mathscr{F}(\mathscr{C})$.

Theorem 11.13. *If $\mathscr{P}$ is a projective plane generated by an openly finite configuration, then either $\mathscr{P}$ is its own core or there is an o.f. configuration $\mathscr{C} \neq \mathscr{P}$ such that $\mathscr{P} = \mathscr{F}(\mathscr{C})$.*

Corollary. *An o.f. generated projective plane is a free closure of an o.f. configuration unless every element of the plane is contained in a finite confined sub-configuration.*

We finish this section with some results about homomorphisms and configurations.

Theorem 11.14. *Let $\mathscr{A}$ and $\mathscr{B}$ be non-degenerate configurations such that $\mathscr{B}$ is contained in a closed configuration $\mathscr{P}$ and suppose β is a homomorphism of $\mathscr{A}$ onto $\mathscr{B}$. Then β extends to a homomorphism $\bar{\beta}$ of $\mathscr{F}(\mathscr{A})$ onto $\langle \mathscr{B} \rangle_\mathscr{P}$.*

Proof. First we show how to extend a homomorphism from $\mathscr{A}$ onto $\mathscr{B}$ to a homomorphism from $\mathscr{A}_1$ into $\langle \mathscr{B} \rangle_\mathscr{P}$, where $\mathscr{A}_1$ is a full free one-step extension of $\mathscr{A}$. For any element P in $\mathscr{A}$ define $P^{\bar{\beta}} = P^\beta$. If X is in $\mathscr{A}_1$ but not in $\mathscr{A}$, then $X = yz$, for a unique pair of elements y and z of $\mathscr{A}$. If $y^\beta \neq z^\beta$, then $X^{\bar{\beta}} = y^\beta z^\beta$ in $\langle \mathscr{B} \rangle_\mathscr{P}$ is forced upon us, while if $y^\beta = z^\beta$, then we choose $X^{\bar{\beta}}$ to be some point on y^β which is also in $\langle \mathscr{B} \rangle_\mathscr{P}$. The non-degeneracy of $\mathscr{B}$ assures us that such a point exists. Thus we can extend β to a homomorphism of $\mathscr{F}(\mathscr{A})$ into $\langle \mathscr{B} \rangle_\mathscr{P}$; but the homomorphic image of a closed configuration is obviously closed, so the extended homomorphism must be onto. $\square$

Corollary. *Every projective plane is the homomorphic image of a free closure. A finite projective plane of order n is the homomorphic image of a free closure $\mathscr{F}(\mathscr{A})$, where $\mathscr{A}$ is open and can be chosen to have open rank $\leq \sqrt{n} + 1$.*

Proof. A proof is wanted only for the last sentence. But if $\mathscr{A}$ is any configuration $\mathscr{W}_t$, where $n+1 \geq t \geq \sqrt{n} + 1$, then it is not difficult to see that a configuration isomorphic to $\mathscr{W}_t$ inside a projective plane $\mathscr{P}$ of order n must generate all of $\mathscr{P}$, since $\mathscr{W}_t$ generates a subplane of $\mathscr{P}$ with at least $t+1$ points on a line, and then (see Theorem 3.7) the subplane must be $\mathscr{P}$. $\square$

Exercise 11.11. Show that if $\mathscr{A}$ is any configuration which is not closed, then $\mathscr{F}(\mathscr{A})$ admits no central collineations at all. (Hint: a central collineation would imply the existence of desargues' configurations outside of $\mathscr{A}$. But a desargues' configuration is confined.)

5. Collineations

We shall now use the machinery developed in the previous sections to prove a number of interesting results about collineations. As always, we write $\operatorname{Aut}\mathscr{A}$ for the group of collineations of the configuration $\mathscr{A}$.

Lemma 11.15. *If $\mathscr{B}$ is a full free extension of $\mathscr{A}$ then $\operatorname{Aut}\mathscr{B}$ has a subgroup isomorphic to $\operatorname{Aut}\mathscr{A}$.*

Proof. Obviously it is only necessary to prove the lemma when $\mathscr{B}$ is a full free one-step extension of $\mathscr{A}$. Let β be a collineation of $\mathscr{A}$ and let P be any element of $\mathscr{B}$ which is not in $\mathscr{A}$. Since $\mathscr{B}$ is a free one-step extension of $\mathscr{A}$, there are exactly two elements x, y of $\mathscr{A}$ which are incident with P in $\mathscr{B}$. The fullness of the extension implies that x^β and y^β, which clearly have no common element in $\mathscr{A}$, meet in $\mathscr{B}$. Defining $P^{\bar\beta}$ as the intersection of x^β with y^β and defining $X^{\bar\beta} = X^\beta$ for all X in $\mathscr{A}$ extends β to be a collineation $\bar\beta$ of $\mathscr{B}$, and the lemma is proved. (Note that we need both fullness and freeness to prove this lemma; it is easy to give examples where the lemma is false without both of these assumptions.) []

Lemma 11.16. *If $\mathscr{C} < \mathscr{P}$ and α is in $\operatorname{Aut}\mathscr{P}$, then $\mathscr{C}$ is confined if and only if $\mathscr{C}^\alpha$ is.*

Proof. Trivial. []

Corollary. *For any α in $\operatorname{Aut}\mathscr{P}$, $[\mathscr{K}(\mathscr{P})]^\alpha = \mathscr{K}(\mathscr{P})$.*

Lemma 11.17. *If $\mathscr{P} = \langle\mathscr{C}\rangle$, and if α and β are elements of $\operatorname{Aut}\mathscr{P}$ such that $X^\alpha = X^\beta$ for all elements X in $\mathscr{C}$, then $\alpha = \beta$.*

Proof. Basically trivial, since if two collineations agree on a set of elements that define a new element in the next extension, then they must agree on that new element too. []

This corollary has the following important theorem as an almost immediate consequence.

Theorem 11.18. *If $\mathscr{C}$ is a confined configuration and $\mathscr{P} = \mathscr{F}(\mathscr{C})$, then $\operatorname{Aut}\mathscr{C} \cong \operatorname{Aut}\mathscr{P}$.*

Proof. Every collineation of $\mathscr{P}$ must send $\mathscr{C}$ to $\mathscr{C}$, since $\mathscr{C}$ is the confined core of $\mathscr{P}$; the kernel of the representation of $\operatorname{Aut}\mathscr{P}$ on $\mathscr{C}$ is the identity, from Lemma 11.17, and this proves the theorem. []

Theorem 11.19. *Let $\mathscr{A}$ be an o.f. configuration and $\mathscr{P} = \mathscr{F}(\mathscr{A})$, and let Γ be a finite subgroup of $\operatorname{Aut}\mathscr{P}$. Then there is an o.f. configuration $\mathscr{B}$ which is a free extension of $\mathscr{A}$ such that Γ fixes $\mathscr{B}$.*

Exercise 11.12. Using the corollary to Lemma 11.7, prove Theorem 11.19.

Exercise 11.13. Let $\mathscr{A}$ be an o.f. configuration and $\mathscr{P} = \mathscr{F}(\mathscr{A})$; let θ be a polarity of $\mathscr{P}$. Show that the number of absolute elements of θ not in $\mathscr{K}(\mathscr{A})$ is finite.

Now we exhibit a confined configuration which has no non-identity collineation, and thus we will have a projective plane whose full group of collineations is the identity; this is, in some sense, as far from the desarguesian case as one might imagine. Let $\mathscr{C}$ be the configuration with the ten points $0, 1, 2, \ldots, 9$ and the ten lines whose points are:

0	1	2	3		2	5	8
0	4	5			2	6	9
0	6	7			3	6	8
1	4	6			3	7	9
1	5	7			4	8	9 .

Then we can argue as follows to show that $\mathscr{C}$ has only the identity collineation: the line 0123 is the only one with four points, so is fixed by any collineation α, and since 489 is the only line not meeting 0123, it is fixed as well. Dually, 6 occurs on four lines, and is the only point with this property, so $6^\alpha = 6$, and 5 is the only point not joined to 6, so $5^\alpha = 5$. Now 067 and 157 are the only lines (other than 0123) not meeting the fixed line 489, so are fixed or interchanged; since each contains a different fixed point (5 or 6), each is fixed, so their intersection 7 is also fixed, and then the third point on each is fixed; that is, $0^\alpha = 0$ and $1^\alpha = 1$. Similarly 3 and 9 do not meet 5, so they are fixed or interchanged. But, since 3 occurs on the fixed line 0123 and 9 does not, we have $3^\alpha = 3$ and $9^\alpha = 9$. It is easy to finish the argument and show that $\alpha = 1$.

For any configuration $\mathscr{C}$ a *polarity* is defined in the obvious way; a one-to-one incidence preserving mapping of the points onto lines and lines onto points. An *absolute element* X of a polarity θ is an element such that X is incident with X^θ.

Exercise 11.14. Show that the configuration above has a unique polarity, and find its absolute points.

Exercise 11.15. If $\mathscr{C}$ is a configuration with a polarity θ show that θ can be extended to a polarity of $\mathscr{F}(\mathscr{C})$. Show also that any polarity of $\mathscr{F}(\mathscr{C})$ must be an extension of a polarity of $\mathscr{C}$, if $\mathscr{C}$ is confined.

Exercise 11.16. Show that the configuration formed from the one above by adding the line 2 4 7 has no collineations and no polarities. (Note: a new proof must be given that this configuration has no collineations!)

The configurations given above lead to examples of projective planes "without collineations" but with a polarity, and "without collineations" and also without a polarity (hence without a correlation).

Note that since the product of any two correlations is a collineation, any projective plane without non-trivial collineations can admit at most one correlation, which must be a polarity.

Exercise 11.17. If $\mathscr{C}$ is a configuration with a polarity θ and if $\bar{\theta}$ is the extension of θ to $\mathscr{C}_1$, the full free one-step extension of $\mathscr{C}$, show that all the absolute elements of $\bar{\theta}$ are in $\mathscr{C}$ and are absolute elements of θ. Hence the extension of θ to the free closure of $\mathscr{C}$ has all its absolute elements in $\mathscr{C}$.

Exercise 11.17 illustrates the sharp contrast between polarities of free closures and the situation in desarguesian planes, or for that matter in finite planes (see Chapters II and XII). We can certainly construct a finite configuration $\mathscr{C}$ which admits polarities whose absolute points do not in any way resemble an oval and then, by Exercises 11.15 and 11.17, $\mathscr{F}(\mathscr{C})$ will also admit polarities with the same absolute points.

We have already proved enough to see that collineations and polarities of free closures act in a somewhat surprising way. However, whenever certain finiteness conditions are put on the generating configuration, the situation is not completely chaotic. We now sketch some very nice structure theorems about finite subgroups of the collineation groups of free closures of o.f. configurations which are not in fact open. The situation for free closures of finite open configurations is not dissimilar, but the proofs are considerably harder, so we deal with the case of an o.f. but not open, configuration. (Many of these results were proved by O'Gorman and a more complete study of these groups can be found in [2].) Let $\mathscr{A}$ be an o.f. configuration which is not open, $\mathscr{K} = \mathscr{K}(\mathscr{A})$, $\mathscr{P} = \mathscr{F}(\mathscr{A})$. Since every element of $\operatorname{Aut}\mathscr{P}$ fixes the configuration $\mathscr{K}$, $\operatorname{Aut}\mathscr{P}$ has a representation on $\mathscr{K}$ as $\operatorname{Aut}\mathscr{P}/\Gamma$, where Γ is the normal subgroup of $\operatorname{Aut}\mathscr{P}$ consisting of those collineations which fix every element of $\mathscr{K}$. Obviously the factor group $\operatorname{Aut}\mathscr{P}/\Gamma$ is isomorphic to some subgroup of $\operatorname{Aut}\mathscr{K}$; in fact $\operatorname{Aut}\mathscr{P}$ splits over Γ and we can be more precise about the factor group.

Theorem 11.20. $\operatorname{Aut}\mathscr{P} = \Gamma \cdot \Sigma$, where Γ is the normal subgroup of $\operatorname{Aut}\mathscr{P}$ fixing every element of $\mathscr{K}$ and Σ is isomorphic to $\operatorname{Aut}\mathscr{K}$ and acts like $\operatorname{Aut}\mathscr{K}$ on $\mathscr{K}$; clearly $\Gamma \cap \Sigma = 1$.

Proof. We must show that $\operatorname{Aut}\mathscr{K}$ can be extended to a collineation group of $\mathscr{P}$, and this will prove the theorem. (Note that this theorem is also true, trivially, when $\mathscr{A}$ is open.) Suppose α is an element of $\operatorname{Aut}\mathscr{K}$, and let us use the canonical representation of $\mathscr{A}$: besides $\mathscr{K}$, $\mathscr{A}$ consists of $r = r_0(\mathscr{A})$ points $X_1, X_2, \ldots, X_r$ on a certain line k of $\mathscr{K}$ (actually k is arbitrary as well). Let $m = k^\alpha$, and choose a point P in $\mathscr{K}$ not on either k or m (this is always possible: why?). Let $\mathscr{A}$ be freely extended by adding the r lines $w_i = P X_i$, then the r points $Y_i = w_i m$ (if $k = m$, then each Y_i is X_i). Now deleting each w_i (for each one has just two incidences) and then deleting X_i, we have a new configuration $\mathscr{A}'$ which is free equivalent to $\mathscr{A}$. We map $\mathscr{A}$ onto $\mathscr{A}'$ by sending every element of $\mathscr{K}$ onto its image under α, and sending X_i to Y_i; this mapping, which we will call α, is an isomorphism from $\mathscr{A}$ to $\mathscr{A}'$, and extends to a isomorphism of $\mathscr{F}(\mathscr{A})$ onto $\mathscr{F}(\mathscr{A}')$; since these two free closures are equal, we have constructed a collineation of $\mathscr{P}$ which acts as α on $\mathscr{K}$. ☐

Now let Λ be a finite subgroup of Γ. By Theorem 11.19 there is an o.f. configuration $\mathscr{B}$, free equivalent to $\mathscr{A}$, such that Λ fixes $\mathscr{B}$. We now reduce $\mathscr{B}$, with respect to Λ as it were, in a manner very similar to that used in the proof of Lemma 11.10. If $\mathscr{B}$ has any 2-incident elements, then Λ sends 2-incident elements to 2-incident elements, so Λ equally fixes the configuration obtained from $\mathscr{B}$ by deleting all its 2-incident elements. Any 1-incident element is sent by Λ to another, and so we can deal with these as before, making sure that in $\mathscr{B}$ they are all incident with an element of $\mathscr{K}$. Then we show, just as before, that all 3-incident elements are in $\mathscr{K}$ itself, and then move all the 1-incident elements (but instead of doing it one at a time, we have to move whole Λ-orbits at a time) in such a way that each "becomes" a point on some chosen line k of K. Finally the 0-incident elements are moved to k, each one of course leading to two points on k. Thus we will have constructed a configuration on which Λ acts as a collineation group, fixing every element of $\mathscr{K}$ (note that this is crucial in the proof!) and permuting somehow the $r_0(\mathscr{A})$ $= r_0(\mathscr{B})$ points not in $\mathscr{K}$ on the line k of $\mathscr{K}$. Thus Λ is a subgroup of the symmetric group of degree $r_0(\mathscr{A})$ acting on these points. Hence:

Theorem 11.21. *Let $\mathscr{A}$ be an o.f. configuration with $\mathscr{K}(\mathscr{A}) \neq \emptyset$ and $r = r_0(\mathscr{A})$, and let Γ be the subgroup of $\operatorname{Aut}\mathscr{F}(\mathscr{A})$ fixing every element of $\mathscr{K}(\mathscr{A})$. Let Λ be a finite subgroup of Γ; then there exists a symmetric group of degree r in Γ such that Λ is a subgroup of this symmetric group.*

To completely determine the finite subgroups of $\operatorname{Aut}\mathscr{F}(\mathscr{A})$ we need to know the finite subgroups of $\operatorname{Aut}\mathscr{K}$ of course. In general this problem is unsolved, as indeed are many of the problems about the group $\operatorname{Aut}\mathscr{F}(\mathscr{A})$. But we have one more nice result about the finite subgroups of Γ:

Theorem 11.22. *Let Σ_1 and Σ_2 be symmetric groups of degree $r = r_0(\mathscr{A})$ contained in the subgroup Γ of $\operatorname{Aut}\mathscr{F}(\mathscr{A})$ (here Γ is the subgroup fixing the core, which is not empty, element-wise and $\mathscr{A}$ is o.f.). Then Σ_1 and Σ_2 are conjugate in Γ.*

Proof. From the remarks preceding Theorem 11.21 we know that there is a set of points $\{P_1, \ldots, P_r\}$ on some line l of $\mathscr{K}$ such that $\mathscr{F}(\mathscr{A}) = \mathscr{F}(\mathscr{K} \cup \{P_1, \ldots, P_r\})$ and Σ_1 induces S_r on $\{P_1, \ldots, P_r\}$. Similarly there is a set of points $\{Q_1, \ldots, Q_r\}$ on some line m of $\mathcal{K}$ such that $\mathcal{F}(\mathcal{A}) = \mathcal{F}(\mathcal{K} \cup \{Q_1, \ldots, Q_r\})$ and Σ_2 induces S_r on $\{Q_1, \ldots, Q_r\}$.

(i) Suppose $l = m$. Let α be the mapping from $\mathcal{K} \cup \{P_1, \ldots, P_r\}$ to $\mathcal{K} \cup \{Q_1, \ldots, Q_r\}$ given by $X^\alpha = X$ for all X in $\mathcal{K}$ and $P_i^\alpha = Q_i$ for $i = 1, \ldots, r$; then it is clear that α is an isomorphism between the two given generating configurations. As an interesting exercise, the reader should now mimic the proof of Lemma 11.15 to show that α extends to a collineation α of $\mathcal{F}(\mathcal{A})$ and that $\alpha^{-1}\Sigma_1\alpha = \Sigma_2$.

(ii) Suppose $l \neq m$. In this case we shall show Σ_2 also induces S_r on a set $\{R_1, \ldots, R_r\}$ of points on l such that $\mathcal{F}(\mathcal{A}) = \mathcal{F}(\mathcal{K} \cup \{R_1, \ldots, R_r\})$ and then use (i).

Let X be any point of $\mathcal{K}$ which is not incident with either l or m and define $R_i = Q_i X \cap l$ for $i = 1, \ldots, r$. Since Σ_2 fixes l, m and X, Σ_2 must induce the same group on $\{Q_1, \ldots, Q_r\}$ and $\{R_1, \ldots, R_r\}$, i.e. Σ_2 induces S_r on $\{R_1, \ldots, R_r\}$. From the definition of the R_i it is clear that they are all contained in the full free one-step extension of $\mathcal{K} \cup \{Q_1, \ldots, Q_r\}$ and, similarly, that the Q_i are all in the full free one-step extension of $\mathcal{K} \cup \{R_1, \ldots, R_r\}$. Thus $\mathcal{F}(\mathcal{K} \cup \{R_1, \ldots, R_r\}) = \mathcal{F}(\mathcal{K} \cup \{Q_1, \ldots, Q_r\}) = \mathcal{F}(\mathcal{A})$ and the theorem is proved. □

References

The original paper of Hall [1] contains an early treatment of free closures. For further results and bibliography on polarities and collineations the reader should consult [2].

1. Hall, M., Jr.: Projective planes. Trans. Am. Math. Soc. **80**, 502–527 (1955).
2. O'Gorman, S. P.: On the generation and automorphisms of projective planes. Ph.D. Thesis. London University 1971.

XII. Polarities

1. Introduction

We saw in Chapter II that polarities in pappian projective planes are of two sorts: orthogonal or unitary. These were different in the first place because one was linear and the other was semi-linear; but when they actually possessed absolute points, they could also be characterized by the structure of these absolute points inside the plane. In the finite case the absolute points told the whole story: if a polarity is orthogonal, then its number of absolute points is $n+1$, where n is the order of the plane, and in addition, these points are collinear when n is even, and form a conic when n is odd; while for a unitary polarity, n must be a square and the number of absolute points is $n^{3/2}+1$, and in addition the set of absolute points and non-absolute lines form a combinatorial structure called a unital. In this chapter we shall generalise these results by showing that if α is a polarity of a finite projective plane of order n, and if $a(\alpha)$ represents the number of absolute points of α, then always $n+1 \leqq a(\alpha) \leqq n^{3/2}+1$ (Theorems of Baer and Seib), and in addition: if $a(\alpha)=n+1$, then the absolute points are collinear when n is even and form an *oval* when n is odd; if $a(\alpha)=n^{3/2}+1$, then the absolute points and non-absolute lines form a unital. We shall see that an oval is exactly the right "non-desarguesian" generalization of a conic (Segre's Theorem). There are some other interesting results, such as the theorem that when n is not a square, then $a(\alpha)=n+1$. We give some examples of polarities in non-desarguesian planes as well.

In view of the results of Chapter XI, where we can construct infinite planes with polarities whose number of absolute points is finite but arbitrary, it is clear that polarities in arbitrary infinite planes must be very different from those in finite planes, and we will have nothing to say about the infinite case in this chapter. It would be most interesting to be able to say more about the structure of unitals however: they have within themselves a rich and complicated structure, and the existence of "non-desarguesian" unitals has been proved, both as objects in their own right, and as designs arising from unitary polarities. For it must be noted that a unital can exist independently of any projective plane, and the conditions under which an abstractly defined unital arise from a unitary polarity of a projective plane, or can even be embedded in a projective plane, are not very well understood at this time.

2. Absolute Elements of Polarities

We recall that a correlation α of a projective plane $\mathscr{P}$ is a one-to-one mapping of the points onto the lines and the lines onto the points such that A is on m if and only if m^{α} is on A^{α}. Thus a projective plane admits a correlation if and only if it is self-dual. A polarity is a correlation of order two and it is not known whether a self-dual plane must admit a polarity.

If α is a polarity of a projective plane $\mathscr{P}$ then a point A (line m), is called *absolute* if A is on A^{α} (m^{α} is on m). We begin our discussion of polarities by looking at the possibilities for the number of absolute elements and the configurations which they may form. Most of the results are due to Baer although some of the proofs are very different to those which he gave.

Lemma 12.1. *Let α be a polarity of a projective plane $\mathscr{P}$. Then every absolute point of α is on a unique absolute line and, dually, every absolute line contains a unique absolute point.*

Proof. We first make the trivial observation that, since α preserves incidence, for any pair of points A and B, A is on B^{α} if and only if B is on A^{α}.

Since the two claims of the lemma are self dual we need only prove one of them. If m is any absolute line of α then, by definition, m^{α} is on m and so m contains at least one absolute point. Put $A = m^{\alpha}$ and suppose there is a second absolute point B on m. Since B is on A^{α}, A must be on B^{α} and thus, since B is also on B^{α} (by the assumption that B is absolute), $B^{\alpha} = AB = m$. But this implies $m^{\alpha} = B$ and contradicts the assumption $B \neq A$. Thus the lemma is proved. $\square$

As an immediate consequence of Lemma 12.1 we see that a polarity of a projective plane cannot possibly have every point as an absolute point; this is in contrast to projective spaces of higher dimension (see the discussion following Theorem 2.28). In the finite case the other extreme, i.e. that a polarity has no absolute elements, is also impossible. Here the finiteness is essential; an example of a polarity of an infinite plane with no absolute points is given by the mapping α of the real plane which sends $\langle (x, y, z) \rangle$ onto the line $\langle (x, y, z)' \rangle$ (see Chapter II). A point $\langle (a, b, c) \rangle$ is absolute if and only if $a^2 + b^2 + c^2 = 0$ which, of course, has no non-trivial solution. In order to prove that a polarity of a finite plane must have absolute elements we first need to solve Exercise 3.12.

Lemma 12.2. *Let I be the $v \times v$ identity matrix, let J be the $v \times v$ matrix with each entry 1 and let $C = nI + J$, where n is a positive integer.*

The C has $n + v$ as a simple eigen root and n as an eigen root with multiplicity $v - 1$.

Proof. For $i = 1, 2, \ldots, v - 1$ define a $1 \times v$ vector $\mathbf{x}_i = (0, 0, \ldots, 1, -1, \ldots, 0)$ where the ith entry is 1, the $i + 1^{\text{st}}$ entry is -1 and all other entries are 0. Then, since $\mathbf{x}_i J = 0$ for all i, $\mathbf{x}_i C = \mathbf{x}_i n I + \mathbf{x}_i J = n \mathbf{x}_i$, so that n is an eigen root with at least $v - 1$ independent eigen vectors. Now if $\mathbf{y}$ is the $1 \times v$ vector with each entry $+1$ then $\mathbf{y} C = n \mathbf{y} + \mathbf{y} J = n \mathbf{y} + v \mathbf{y} = (n + v) \mathbf{y}$, so that $n + v$ is also an eigen root. Since the total number of eigen roots of C is at most v, C must have n as an eigen root of multiplicity $v - 1$ and $n + v$ with multiplicity 1. $\square$

Lemma 12.3. *Let α be a polarity of a finite projective plane $\mathcal{P}$ of order n. Then α has at least one absolute point. Furthermore, if n is not a square α has exactly $n + 1$ absolute points.*

Proof. Write $v = n^2 + n + 1$, label the points $X_1, X_2, \ldots, X_v$ in an arbitrary manner and then label the lines $m_1, m_2, \ldots, m_v$ so that $m_i = X_i^\alpha$ for $i = 1, 2, \ldots, v$. If A is the incidence matrix of $\mathcal{P}$ given by this labelling (i.e. $a_{ij} = 1$ if and only if X_i is on m_j), then, since X_i is on $X_j^\alpha = m_j$ precisely when X_j is on $X_i^\alpha = m_i$, A is symmetric. Furthermore the point X_i is absolute if and only if $a_{ii} = 1$ which means that the number of absolute points of α is equal to the trace of A.

By Theorem 3.12, $A A' = n I_v + J_v$ and so, by the symmetry of A, $A^2 = n I_v + J_v$. Thus, by Lemma 12.2, A^2 has n as an eigen root with multiplicity $v - 1 = n^2 + n$ and $n + v = (n + 1)^2$ as a simple eigen root. Clearly, since $|A^2 - \lambda^2 I| = |A - \lambda I| |A + \lambda I|$, any eigen root of A is a square root of an eigen root of A^2; so the possibilities for the eigen roots of A are $n + 1$, $-(n + 1)$, $\sqrt{n}$, $-\sqrt{n}$. If $\sqrt{n}$ has multiplicity r and $-\sqrt{n}$ has multiplicity s then $r + s = n^2 + n$; also A has one other simple eigen root, either $n + 1$ or $-(n + 1)$. However, since each column of A has $n + 1$ non zero entries, straightforward verification gives $n + 1$ as an eigen root of A with the $1 \times v$ vector $\mathbf{y} = (1, 1, \ldots, 1)$ as eigen vector. Since the trace of A is the sum of its eigen roots we have that the trace of A is $n + 1 + (r - s) \sqrt{n}$ which, as every entry of A is ether 0 or 1, must be an integer. If n is not a square, $n + 1 + (r - s) \sqrt{n}$ is an integer if and only if $r = s$ and so, in this case, α has exactly $n + 1$ absolute points. If n is a square then any prime divisor of $\sqrt{n}$ is also a divisor of n and, consequently, $n + 1 + (r - s) \sqrt{n} \neq 0$. Hence α always has at least one absolute point. $\square$

This lemma can be strengthened to state that α must always have at least $n + 1$ absolute points. To do this we first need to prove;

Lemma 12.4. *Let α be a polarity of a finite projective plane of order n. If m is any non-absolute line of α then the number of absolute points on m is congruent to $n + 1$ (mod 2).*

Proof. If X is any point of m then, since m is not absolute, $X^\alpha \neq m$. For any X on m define X^θ by $X^\theta = m \cap X^\alpha$; then $(X^\theta)^\theta = (X^\theta)^\alpha \cap m = (X^\alpha \cap m)^\alpha \cap m = X \cdot m^\alpha \cap m = X$. Thus θ is an involutory permutation of the points of m and, if t is the number of points fixed by θ, then $n + 1 - t \equiv 0 \pmod 2$ or $t \equiv (n + 1) \pmod 2$. But $A^\theta = A$ if and only if A is on A^α, so that t is also equal to the number of absolute points of α on m and the lemma is proved. $\square$

Corollary. *If α is a polarity of a finite projective plane $\mathscr{P}$ of even order n then every line of $\mathscr{P}$ contains at least one absolute point.*

Theorem 12.5. *Let α be a polarity of a finite projective plane $\mathscr{P}$ of order n. Then α has at least $n + 1$ absolute points.*

Proof. If $a(\alpha)$ is the number of absolute points of α then, by Lemma 12.3, $a(\alpha) \geq 1$.

Case (a). n is even.
By the corollary to Lemma 12.4, each line contains an absolute point of α. If t is the number of flags such that the point is an absolute point of α then, clearly, $t = a(\alpha)(n + 1)$. However, as every line is in at least one such flag, $t \geq n^2 + n + 1$ which gives $a(\alpha)(n + 1) \geq n^2 + n + 1 > n^2 + n$, i.e. $a(\alpha) > n$. Thus, since $a(\alpha)$ is an integer, $a(\alpha) \geq n + 1$.

Case (b). n is odd.
Since $a(\alpha) \geq 1$, let X be an absolute point of α. By Lemma 12.1 X is incident with n non-absolute lines. Since $n + 1$ is even, Lemma 12.4 implies that each of these lines contains an even number of absolute points; hence each of these n lines contains at least one further absolute point other than X and α has at least $n + 1$ absolute points. $\square$

As we saw in Chapter II, $\mathscr{P}_2(q)$ has polarities with exactly $q + 1$ absolute points so that Theorem 12.5 certainly gives the best possible lower bound for $a(\alpha)$. We now consider the situation where this lower bound is attained and show that, for arbitrary planes, the situation is very similar to the desarguesian case. In $\mathscr{P}_2(q)$ we know that if a polarity α has exactly $q + 1$ absolute points then the absolute points are either collinear (if q is even) or form a conic (if q is odd). Defining an *oval* in a finite projective plane of order n as a set of $n + 1$ points no three of which are collinear, we shall show that for an arbitrary plane of order n then, if α has exactly $n + 1$ absolute points, the absolute points are either collinear (if n is even) or form an oval (if n is odd). Clearly the points of any conic of $\mathscr{P}_2(q)$ form an oval but we shall also prove the remarkable fact that, for q odd, the points in any oval of $\mathscr{P}_2(q)$ are the points of a conic.

Theorem 12.6. *Let α be a polarity of a finite projective plane of order n. If α has exactly $n + 1$ absolute points then*
(a) *if n is even the absolute points of α are collinear,*
(b) *if n is odd the absolute points of α form an oval.*

Proof. *Case (a).* Suppose n is even.

If m is a line joining two absolute points of α then we must show that every point of m is an absolute point. We shall suppose that X is a non-absolute point on m and show that this leads to a contradiction. Let $m_1, m_2, \ldots, m_n$ be the n lines through X distinct from m and let a_i be the number of absolute points of α on m_i. Then, by the corollary to Lemma 12.4, $a_i \geq 1$ for all i. Furthermore, since X is non-absolute, the a_i absolute points on m_i are distinct from the a_j absolute points on m_j for $i \neq j$, and so the total number of absolute points not on m is $\sum_{i=1}^{n} a_i \geq n$. But there are at least two absolute points on m, and this implies that α has at least $n+2$ absolute points and gives the required contradiction.

Case (b). Suppose n is odd.

If m is a line joining two absolute points A and B we must show that m cannot contain a third absolute point. So we assume m contains a third absolute point C and show that this gives a contradiction. By Lemmas 12.1 and 12.4, A is incident with only one absolute line and each of the n non-absolute lines through A contains a further absolute point distinct from A. Thus each of the $n-1$ non-absolute lines other than m through A contains an absolute point distinct from A, B and C. But now α has at least $(n-1)+3 = n+2$ absolute points and the theorem is proved. ☐

As a direct corollary to Lemma 12.3 and Theorem 12.6 we have the following theorem which gives precise information about polarities in planes of non-square order.

Theorem 12.7. *Let α be a polarity of a finite projective plane of non-square order n. Then α has $n+1$ absolute points and*
(a) *if n is even the absolute points are collinear,*
(b) *if n is odd the absolute points form an oval.*

Clearly, by the dual arguments, the absolute lines of α must be concurrent if n is even and must be such that no three are concurrent if n is odd. When n is odd we shall show the absolute lines of α are geometrically determined by the oval. This is just one consequence of the following discussion of ovals.

3. Ovals

Let $\mathcal{O}$ be an oval in a finite projective plane $\mathcal{P}$ of order n. A line m of $\mathcal{P}$ is called a *secant, tangent* or *exterior line* of $\mathcal{O}$ if m contains 2, 1 or 0 points of $\mathcal{O}$ respectively. Note that any line of $\mathcal{P}$ is of one of these three types.

Lemma 12.8. *If A is a point of an oval $\mathcal{O}$ in a finite projective plane $\mathcal{P}$ then there is a unique tangent m to $\mathcal{O}$ such that A is on m.*

Proof. Let $\mathcal{O} = \{A, A_1, \ldots, A_n\}$. Then each of the lines $m_i = AA_i$ is a secant through A and, clearly, $m_i \neq m_j$ if $i \neq j$. Thus A has exactly n secants through it. If m is the remaining line through A then, as A is on m, m is not an exterior line and, as $m \neq m_i$ for any i, is not a secant; thus m is a tangent. $\square$

Exercise 12.1. If α is a polarity of a finite projective plane whose absolute points form an oval $\mathcal{O}$, show that the absolute lines of α are the tangents to $\mathcal{O}$.

Exercise 12.2. If $\mathcal{O}$ is an oval of a finite projective plane $\mathcal{P}$ show that $\mathcal{P}$ admits at most one polarity whose absolute points are the points of $\mathcal{O}$.

Exercise 12.3. If $\mathcal{O}$ is an oval of a finite projective plane $\mathcal{P}$ of order n, show that $\mathcal{O}$ has $n+1$ tangents, $\frac{1}{2}n(n+1)$ secants and $\frac{1}{2}n(n-1)$ exterior lines.

We have already seen that in $\mathcal{P}_2(q)$ if there are $q+1$ absolute points then they are either collinear or lie on a conic. In fact, as we claimed earlier, conics provide the only examples of ovals in $\mathcal{P}_2(q)$ for q odd.

Theorem 12.9 (Segre's theorem). *Every oval in a desarguesian plane of odd order is a conic.*

Proof. This theorem has a very elegant and simple proof which we include in the first appendix at the end of the chapter. The proof given is essentially the original one of Segre's and could have been given in Chapter 2. Its "flavour" is very different to the rest of the other chapters, in that it uses homogeneous coordinates and the approach typical of "classical" projective geometry. $\square$

Corollary. *If $\mathcal{O}$ is any oval in a desarguesian projective plane $\mathcal{P}_2(q)$ of odd order then $\mathcal{P}_2(q)$ admits a unique polarity whose absolute points are the points of $\mathcal{O}$.*

This corollary is definitely false for non-desarguesian planes. In Section 5 we shall show that any translation plane which admits a polarity must be a division ring plane. Thus any oval in a translation plane which is not a division ring plane cannot be the absolute points of a polarity. It is also well known that the Hughes planes contain ovals which are not the absolute points of a polarity.

Exercise 12.4. Find an oval in the Hughes plane of order 9.

Exercise 12.5. Try to find ovals in some of the translation planes, but not division ring planes, of small order given in Chapter VII.

From Theorem 12.6 we know that a plane of even order cannot admit a polarity whose absolute points form an oval. While it is true that conics in desarguesian planes of even order are ovals, there are ovals in these planes which are not conics (see below); thus Segre's theorem is the best possible result.

Lemma 12.10. *Let $\mathcal{O}$ be an oval in a projective plane of even order n. Then all the tangents to $\mathcal{O}$ are concurrent at a point called the knot (or nucleus) of $\mathcal{O}$.*

Proof. Let (X, m) be any flag such that X is not in $\mathcal{O}$ and m is a secant line. Since every line through X contains either 0, 1 or 2 points of $\mathcal{O}$ and the number of points of $\mathcal{O}$ is odd, at least one line through X is a tangent to $\mathcal{O}$. Let $X_1, X_2, \ldots, X_{n-1}$ be the points of m not on $\mathcal{O}$ and let h_i be the number of tangents through X_i; then we have just shown $h_i \geq 1$ for all i.

If A and B are the points of m on $\mathcal{O}$ then, by Lemma 12.8, A and B are on unique tangents a and b respectively. Each of the $n-1$ tangents to $\mathcal{O}$ other than a or b must intersect m in one of the X_i. Thus $n-1 = \sum_{i=1}^{n} h_i$ and so, since $h_i \geq 1$, we have $h_i = 1$ for $i = 1, 2, \ldots, n-1$. Since m was any secant this implies that no two tangents can intersect at a point on a secant.

Let $C = ab$. Since C cannot lie on a secant, the join of C to any point of $\mathcal{O}$ must be a tangent. Hence all the tangents to $\mathcal{O}$ are concurrent at C and, furthermore, every line through C is a tangent to $\mathcal{O}$. □

Clearly $\mathcal{O}$ together with its knot form a set of $n+2$ points such that no three are collinear and any subset of $n+1$ of these $n+2$ points form an oval. Since no two conics of a desarguesian plane can have more than four common points, it is easy to see that for $s \geq 3$ we can find an oval in $\mathscr{P}_2(2^s)$ which is not a conic.

Exercise 12.6. Let $\mathscr{P}$ be a finite projective plane of even order n. If $\mathcal{O}$ is any set of k points such that no three are collinear show that $k \leq n+2$. If $k = n+2$ show that any subset of $n+1$ of these points is an oval with the remaining one as knot.

4. Polarities of Planes of Square Order

In Theorem 12.7 we saw that polarities in non-desarguesian planes of non-square order behave in a similar manner to those of desarguesian planes. We now consider the case where the planes have square order and here very different situations may occur. We begin, however, by showing that if $\mathscr{P}$ is any projective plane of order $n = s^2$ then the number $a(\alpha)$ of absolute points of a polarity α is bounded above by $s^3 + 1$. Since in the desarguesian case $a(\alpha)$ is always $s^2 + 1$ or $s^3 + 1$, this upper bound is again the best possible. We shall show that when the upper bound is actually attained then the situation is very similar to the desarguesian one. (Most of the results here are due to Seib. However Seib's proofs use results not established in this book; the proofs here are due to Kimberley.)

Theorem 12.11. *Let* α *be a polarity of a finite projective plane* $\mathscr{P}$ *of order* $n = s^2$. *Then* $a(\alpha) \leqq s^3 + 1$.

Proof. From the proof of Lemma 12.3, $a(\alpha)$ is of the form $1 + s^2 + (k_1 - k_2)s$ where $k_1 + k_2 = s^4 + s^2$. Furthermore, by Theorem 12.5, $k_1 - k_2$ is non-negative and, since $k_1 + k_2$ is even, $k_1 - k_2$ is even. Thus $a(\alpha)$ is of the form $1 + 2rs + s^2$ where r is a non-negative integer.

Putting $w = 1 + 2rs + s^2$ then the number b of non-absolute lines is given by $b = n^2 + n + 1 - (1 + 2rs + s^2) = s^4 - 2rs$. Label the non-absolute lines $l_1, l_2, \ldots, l_b$ and let $k(l_i)$ be the number of absolute points on l_i. Then clearly, the number of absolute point and non-absolute line flags is given by $\sum_{i=1}^{b} k(l_i)$. But there are w absolute points and, by Lemma 12.1, each is incident with s^2 non-absolute lines. This gives the total number of absolute point and non-absolute line flags as ws^2 and so

$$\sum_{i=1}^{b} k(l_i) = ws^2. \tag{1}$$

Let P be a fixed absolute point; then, by Lemma 12.1, every other absolute point is joined to P by a non-absolute line and, furthermore, P is on s^2 non-absolute lines and thus

$$\sum_{P \in l_i} k(l_i) = (w - 1) + s^2. \tag{2}$$

If we now let P vary over the absolute points then $k(l_i)$ will be counted once for each absolute point on l_i. Hence, summing (2) over P gives

$$\sum_{i=1}^{b} k(l_i)^2 = w\big((w - 1) + s^2\big). \tag{3}$$

But[4] if $a_1, a_2, \ldots, a_m$ is any set of positive integers $m \sum_{i=1}^{m} a_i^2 \geqq \left(\sum_{i=1}^{m} a_i\right)^2$.

Thus, from (1) and (3), we have

$$bw(w - 1 + s^2) \geqq w^2 s^4,$$
$$b(w - 1 + s^2) \geqq ws^4,$$
$$(s^4 - 2rs)(2s^2 + 2rs) \geqq s^4(s^2 + 2rs + 1),$$
$$2(s^3 - 2r)(s + r) \geqq s^2(s^2 + 2rs + 1).$$

This simplifies to $s^4 - (2r + s)^2 \geqq 0$; i.e. $r \leqq (s^2 - s)/2$ and so $w = 1 + 2rs + s^2 \leqq 1 + s(s^2 - s) + s^2 = 1 + s^3$. ☐

---

[4] This simple inequality is proved as follows. For any x, $\sum_{i=1}^{m} (a_i + x)^2 = \Sigma a_i^2 + 2x\Sigma a_i + mx^2 \geqq 0$. Thus $(mx + \Sigma a_i)^2 + m\Sigma a_i^2 - (\Sigma a_i)^2 \geqq 0$ for all x, which implies the given inequality.

Although in general polarities may behave very differently from the desarguesian situation, if the number of absolute points is the maximum possible then the absolute points form the same "type" of configuration as in the desarguesian case. In order to clarify this statement we define a special type of "design":

A $2-(v,k,1)$ *design* is a set of v points with distinguished subsets called *lines* such that each line contains k points and any two distinct points are contained in a unique line.

Clearly a finite projective plane of order n is a $2-(n^2+n+1, n+1, 1)$ design.

Exercise 12.7. Show that a $2-(n^2+n+1, n+1, 1)$ design is a projective plane, if $n > 2$.

Exercise 12.8. If α is a unitary polarity of $\mathscr{P}_2(q^2)$ show that the set of absolute points and non-absolute lines form a $2-(q^3+1, q+1, 1)$ design.

In view of Exercise 12.8 we call a $2-(s^3+1, s+1, 1)$ design a *unital*. We now show that if the number of absolute points of a polarity of any plane of square order is a maximum, then the absolute points and non-absolute points form a unital, thus exhibiting the similarity with the desarguesian case.

Theorem 12.12. *Let α be a polarity of a finite projective plane of order s^2. If $a(\alpha) = s^3 + 1$ then the set of absolute points and non-absolute lines forms a unital.*

Proof. Using the notation of the proof of Theorem 12.11 and putting $v = s^3 + 1$ and $b = s^4 - s^3 + s^2$ in (1) and (3) we have

$$\sum_{l=1}^{b} k(l_i) = s^2(s^3+1), \tag{4}$$

$$\sum_{i=1}^{b} k(l_i)^2 = (s^3+s^2)(s^3+1). \tag{5}$$

We now consider $\sum_{i=1}^{b} \left(k(l_i) - (s+1) \right)^2$.

$$\sum_{i=1}^{b} \left(k(l_i) - (s+1) \right)^2 = \sum_{i=1}^{b} k(l_i)^2 - 2(s+1) \sum_{i=1}^{b} k(l_i) + b(s+1)^2.$$

Substituting from (4) and (5) we have

$$\sum_{i=1}^{b} \left(k(l_i) - (s+1) \right)^2 = (s^3+s^2)(s^3+1) - 2(s+1)s^2(s^3+1) + (s+1)^2(s^4-s^3+s^2)$$

$$= (s^3+1)(s^3+s^2-2s^3-2s^2) + (s+1)^2 s^2(s^2-s+1)$$

$$= s^2 \left[(s+1)^2(s^2-s+1) - (s^3+1)(s+1) \right]$$

$$= 0.$$

But $\sum_{i=1}^{b} (k(l_i) - (s+1))^2$ is the sum of squares of integers and, consequently, can only be zero if each summand is zero. Thus $k(l_i) = s+1$ for each i and the theorem is proved. $\square$

If $\mathcal{P}$ is a finite projective plane of order n then a polarity α of $\mathcal{P}$ is called *orthogonal* if $a(\alpha) = n+1$ and *unitary* if $a(\alpha) = n^{3/2} + 1$. As we have already seen, any polarity of a desarguesian plane is either orthogonal or unitary. Not much is known about polarities with $n+1 < a(\alpha) < n^{3/2} + 1$, nor are many examples known. Each of the known examples is *irregular* in the following sense: a polarity α is called *regular* if there is an integer t such that the number of absolute points on a non-absolute line (absolute lines through a non-absolute point) is 0, 1 or $t+1$. Note that all polarities of finite desarguesian planes are regular.

We shall give examples of irregular polarities and of unitary and orthogonal polarities in non-desarguesian planes. As we noted earlier, there are no known regular polarities which are not orthogonal or unitary and it has been conjectured that any regular polarity must be either unitary or orthogonal.

5. Polarities in Translation Planes

If ϕ is a polarity and α a collineation then, clearly, $\phi\alpha\phi$ is a collineation.

Exercise 12.9. If α is a (P, m)–central collineation and ϕ is a polarity show that $\phi\alpha\phi$ is a (m^ϕ, P^ϕ)–central collineation.

Exercise 12.10. If m is a translation line of a projective plane $\mathcal{P}$ and ϕ is a polarity of $\mathcal{P}$ show that $\mathcal{P}$ is a dual translation plane with respect to m^ϕ.

As an immediate corollary to the last exercise we have

Theorem 12.13. *If a translation plane admits a polarity then it is a division ring plane.*

Our main aim in this section is to give examples of polarities in non-desarguesian division ring planes. However before we can actually exhibit the examples we give a brief discussion of polarities in division ring planes. Suppose that θ is a polarity of a proper division ring plane $\mathcal{P}$ then, by Exercise 12.10, θ interchanges (∞) and $[\infty]$. If $Q \neq (\infty)$ is any absolute point with $Q^\theta = l$ then $\mathcal{P}$ may be coordinatized so that $Q = (0, 0)$ and $l = [0, 0]$. Then $l \cap [\infty]$ is (0), (∞) is joined to $(0, 0)$ by the line $[0]$ and θ must interchange (0) with $[0]$. If D is a division ring coordinatizing $\mathcal{P}$ with this choice of $(0, 0)$ and (0), then θ determines two permutations

α, β of D as follows:

$$(m, 0)^\theta = [m^\beta, 0] , \tag{1}$$

$$(0, y)^\theta = [0, -y^\alpha] . \tag{2}$$

Note that $0^\alpha = 0^\beta = 0$.

Exercise 12.11. Show that θ may be described as follows

$$(\infty) \leftrightarrow [\infty], \quad (m)^\theta = [m^{\beta^{-1}}], \quad [k]^\theta = (k^\beta)$$
$$(x, y)^\theta = [x^\beta, -y^\alpha], \quad [m, k]^\theta = (m^{\beta^{-1}}, -k^\alpha) .$$

Since θ preserves incidence, (x, y) is on $[m, k]$ if and only if $(m^{\beta^{-1}}, -k^\alpha)$ is on $[x^\beta, -y^\alpha]$. Thus $mx + y = k$ if and only if $x^\beta m^{\beta^{-1}} - k^\alpha = -y^\alpha$ which gives

$$y^\alpha + x^\beta m^{\beta^{-1}} = k^\alpha = (mx + y)^\alpha . \tag{1}$$

Putting $y = 0$ in (1) and using $0^\alpha = 0$ we have

$$(mx)^\alpha = x^\beta m^{\beta^{-1}} . \tag{2}$$

Thus

$$(mx + y)^\alpha = (mx)^\alpha + y^\alpha \tag{3}$$

so that α is additive.

Since D is a division ring $(x(y + z))^\alpha = (xy + xz)^\alpha$ and so, by (1) and (2)

$$(y + z)^\beta x^{\beta^{-1}} = (xy)^\alpha + (xz)^\alpha = y^\beta x^{\beta^{-1}} + z^\beta x^{\beta^{-1}}$$

which simplifies to

$$(y + z)^\beta = y^\beta + z^\beta . \tag{4}$$

From (2) $(mx)^{\alpha^2} = (x^\beta m^{\beta^{-1}})^\alpha = (m^{\beta^{-1}})^\beta \cdot (x^\beta)^{\beta^{-1}} = mx$ so that $\alpha^2 = 1$. This discussion proves the following theorem.

Theorem 12.11. *Given a polarity of a finite division ring plane $\mathscr{P}(D)$, there exist two permutations α and β of D such that*
  (i) $0^\alpha = 0 = 0^\beta$,
  (ii) $(xy)^\alpha = y^\beta x^{\beta^{-1}}$ *for any x, y in D,*
  (iii) $(x + y)^\alpha = x^\alpha + y^\alpha$ *for any x, y in D,*
  (iv) $(x + y)^\beta = x^\beta + y^\beta$ *for any x, y in D,*
  (v) $\alpha^2 = 1$.
Conversely any pair of permutations α, β of D satisfying (i)–(v) determine a polarity of $\mathscr{P}(D)$ (by the rules given in Exercise 12.11).

Our method of giving the examples will be to give the two mappings α and β. But first we must establish a method of determining the "type" of polarity determined by α and β. Since we are only interested in exhibiting polarities, rather than proving best possible results, we may assume that the characteristic of D is odd.

Given the polarity θ, α acts in a natural way as a mapping $\bar{\alpha}$ on the points of $[0] \backslash (\infty)$; $\bar{\alpha}$ is given by $(0, y)^{\bar{\alpha}} = [0] \cap (0, y)^\theta = (0, -y^\alpha)$. Clearly a point $(0, y)$ is absolute if and only if $y = -y^\alpha$ and, thus, the number of absolute points on $[0] \backslash (\infty)$ is $|\{y \in D \mid y = -y^\alpha\}|$.

Exercise 12.12. If $D_1 = \{x \in D \,|\, x^\alpha = x\}$ and $D_2 = \{x \in D \,|\, x^\alpha = -x\}$, show that, considered as additive groups, $D = D_1 \oplus D_2$.

Theorem 12.15. *If $\mathscr{P}(D)$ has odd order n and θ is a polarity determined by α and β as in Exercise 12.11 then*

$$a(\theta) = n|D_2| + 1$$

Proof. We have already shown that the number of absolute points on $[0]\backslash(\infty)$ is $|D_2|$. If k is any element of D then any point of $[k]\backslash(\infty)$ is of the form (k, y). If (k, y) is an absolute point, then (k, y) is on $[k^\beta, -y^\alpha]$ so that $k^\beta k + y = -y^\alpha$. If we write $y = d_1 + d_2$, where $d_i \in D_i$ $(i = 1, 2)$, then $k^\beta k + d_1 + d_2 = -d_1 + d_2$ and, hence, $2d_1 = -k^\beta k$. Thus if a point of $[k]\backslash(\infty)$ is absolute, then it is of the form $(k, -\frac{1}{2}k^\beta k + d_2)$; but clearly any point of this form is an absolute point of θ. This shows that the number of absolute points on $[k]\backslash(\infty)$ is $|D_2|$ and proves the theorem. ◻

Exercise 12.13. Show that θ is orthogonal if and only if $\alpha = 1$, and is unitary if and only if $|D_2| = \sqrt{n}$.

Example. In order to exhibit any examples it is natural to look for the simplest possible α and β. If we try to choose $\alpha = \beta$ then, by Theorem 12.14, α must be an involutory anti-automorphism of D, and so, if we choose D to be commutative, any involutory automorphism of D will give a polarity of $\mathscr{P}(D)$. Immediately this gives that for any commutative division ring D, $\mathscr{P}(D)$ admits an orthogonal polarity (put $\alpha = 1$ and appeal to Exercise 12.13).

Let F be a finite field and let D be the Dickson commutative division ring which is a two-dimensional vector space over F (see Chapter IX). The following exercises show that $\mathscr{P}(D)$ admits a unitary polarity.

Exercise 12.14. Show that the mapping α given by $(x + \lambda y)^\alpha = (x - \lambda y)$ is an automorphism of D, whose fixed elements are the elements of F.

Exercise 12.15. Show that the polarity θ of $\mathscr{P}(D)$ defined by taking α as in Exercise 12.14 and $\beta = \alpha$ is unitary.

The results in this section are due to Ganley. Since we were mainly interested in giving examples we have not stated all his results in their best forms. For a detailed discussion of polarities in division ring planes see [1].

As a final example, we give an exercise which gives a polarity of a division ring plane of order n having $n^{5/4} + 1$ absolute points. This shows that division ring planes can admit polarities which are neither orthogonal nor unitary.

*Exercise 12.16. Let $F = GF(q)$ and let D be a division ring whose elements are $x + \lambda y$ for all x, y in F where addition in D is componentwise and multiplication is given by $(x + \lambda y)(u + \lambda v) = xu + \delta y^\sigma v + \lambda(yu + x^\sigma v)$ where σ is in $\operatorname{Aut} F$, $\sigma^2 = 1 \neq \sigma$ and $\delta \in F$, $\delta^\sigma \neq \delta$ (see Chapter IX, Section 4).

If α is a permutation of the elements of D given by $(x + \lambda y)^\alpha = x + \lambda b y^\sigma$, where $b^{1+\sigma} = 1$, show that α is an involutory automorphism of D and that the polarity θ of $\mathscr{P}(D)$ given by taking $\beta = \alpha$ has $|D|^{5/4} + 1$ absolute points.

*Exercise 12.17. Show that the polarity of Exercise 12.16 is irregular, if $q = 9$ or $q = 25$.

We conclude this section by remarking that the Hughes planes (which are not translation planes) also admit polarities. The polarities of the Hughes plane have been completely determined and are all irregular.

6. Appendix (Proof of Segre's Theorem)

Let $F = GF(q)$ and let $\mathcal{O}$ be any oval in $\mathscr{P}_2(q)$. If a triangle has its vertices on $\mathcal{O}$ then we say it is an *inscribed* triangle of $\mathcal{O}$; a *circumscribed* triangle is one whose sides are tangents to $\mathcal{O}$. The points of contact of the sides of a circumscribed triangle are the vertices of an inscribed triangle, and we call these two triangles *mates*.

Lemma 12.16. *Every inscribed triangle of $\mathcal{O}$ and its mate are in perspective.*

Proof. We may take the inscribed triangle as the triangle of reference for a homogeneous coordinatization of $\mathscr{P}_2(q)$. Thus the triangle is

$$A_1 = \langle (1, 0, 0) \rangle, \quad A_2 = \langle (0, 1, 0) \rangle, \quad A_3 = \langle (0, 0, 1) \rangle.$$

If a_1, a_2, a_3 are the tangents to $\mathcal{O}$ at A_1, A_2, A_3 respectively then the equations of a_1, a_2, a_3 are of the form

$$a_1 : x_2 = k_1 x_3, \quad a_2 : x_3 = k_2 x_1, \quad a_3 : x_1 = k_3 x_2$$

where k_1, k_2, k_3 are non-zero elements of F. If $B = \langle c_1, c_2, c_3 \rangle$ is any of the $q - 2$ points of $\mathcal{O}$ distinct from A_1, A_2, A_3 then $c_1 c_2 c_3 \neq 0$ and the lines $A_1 B$, $A_2 B$, $A_3 B$ have respective equations $x_2 = h_1 x_3$, $x_3 = h_2 x_1$, $x_1 = h_3 x_2$ where $h_1 = c_2 c_3^{-1}$, $h_2 = c_3 c_1^{-1}$ and $h_3 = c_1 c_2^{-1}$. Clearly the h_i satisfy

$$h_1 h_2 h_3 = 1. \tag{1}$$

Conversely, if h_1 is any of the $q - 2$ elements of F distinct from zero and k_1, the line $x_2 = h_1 x_3$ meets $\mathcal{O}$ at A_1 and at a further point, B say, distinct from A_1, A_2, A_3; hence the coefficients h_2, h_3 in the equations

$x_3 = h_2 x_1, x_1 = h_3 x_2$ of the lines $A_2 B$, $A_3 B$ are functions of h_1, connected by (1), which take once each of the non-zero values of F distinct from k_2, k_3 respectively. On multiplying together the $q-2$ equations for $h_1 h_2 h_3$ thus obtained, we have

$$\Pi^3 = k_1 k_2 k_3 , \tag{2}$$

where Π denotes the product of the $q-1$ non-zero elements of F. (Note that (1) holds if $h_i \neq k_i$ $(i = 1, 2, 3)$. If $h_i = k_i$ then $h_j = 1$ for $i \neq j$ so that $h_1 h_2 h_3 = k_i$.) But $\Pi = -1$ (see Exercise 12.18, at the end of this appendix) and thus

$$k_1 k_2 k_3 = -1 . \tag{3}$$

The points $a_2 a_3 = (\langle k_3, 1, k_2 k_3 \rangle)$, $a_3 a_1 = \langle (k_3 k_1, k_1, 1) \rangle$, $a_1 a_2 = \langle (1, k_1 k_2, k_2) \rangle$ are joined to A_1, A_2, A_3 respectively by the lines:

$$x_3 = k_2 k_3 x_2, \qquad x_1 = k_3 k_1 x_3, \qquad x_2 = k_1 k_2 x_1 .$$

But, by (3), these lines are concurrent at $K = \langle (1, k_1 k_2, -k_2) \rangle$ and the lemma is proved. $\square$

We now prove Segre's theorem and, without any loss of generality, assume that K is the unit point $\langle (1, 1, 1) \rangle$ i.e. $k_1 = k_2 = k_3 = -1$.

If $B = \langle (c_1, c_2, c_3) \rangle$ is any of the $q-2$ points of $\mathcal{O}$ distinct from A_1, A_2, A_3, we denote by $b\colon x_1 b_1 + x_2 b_2 + x_3 b_3 = 0$ the tangent to $\mathcal{O}$ at B. Since B is on b we have

$$c_1 b_1 + c_2 b_2 + c_3 b_3 = 0 . \tag{4}$$

Furthermore none of A_1, A_2, A_3 is on B and so if we put $d_1 = b_1 - b_2 - b_3, d_2 = -b_1 + b_2 - b_3, d_3 = -b_1 - b_2 + b_3$ we have

$$b_1 b_2 b_3 d_1 d_2 d_3 \neq 0 \tag{5}$$

By the lemma, the triangles $BA_2 A_3$ and $ba_2 a_3$ are in perspective. Straightforward calculation shows that this implies

$$\det \begin{vmatrix} c_3 - c_1 & c_1 - c_3 & -c_1 - c_2 \\ b_1 - b_3 & b_2 & 0 \\ b_1 - b_2 & 0 & b_3 \end{vmatrix} = 0,$$

which simplifies to $b_2(c_1 + c_2) = b_3(c_1 + c_3)$.

A similar consideration for the triangles $BA_3 A_1$, $BA_1 A_2$ and their mates gives $b_3(c_2 + c_3) = b_1(c_2 + c_1)$ and $b_1(c_3 + c_1) = b_2(c_3 + c_2)$. But these last three equations imply $b_1 : b_2 : b_3 = (c_2 + c_3):(c_3 + c_1):(c_1 + c_2)$; substituting for b_1, b_2, b_3 in (4) this gives

$$c_1(c_2 + c_3) + c_2(c_3 + c_1) + c_3(c_1 + c_2) = 0$$

which, since q is odd, gives

$$c_1 c_2 + c_2 c_3 + c_3 c_1 = 0 . \tag{6}$$

By (6), each of the $q-2$ points of $\mathcal{O}$ distinct from A_1, A_2, A_3 lies on the conic with equation $x_2 x_3 + x_3 x_1 + x_1 x_2 = 0$ which, clearly, also contains A_1, A_2, A_3. Thus since the conic also has exactly $q+1$ points, the points of $\mathcal{O}$ are the points of the conic with equation $x_2 x_3 + x_3 x_1 + x_1 x_2 = 0$. $\square$

Exercise 12.18. If Π denotes the product of the $q-1$ non-zero elements of $GF(q)$, q odd, show that $\Pi = -1$. (Hint: Use the fact that the multiplicative group of $GF(q)$ is cyclic and, hence, contains only one element of order 2.)

References

Since the chapter is self-contained no references (except to Chapter II) are needed for the understanding of the material. However the reader may feel that he wishes to know a little more about the known examples of polarities. Much more is known about polarities of translation planes than we have included here and the interested student should see [1]. The polarities of the Hughes planes have been completely determined and can be found in [2].

1. Ganley, M. J.: Polarities in Translation Planes. Geometriae Dedicata 1, 103–116 (1972).
2. Piper, F. C.: Polarities in the Hughes plane. Bull. London Math. Soc. 2, 209–213 (1970).

XIII. Collineation Groups

1. Introduction

This chapter takes up again the theme of collineations and collineation groups, but dealing with somewhat deeper theorems than we found in Chapter IV or in the various sections of the other chapters that have touched on collineations. We shall be concerned almost exclusively with finite planes, and our first concern will be with orbits, fixed points, and so on; Section 2 contains the very important Orbit Theorem, which is crucial to any work on collineation groups of finite planes: *a collineation group of a finite projective plane has equally many point and line orbits.* Section 3 develops a number of results about perspectivities, which besides their intrinsic interest, enable us to give interesting and useful characterizations of finite desarguesian planes in Section 4. In Section 5 we discuss Singer groups, already studied in Chapter II, and introduce difference sets. Hall's multiplier theorem is stated, but its proof is postponed to the Appendix (since the proof is more than usually unrelated to the geometric ideas of the theorem itself, involving as it does group algebras). It is worth commenting that the subject of Singer groups and difference sets was once one of the most active branches in the study of projective planes, but the difficulty in proving the central conjecture (a finite plane with a Singer group is desarguesian), or of finding a counter-example, has been so great that it has almost lain in abeyance for many years. Finally in Section 6 we consider some non-existence theorems, which are very like the Bruck-Ryser theorem, and include it as a special case. These are non-existence theorems for planes with collineation groups of specified types (e.g. of prime order, fixing a certain number of points); the proofs are extremely long and detailed, and involve a knowledge of number theory far beyond the level of this book, so they are mostly only quoted. But one result has an accessible proof, which is given here: *a plane of order $n \equiv 2$* (mod 4), $n > 2$, *can have no collineations of even order.*

2. Some Orbit Theorems

If Π is any collineation group of a projective plane $\mathscr{P}$ then Π induces a permutation group, $\Pi(P)$ say, on the points of $\mathscr{P}$ and a second permutation group $\Pi(l)$ on the lines of $\mathscr{P}$. It is natural to look for properties of these two representations of Π.

If $\mathscr{P}^{l_\infty}$ is a translation plane of order n and if $\Pi = \operatorname{Aut}\mathscr{P}$ and $\Lambda = \Pi_{(l_\infty, l_\infty)}$, then $|\Lambda| = n^2$. The group $\Lambda(P)$ has $n+1$ point orbits of length 1, namely the points of l_∞, and one point orbit of length n^2. For any point X on l_∞, Λ is transitive on the n lines through X and distinct from l_∞, so $\Lambda(l)$ has $n+1$ line orbits of length n and one of length 1. This example illustrates that $\Pi(P)$ and $\Pi(l)$ need not have the same orbit structure. In fact in this example $\Lambda(P)$ acts faithfully and regularly on the affine points of $\mathscr{P}^{l_\infty}$ whereas $\Lambda(l)$ is not faithful on any of its orbits. However if Π is any finite collineation group then it is impossible for both $\Pi(P)$ and $\Pi(l)$ to be non-faithful on all of their orbits.

Theorem 13.1. *If Π is a finite collineation group of a projective plane $\mathscr{P}$ then Π induces a faithful permutation group on every orbit of maximal length.*

Proof. Since Π is finite, all orbits have finite length and, consequently, there is certainly an orbit of maximal length which, without any loss of generality, we may assume to be a point orbit. (Otherwise the dual argument will prove the theorem.)

Let P and Q be any pair of distinct points in an orbit $\mathcal{O}$ of maximal length and let $\Pi_\mathcal{O}$ be the subgroup of Π fixing every point of $\mathcal{O}$. Then Π induces a faithful permutation group on $\mathcal{O}$ if and only if $\Pi_\mathcal{O} = 1$. We shall assume that $\Pi_\mathcal{O} \neq 1$ and then show that this assumption enables us to find an orbit whose length is greater than that of $\mathcal{O}$. This contradiction will then prove the theorem.

Since $\mathcal{O}$ is an orbit of Π, $\Pi_\mathcal{O} \trianglelefteq \Pi$ so that Π permutes the fixed elements of $\Pi_\mathcal{O}$. If $\alpha \in \Pi$ is such that $P^\alpha = Q$ (such an α certainly exists since P and Q are in the same orbit $\mathcal{O}$), then for any line l through P the line l^α passes through Q. Furthermore, as the fixed elements of $\Pi_\mathcal{O}$ are permuted by Π, l is fixed by $\Pi_\mathcal{O}$ if and only if l^α is. In other words α induces a one-to-one mapping of the lines through P fixed by $\Pi_\mathcal{O}$ onto the fixed lines of $\Pi_\mathcal{O}$ through Q and, consequently, P is fixed linewise by $\Pi_\mathcal{O}$ if and only if Q is. Hence, since we are assuming $\Pi_\mathcal{O} \neq 1$, P is not fixed linewise by $\Pi_\mathcal{O}$ and there is a line m through P and a collineation $\beta \in \Pi_\mathcal{O}$ such that $m^\beta \neq m$. Clearly any line containing two points of $\mathcal{O}$ is fixed by $\Pi_\mathcal{O}$ and so P is the only point of $\mathcal{O}$ incident with either m or m^β. For any $\pi \in \Pi$, m^π and $m^{\beta\pi}$ are two distinct lines in the orbit of m such that P^π is the only point of $\mathcal{O}$ on either m^π or $m^{\beta\pi}$. Hence each point of $\mathcal{O}$ is incident with at least two lines in the orbit of m while each line in this orbit contains only one point of $\mathcal{O}$. This shows that the length of the orbit of m is at least twice that of $\mathcal{O}$ which contradicts the choice of $\mathcal{O}$ and proves the theorem. $\quad\square$

Corollary 1. *Any collineation group of a finite projective plane induces a faithful permutation group on at least one orbit.*

If a permutation group acts faithfully on a given orbit than we refer to that orbit as a *faithful orbit*. If Γ is an abelian group on a set $\mathscr{S}$ then, since every subgroup of Γ is normal, Γ acts regularly on each of its orbits. Thus an orbit of Γ is faithful if and only if its length is equal to the order of $|\Gamma|$. This simple observation leads to another corollary to Theorem 13.1:

Corollary 2. *If α is a collineation of a finite projective plane of order n and if the order of α is t than $t \leq n^2 + n + 1$.*

We have already seen in Chapter II that $\mathscr{P}_2(q^a)$ admits a collineation of order $q^{2a} + q^a + 1$, so that Corollary 2 gives a best possible upper bound.

Exercise 13.1. Let $\mathscr{P}$ be a projective plane with a collineation group Π. If Π has a non-faithful orbit which contains a quadrangle show that Π must leave some subplane of $\mathscr{P}$ invariant.

Exercise 13.2. If Π is an abelian collineation group of a finite projective plane of order n such that $|\Pi| \neq n^2 + n + 1$, show that $|\Pi| \leq n^2$.

As we saw by considering the translation group of a translation plane, it is not in general true that, for a collineation group Π, $\Pi(P)$ and $\Pi(l)$ have the same orbit structure. It is, however, true in the special case where Π is cyclic, and to prove this we use the incidence matrix of the plane.

An $n \times n$ matrix A of 0's and 1's is called a *permutation matrix* if each row of A and each column of A has exactly one entry 1.

Exercise 13.3. If A is any permutation matrix show that $A A' = I = A' A$.

Lemma 13.2. *If A is an incidence matrix for a finite projective plane $\mathscr{P}$ then any collineation of $\mathscr{P}$ may be represented by a pair of permutation matrices P, Q with $P A = A Q$.*

Proof. Let α be any collineation of $\mathscr{P}$ and let $v = n^2 + n + 1$. Let $P_1, \ldots, P_v, l_1, \ldots, l_v$ be a labelling of the points and lines of $\mathscr{P}$ such that $A = (a_{ij})$ with $a_{ij} = 1$ if and only if $P_i \in l_j$ and define two $v \times v$ matrices P and Q by $P = (p_{ij})$ where $p_{ij} = 1$ if and only if $P_i^\alpha = P_j$, otherwise $p_{ij} = 0$; $Q = (q_{ij})$ where $q_{ij} = 1$ if and only if $l_i^\alpha = l_j$, otherwise $q_{ij} = 0$.

Then since α is a permutation on the points and a permutation on the lines both P and Q are clearly permutation matrices (check this!).

The (i, k)-entry of $P A = \sum_{j=1}^{v} p_{ij} a_{jk} = a_{xk}$, where x is uniquely determined by $p_{ix} = 1$; i.e. x is uniquely determined by $P_i^\alpha = P_x$.

Similarly the (i, k)-entry of $AQ = \sum\limits_{j=1}^{v} a_{ij}q_{jk} = a_{iy}$ where y is uniquely determined by $l_y^\alpha = l_k$. We must show $a_{xk} = a_{iy}$.

$$a_{xk} = 1 \Leftrightarrow P_x \in l_k$$
$$\Leftrightarrow P_x^{\alpha^{-1}} \in l_k^{\alpha^{-1}}$$
$$\Leftrightarrow P_i \in l_y$$
$$\Leftrightarrow a_{iy} = 1 \,.$$

Thus $a_{xk} = a_{iy}$ and the lemma is proved. □

Theorem 13.3. *A collineation of a finite projective plane has an equal number of fixed points and lines.*

Proof. Let $\mathscr{P}$ be a finite projective plane and let A be an incidence matrix for $\mathscr{P}$. Then, by Lemma 13.2, any collineation α can be represented by two permutation matrices P, Q with $PA = AQ$. If $P = (p_{ij})$ then $p_{ij} = 1$ if and only if $P_i^\alpha = P_j$ so that $p_{ii} = 1$ if and only if $P_i^\alpha = P_i$ and the number of fixed points of α is the trace of P. Similarly the number of fixed lines of α is the trace of Q.

But since $PA = AQ$ and, by Theorem 3.14, A is non-singular, we have $P = AQA^{-1}$ so that the trace of P is equal to the trace of Q. □

Corollary. *Let Π be a cyclic collineation group of a finite projective plane $\mathscr{P}$. Then $\Pi(P)$ has the same cyclic structure as $\Pi(l)$.*

Proof. Let $\Pi = \langle \pi \rangle$. If p_x is the number of point cycles of length x and l_x is the number of line cycles of length x, we have to prove $p_x = l_x$ for all x.

The proof is by induction. By Theorem 13.3 $p_1 = l_1$.

Let t be an integer greater than 1 and suppose $p_x = l_x$ for all $x < t$. Clearly if $x \nmid |\pi|$ then $p_x = l_x = 0$ (see Result 1.13) so we may assume that $t \mid |\pi|$.

Consider π^t. For any s which divides t, π^t fixes all points and lines in cycles of length s. Conversely any point or line fixed by π^t must lie in a cycle whose length divides t. Thus the number of points fixed by π^t is $\sum\limits_{x \mid t} x p_x = \sum\limits_{\substack{x \mid t \\ x \neq t}} x p_x + t p_t$. Similarly the number of lines fixed by π^t is $\sum\limits_{x \mid t} x l_x = \sum\limits_{\substack{x \mid t \\ x \neq t}} x l_x + t l_t$. By Theorem 13.3 these numbers are equal. Thus

$$\sum\limits_{\substack{x \mid t \\ x \neq t}} x p_x + t p_t = \sum\limits_{\substack{x \mid t \\ x \neq t}} x l_x + t l_t \,.$$

But, by the induction hypothesis, $p_x = l_x$ for all $x \mid t$, $x \neq t$. Thus $t p_t = t l_t$ or $p_t = l_t$. □

As we saw from looking at the group of translations of a translation plane, neither Theorem 13.3 nor its corollary is true for arbitrary collineation groups. However we do have

Theorem 13.4. (The orbit theorem)[5]. *Let Π be a collineation group of a finite projective plane $\mathscr{P}$. Then Π has an equal number of point and line orbits.*

Proof. Left t_1 be the number of points orbits of Π and t_2 the number of line orbits. Further, for any $\alpha \in \Pi$, let $\chi(\alpha)$ be the number of fixed points of α which, by Theorem 13.3, is also the number of lines fixed by α. Applying Result 1.14 to $\Pi(P)$ and $\Pi(l)$ respectively we get $|\Pi| t_1 = \sum_{\alpha \in \pi} \chi(\alpha) = |\Pi| t_2$. Hence $t_1 = t_2$. $\square$

Corollary. Π *is transitive on the points of* $\mathscr{P}$ *if and only if* Π *is transitive on the lines of* $\mathscr{P}$.

Exercise 13.4. Let $\mathscr{P}$ be a finite projective plane with a collineation group Π. Show that Π is doubly transitive on the points of $\mathscr{P}$ if and only if Π is transitive on the flags of $\mathscr{P}$ and on the anti-flags of $\mathscr{P}$. (Recall that a *flag* is an incident point-line pair while an *anti-flag* is a non-incident point-line pair.)

Exercise 13.5. Use Exercise 13.4 to show that Π is doubly transitive on the points of $\mathscr{P}$ if and only if Π is doubly transitive on the lines of $\mathscr{P}$.

Exercise 13.6. Let $\mathscr{P}^{l_\infty}$ be a finite affine plane with a collineation group Π. Show that Π is transitive on the affine lines of $\mathscr{P}^{l_\infty}$ if and only if Π is transitive on the affine points of $\mathscr{P}^{l_\infty}$ and on the points of l_∞.

Exercise 13.7. Show that any collineation group Π of a finite affine plane $\mathscr{A}$ has at least as many line orbits as point orbits. Show that the number of line orbits is equal to the number of point orbits if and only if Π is transitive on the special points of $\mathscr{A}$.

3. Collineation Groups Containing Perspectivities

By Theorem 6.20, we know that a finite Moufang plane is desarguesian so that a finite projective plane is desarguesian if and only if it is (P, l) – transitive for all choices of P and l with P on l. In Chapter IV we saw how the existence of a few perspectivities was often sufficient to guarantee many more, We now extend those results to give weaker necessary and sufficient conditions for a finite projective plane to be desarguesian. A crucial tool in our arguments will be the following permutation group lemma.

---

[5] This theorem, often called the Dembowski-Hughes-Parker Theorem, can actually be found in another work due to Brauer [1], although Brauer formulated the result in a very different way.

Lemma 13.5. *Let Γ be a permutation group on a finite set $\mathscr{S}$ and let p be a prime. If $\mathscr{T}$ is a subset of $\mathscr{S}$ such that, for any $t \in \mathscr{T}$, Γ contains an element of order p fixing t and no other element of $\mathscr{S}$, then $\mathscr{T}$ is contained in an orbit of Γ.*

Proof. Let t_1, t_2 be any two distinct elements of $\mathscr{T}$ and suppose that t_1 and t_2 lie in different orbits of Γ. Let $\gamma_i \in \Gamma$, $(i = 1, 2)$, have prime order p and be such that $t_i^{\gamma_i} = t_i$ but $s^{\gamma_i} \neq s$ for all $s \in \mathscr{S}$, $s \neq t_i$. Then, for any $s \in \mathscr{S}$, $s \neq t_i$, the length of the orbit of s under $\langle \gamma_i \rangle$ is p. However, since $\langle \gamma_i \rangle$ is a subgroup of Γ, any orbit of Γ must be a union of orbits of $\langle \gamma_i \rangle$. Putting $i = 1$ we see that, since t_1 is the only orbit of length 1 under $\langle \gamma_1 \rangle$ and since $t_2 \neq t_1$, the orbit of t_2 under Γ is partitioned into disjoint cycles of length p by $\langle \gamma_1 \rangle$ so that $|t_2 \Gamma| \equiv 0 \pmod{p}$ whereas $|t_1 \Gamma| \equiv 1 \pmod{p}$ (because $\langle \gamma_1 \rangle$ partitions the points of $t_1 \Gamma \setminus t_1$ into cycles of length p). However the same argument with $i = 2$ gives $|t_1 \Gamma| \equiv 0 \pmod{p}$ and $|t_2 \Gamma| \equiv 1 \pmod{p}$ and this contradiction proves the lemma. ☐

The following lemma gives a simple geometric application of Lemma 13.5.

Lemma 13.6. *Let $\mathscr{P}$ be a finite projective plane. If α is an (A, a)-elation of prime order p and β is a (B, b)-elation of the same prime order p with B not on a and b not passing through A then $\langle \alpha, \beta \rangle$ contains an element γ with $A^\gamma = B$ and $a^\gamma = b$.*

Proof. Let $Q = ab$ and $l = AB$, whence by hypothesis, $Q \neq A$, B and $l \neq a, b$ and, since α and β both fix Q and l, $\langle \alpha, \beta \rangle$ fixes both Q and l. Thus $\langle \alpha, \beta \rangle$ induces a permutation group Γ on the points of l. Since α, β are elations of $\mathscr{P}$, neither with l as axis, the only point of l fixed by α is A and the only point of l fixed by β is B so that taking $\mathscr{S}$ as the points of l and $\mathscr{T}$ as $\{A, B\}$, we can apply Lemma 13.5 to show that Γ contains an element γ' with $A^{\gamma'} = B$. But γ' must be induced by a collineation $\gamma \in \langle \alpha, \beta \rangle$ and thus, since $\langle \alpha, \beta \rangle$ fixes Q, we have $A^\gamma = B$ and $a^\gamma = (QA)^\gamma = QB = b$. ☐

If Π is any collineation group of $\mathscr{P}$ such that $\langle \alpha, \beta \rangle \subset \Pi$ then, by Corollary 2 of Lemma 4.11, $\Pi_{(B,b)} = \gamma^{-1} \Pi_{(A,a)} \gamma$ so that $|\Pi_{(B,b)}| = |\Pi_{(A,a)}|$.

In Theorem 4.14 we saw a situation under which elations of a finite projective plane are forced to have the same prime order. We shall now extend that result. If A is the centre of an elation in a given collineation group Π and if l is an axis of an elation in Π such that A is on l then we shall call the flag (A, l) an *elation flag* of Π. Note that by Lemma 4.15 (A, l) is an elation flag of Π if and only if $\Pi_{(A,l)} \neq 1$. Two elation centres are said to be *connected* in Π if they are joined by a sequence of elation flags of Π. Connectivity between elation axes or between an elation centre and an elation axis is defined in an analogous way. Clearly "being connected in Π" is an equivalence relation; an equivalence class under this relation is called a *connected set of Π* and a connected set is called *trivial* if it consists of a flag only. (Often we shall talk about two centres being connected etc. without referring to a specific group. This may be for one of two reasons; either it will be clear which group we are discussing or else the group will be unimportant and may be taken as the full collineation group of the plane.) The motivation behind the preceding definitions is given by the following lemma which is an extension of Theorem 4.14.

Lemma 13.7. *Let $\mathscr{P}$ be a finite projective plane. If α and β are elations whose centres A and B belong to a non-trivial connected set $\mathscr{S}$ then α and β have the same prime order.*

Proof. We first note, although the observation is unnecessary for the proof, that the axes of α and β must also belong to $\mathscr{S}$. Since A and B are in $\mathscr{S}$ they are connected by a sequence of elation flags. But, by Theorem 4.14 and its dual, any two elations whose centre-axis flags have an element in common have the same prime order. Thus repeated use of Theorem 4.14 and its dual proves the lemma. □

So any non-trivial connected set $\mathscr{S}$ of a finite projective plane has associated with it a unique prime p, namely the order of any elation whose centre belongs to $\mathscr{S}$. We shall call p the *characteristic* of $\mathscr{S}$.

We now combine Lemmas 13.6 and 13.7 with Corollary 2 of Theorem 4.26 to show how the existence of seemingly few elations with centre on a given line l is sufficient to make l a translation line.

Lemma 13.8. *Let $\mathscr{P}$ be a finite projective plane and let Γ be a collineation group of $\mathscr{P}$. If l is a line of $\mathscr{P}$ such that* (i) $\Gamma_{(l,l)} \ne 1$ *and* (ii) *for any A on l there is a line a through A, $a \ne l$, such that $\Gamma_{(A,a)} \ne 1$, then $\Gamma_{(l,l)}$ is transitive on the affine points of $\mathscr{P}^l$, i.e. l is a translation line.*

Proof. Let $A_1, A_2, \ldots, A_{n+1}$ be the points of l and for each A_i let $a_i \ne l$ be a line through A_i such that $\Gamma_{(A_i, a_i)} \ne 1$. Clearly the A_i, a_i and l $(i = 1, 2, \ldots, n+1)$ are connected so that, by Lemma 13.7, if $\alpha_i \in \Gamma_{(A_i, a_i)}$ and $\alpha_j \in \Gamma_{(A_j, a_j)}$, $\alpha_i \ne 1 \ne \alpha_j$, then α_i and α_j have the same prime order p. For each i pick a non-trivial $\alpha_i \in \Gamma_{(A_i, a_i)}$ and let $\Sigma = \langle \alpha_1, \alpha_2, \ldots, \alpha_{n+1} \rangle$. Since each of the α_i fixes l, Σ induces a permutation group on the points of l and, for each A_i on l, α_i has prime order p and fixes A_i but no other point on l. Thus, by Lemma 13.5, Σ is transitive on the points of l and Corollary 2 of Lemma 4.26 completes the proof. □

Unfortunately homologies do not behave quite so uniformly or simply as elations. We have already seen in Chapter VII that in any quasifield the group of $((0, 0), [\infty])$-homologies is isomorphic to the multiplicative group of a skewfield. Hence if finite it is cyclic and, in general, contains elements of different prime orders. However, as we saw in Corollary 2 of André's Theorem (Theorem 4.25), if the plane is finite and a line is a homology axis for at least two distinct centres then it is also an elation axis. Thus we see that the existence of sufficient homologies guarantees the existence of elations and enables us to use the previous lemmas. As an analogue to Lemma 4.15 we have

Lemma 13.9. *If* α *is an* (A, a)*-elation and* β *is a* (B, b)*-homology with A not on b but a passing through B, then* $\langle \alpha, \beta \rangle$ *contains a* (B, a)*-elation.*

Proof. As in the proof of Lemma 4.15 we consider $\alpha^{-1}\beta^{-1}\alpha\beta$. For any X on a, $X^{\alpha^{-1}\beta^{-1}\alpha\beta} = X^{\beta^{-1}\alpha\beta}$. But, since X^β is on a, $X^{\beta^{-1}\alpha} = X^{\beta^{-1}}$ so that $X^{\beta^{-1}\alpha\beta} = X^{\beta^{-1}\beta} = X$ and $\alpha^{-1}\beta^{-1}\alpha\beta$ fixes a pointwise. Similarly $\alpha^{-1}\beta^{-1}\alpha\beta$ fixes B linewise. Finally since $A^\beta \neq A$, $\alpha\beta \neq \beta\alpha$ (by Corollary 1 to Lemma 4.11) so that $\alpha^{-1}\beta^{-1}\alpha\beta$ is a non-trivial (B, a)-elation. □

4. Some Characterizations of Desarguesian Planes

We now combine the lemmas of Section 3 and Chapter IV to give some characterizations of desarguesian planes. The first is due to Gleason.

Theorem 13.10. *Let* $\mathscr{P}$ *be a finite projective plane and let* Π *be a collineation group of* $\mathscr{P}$*. If every point of* $\mathscr{P}$ *is the centre of a non-trivial elation in* Π *and if every line is the axis of a non-trivial elation in* Π *then* $\mathscr{P}$ *is desarguesian and* Π *contains the little projective group of* $\mathscr{P}$*.*

Proof. If l is any line of $\mathscr{P}$ we shall use the lemmas to show that l is a translation line of $\mathscr{P}$ and that Π contains all the elations of $\mathscr{P}$ with axis l. Then, since the choice of l is arbitrary, this will imply that $\mathscr{P}$ is a finite Moufang plane which, by Theorem 6.20, is desarguesian. The claim about Π will follow since Π contains all elations of $\mathscr{P}$ (see Exercise 2.25 and the remarks following it).

Since every point is the centre of a non-trivial elation in Π and every line is the axis of a non-trivial elation in Π we have, by Lemma 4.15, that for any flag (A, a), $\Pi_{(A,a)} \neq 1$. Thus, by Lemma 13.8, $\Pi_{(l,l)}$ is transitive on the affine points of $\mathscr{P}^l$ and the theorem is proved. □

The second characterization is a generalization of Theorem 13.10 and is due to Wagner.

Theorem 13.11. *Let* $\mathscr{P}$ *be a finite projective plane and let* Π *be a collineation group of* $\mathscr{P}$*. If* Π *is transitive on the points of* $\mathscr{P}$ *and contains a perspectivity then* $\mathscr{P}$ *is desarguesian and* Π *contains the little projective group of* $\mathscr{P}$*.*

Proof. By the corollary to Theorem 13.4, Π is transitive on both the points and the lines of $\mathscr{P}$. Suppose Π contains an elation α and has centre A and axis a. For any point B of $\mathscr{P}$ there is a collineation β in Π with $A^\beta = B$. But then, by Lemma 4.11, $\beta^{-1}\alpha\beta$ is an elation with centre B so that every point of $\mathscr{P}$ is the centre of an elation in Π. Similarly every line of $\mathscr{P}$ is the axis of an elation in Π and Theorem 13.10 proves the theorem.

Suppose that every perspectivity in Π is a homology then, by the above argument, every point is the centre of a homology in Π and every line is the axis of a homology in Π. If any line is a homology axis for two centres, or dually if any point is a homology centre for two axes, then by Corollary 2 of Theorem 4.25 (or its dual) Π contains an elation. Thus for each point P of $\mathscr{P}$ there is a unique line P^θ such that P^θ is the axis of all non-trivial homologies with centre P and P is the centre of all non-trivial homologies with axis P^θ.

If θ is the mapping which interchanges P with P^θ then clearly $\theta^2 = 1$. Suppose A and l are such that A is on l but l^θ is not on A^θ. If α is a non-trivial (l^θ, l) homology then $A^\alpha = A$ but $A^{\theta\alpha} \neq A^\theta$ and so if δ is any non-trivial (A, A^θ)-homology, $\alpha^{-1}\delta\alpha$ is a $(A^\alpha, A^{\theta\alpha})$-homology (by Lemma 4.11). Since $A^\alpha = A$ this means that $A^{\theta\alpha}$ and A^θ are distinct homology axes for A and this contradiction shows that A is on l if and only if l^θ is on A^θ; i.e. that θ is a polarity. But this leads to another contradiction since, by the definition of θ, A is not on A^θ for any A in $\mathscr{P}$ so that θ is a polarity without absolute points, contrary to Theorem 12.3. Hence Π must contain an elation and the theorem is proved. □

*Exercise 13.8. Let $\mathscr{P}$ be a finite projective plane and let Π be a collineation group of $\mathscr{P}$. If every point of $\mathscr{P}$ is the centre of a non-trivial elation in Π show that either (a) $\mathscr{P}$ is desarguesian and Π contains its little projective group or (b) $\mathscr{P}$ is the dual of a translation plane and Π contains the dual translation group of $\mathscr{P}$.

*Exercise 13.9. Let $\mathscr{P}$ be a finite projective plane and let Π be a collineation group of $\mathscr{P}$. If every point of $\mathscr{P}$ is the centre of a non-trivial homology in Π show that either conclusions (a) or (b) of Exercise 13.8 hold or (c) $\mathscr{P}$ is a translation plane and Π contains the translation group of $\mathscr{P}$.

**Exercise 13.10. Let $\mathscr{P}$ be a finite projective plane of order n and let Π be a collineation group of $\mathscr{P}$. If every point of $\mathscr{P}$ is the centre of a non-trivial perspectivity in Π and if n is not a square show that one of the conclusions of Exercise 13.9 hold.

(It is still unknown what the conditions of Exercise 13.10 imply if n is a square. Certainly the unitary group of $\mathscr{P}_2(q^2)$ satisfies the conditions of the exercise so that conclusions (a), (b), (c) of Exercise 13.9 do not exhaust all possibilities (see Chapter II, Section 8).)

Exercise 13.11. Let $\mathscr{P}$ be a finite projective plane of non-square order with a collineation group Π doubly transitive on the points of $\mathscr{P}$. Show that $\mathscr{P}$ is desarguesian and Π contains the little projective group of $\mathscr{P}$. (This is a special case of the celebrated Ostrom-Wagner Theorem; see Chapter XIV.)

5. Singer groups and Difference Sets

In Chapter II we saw that many desarguesian projective planes admit collineation groups acting regularly (and hence transitively) on their points and lines. If $\mathscr{P}$ is any projective plane then, (following Chapter II), we call a collineation group Π of $\mathscr{P}$ a *Singer group* for $\mathscr{P}$ if it is regular on both the points and lines of $\mathscr{P}$.

Exercise 13.12. Show that a collineation group Π of a finite projective plane $\mathscr{P}$ is regular on the points if and only if it is regular on the lines of $\mathscr{P}$.

If Π is a Singer group of a projective plane $\mathscr{P}$ then for any given point X and line m of $\mathscr{P}$, the subset $\mathscr{D} = \{\pi \in \Pi \mid X^{\pi} \text{ is on } m\}$ is called a *difference set* of Π. (The terminology is historical and, as we shall see, makes more sense when Π is abelian and is written additively.)

Lemma 13.12. *Let Π be a Singer group with a difference set $\mathscr{D}$. Then for every $\beta \neq 1$ in Π there exists a unique pair δ_1, δ_2 in $\mathscr{D}$ such that $\beta = \delta_1 \delta_2^{-1}$; also there exists a unique pair δ_3, δ_4 in $\mathscr{D}$ such that $\beta = \delta_3^{-1} \delta_4$.*

Proof. Since Π is regular on the points of $\mathscr{P}$, X and X^{β} are distinct points; thus there is a unique line joining them and, by the transitivity of Π, this line is m^{α} for some α in Π. But now $X^{\alpha^{-1}}$ and $X^{\beta\alpha^{-1}}$ are on m so that $\alpha^{-1} = \delta_2$ is in $\mathscr{D}$ and $\beta\alpha^{-1} = \delta_1$ is also in $\mathscr{D}$; solving gives $\beta = \delta_1 \delta_2^{-1}$. Conversely if $\beta = \delta_1' \delta_2'^{-1}$, where δ_1', δ_2' are in D, then $\beta\delta_2' = \delta_1'$ and so $X^{\beta\delta_2'} = X^{\delta_1'}$ is on m. Thus X^{β} is on $m^{\delta_2'^{-1}}$ and clearly since δ_2' is in $\mathscr{D}$, X is also on $m^{\delta_2'^{-1}}$; thus $m^{\delta_2'^{-1}}$ is the line joining X to X^{β}. But, this implies $m^{\delta_2'^{-1}}$ is m^{α} and so, by the regularity of Π, $\alpha = \delta_2'^{-1}$. It is now easy to see $\delta_1 = \delta_1'$, $\delta_2 = \delta_2'$ so that δ_1 and δ_2 are unique.

Similarly, by considering the lines m and m^{β}, we find that there is a unique δ_3, δ_4 in $\mathscr{D}$ such that $\beta = \delta_3^{-1} \delta_4$. $\square$

Lemma 13.12 has a very interesting converse which shows that any group with the appropriate properties is a Singer group for a projective plane.

Theorem 13.13. *Let Γ be a group with a subset $\mathscr{D}$ such that every element of Γ, except the identity, can be uniquely represented as $\delta_1 \delta_2^{-1}$ and as $\delta_3^{-1} \delta_4$, where each δ_i is in $\mathscr{D}$. Then if $|\Gamma| > 3$, there is a (unique) projective plane $\mathscr{P}$ such that Γ is a Singer group for $\mathscr{P}$ with difference set $\mathscr{D}$.*

Proof. For each α, β in Γ, let (α) be a point of $\mathscr{P}$ and let $[\beta]$ be a line of $\mathscr{P}$. Define incidence by: (α) is on $[\beta]$ if and only if $\alpha\beta^{-1}$ is in $\mathscr{D}$.

If (α_1), (α_2) are any two distinct points, then they lie on a line $[\beta]$ if, and only if, $\alpha_1 \beta^{-1}$ and $\alpha_2 \beta^{-1}$ are both in $\mathscr{D}$. Since $\alpha_1 \neq \alpha_2$, $\alpha_1 \alpha_2^{-1} \neq 1$ and so, by assumption, there are unique elements δ_1, δ_2 in $\mathscr{D}$ such that

$\alpha_1 \alpha_2^{-1} = \delta_1 \delta_2^{-1}$. If we put $\beta = \delta_2^{-1} \alpha_2$ then it is clear that $\alpha_1 \beta^{-1} = \delta_1$ and $\alpha_2 \beta^{-1} = \delta_2$ so that the two points both lie on $[\beta]$. The proof that β is unique is left to the reader. A similar argument shows that any pair of distinct lines have a unique common point.

Clearly the set of points on a line, or the set of lines through a point, is in one-to-one correspondence with the elements in $\mathscr{D}$. Since every element of Γ has a representation of the form $\delta_1 \delta_2^{-1}$, $\mathscr{D}$ must have at least three elements, otherwise we would have $|\Gamma| < 3$. Exercise 13.13, at the end of this proof, shows that this is enough to make the set $\mathscr{P}$ of points (α) and lines $[\beta]$ be a projective plane.

For any γ in Γ the mapping which sends the point (α) to $(\alpha\gamma)$ and the line $[\beta]$ to $[\beta\gamma]$ is easily seen to be a collineation of $\mathscr{P}$. Thus Γ acts (faithfully) as a collineation group of $\mathscr{P}$. Obviously this collineation group is regular on the points and lines of $\mathscr{P}$. ☐

Exercise 13.13. Show that if $\mathscr{P}$ is any non-empty set of points and lines with an incidence relation such that any pair of distinct points (lines) are on exactly one line (point), then if every line has more than two points and every point is on more than two lines it follows that $\mathscr{P}$ is a projective plane.

Exercise 13.14. Show that the plane $\mathscr{P}$ constructed in Theorem 13.13 is (up to isomorphism) the only projective plane on which Γ can act as a Singer group with the given difference set $\mathscr{D}$.

Exercise 13.15. If $\mathscr{P}$ is a projective plane with a Singer group Γ and difference set $\mathscr{D}$, show that, for any α, β in Γ, $\alpha^{-1} \mathscr{D} \beta$ is also a difference set for Γ acting on $\mathscr{P}$, and that any difference set for Γ acting on $\mathscr{P}$ is of this form.

Before continuing our discussion of Singer groups we recall the construction of a Singer group for a pappian plane given in Chapter II, Section 10. Let $\mathscr{P} = \mathscr{P}_2(K)$ and let F be a cubic extension of K. Then F is, in a natural way, a three-dimensional vector space over K and the left multiplications $L_a : x \to ax$ are linear transformations which are non-singular provided $a \neq 0$. Then the set of all $L_a, a \neq 0$, induces a collineation group Γ of $\mathscr{P}$ which is a Singer group for $\mathscr{P}$ and is isomorphic to F^*/K^*. Clearly each Γ constructed in this way is abelian and if K is finite then Γ is even cyclic. Furthermore we note that if K is finite then it always possesses a cubic extension so that, in this case, $\mathscr{P}$ always has at least one Singer group. That this group is not unique, even up to isomorphism, is shown by the following exercise.

*Exercise 13.16. If F is a nearfield of dimension 3 over its kernel K show that there is a Singer group isomorphic to F^*/K^* acting on $\mathscr{P}_2(K)$. Show that there are finite André systems F of this sort such that F^*/K^* is not abelian. (Hint: find an associative André quasifield (not a field), of order 64.)

When Γ is cyclic and finite, it is convenient to represent it additively as the cyclic group Z_v, i.e. the integers modulo $v = n^2 + n + 1$, where n is the order of the plane. The actual construction of the difference set $\mathscr{D}$ for a given desarguesian plane is straightforward but labourious, since it involves nothing more than following the recipe in Chapter II. In this case the points of $\mathscr{P}_2(K)$ are identified with the elements of F (and their "scalar" multiples), and then the multiplication of F is used to tell us how to move the points around. We now give a brief sketch of the construction for $K = GF(2)$.

First we construct $F = GF(8)$ by adjoining an element λ to $GF(2)$, where $\lambda^3 = 1 + \lambda$. Then we write down the elements of F in order

$$\lambda^0 = 1$$
$$\lambda^1 = \quad\;\; \lambda$$
$$\lambda^2 = \qquad\quad \lambda^2$$
$$\lambda^3 = 1 + \lambda$$
$$\lambda^4 = \quad\;\; \lambda + \lambda^2$$
$$\lambda^5 = 1 + \lambda + \lambda^2$$
$$\lambda^6 = 1 \qquad\; + \lambda^2,$$

and, of course, $\lambda^7 = 1$. If $\Gamma = \langle\lambda\rangle$ then the seven points of $\mathscr{P}$ are written in the (cyclic) order that Γ induces on them. Since 1, λ and $1 + \lambda = \lambda^3$ are the non-zero vectors of a two-dimensional subspace, we can use "$P = 1$" and "$k = $ line on 1 and λ" as the base point and line to construct our difference set $\mathscr{D}$. It is then easy to see that $\mathscr{D}$ consists of the three elements L_a where $a = 1$, λ, λ^3. Thus if we had chosen to write Γ additively as the integers modulo 7, then $\mathscr{D}$ could be represented by 0, 1, 3.

Exercise 13.17. Repeat the above example for the field $K = GF(3)$.

A difference set for a Singer group often provides a particularly easy way to represent the plane. For instance, in the example above, we can give the seven lines of $\mathscr{P}_2(2)$ by writing down $\mathscr{D}$ and its seven "translates" $\mathscr{D} + a$:

$$
\begin{array}{ccc}
0 & 1 & 3 \\
1 & 2 & 4 \\
2 & 3 & 5 \\
3 & 4 & 6 \\
4 & 5 & 0 \\
5 & 6 & 1 \\
6 & 0 & 2 \,.
\end{array}
$$

Here we think of the integers modulo 7 as the points and the seven sets above as the lines. For many purposes (e.g. in coding theory) this sort of representation is extremely useful.

The following theorem often helps in the complete construction of a difference set, without computing any extension fields at all. In addition it also serves as a powerful non-existence theorem. As the proof has nothing to do with our geometric considerations we consign it as usual to an appendix at the end of the chapter; the theorem is M. Hall's.

Theorem 13.14. *Let $\mathscr{P}$ be a finite projective plane of order n and suppose that Γ is an abelian Singer group for $\mathscr{P}$ with difference set $\mathscr{D}$. Let p be any prime divisor of n and let ϕ be the mapping of Γ given by $\alpha^\phi = \alpha^p$. Then there is an element λ of Γ such that $\mathscr{D}^\phi = \mathscr{D}\lambda$. Furthermore the mapping $\bar{\phi}$ defined below is a collineation of $\mathscr{P}$ which normalizes Γ.*

$$\bar{\phi} : (\alpha) \to (\alpha^\phi),$$

$$[\beta] \to [\lambda\beta^\phi].$$

If n is a prime power then clearly n has only one prime divisor and there is only one such mapping $\bar{\phi}$. In our next theorem we look at the collineation group of $\mathscr{P}$ generated by all possible $\bar{\phi}$ and will establish results which will prove the non-existence of planes with abelian Singer groups for certain non-prime power orders.

Theorem 13.15. *Let $\mathscr{P}$ be a finite projective plane of order n and suppose that Γ is an abelian Singer group for $\mathscr{P}$. Let Φ be the group of all collineations of $\mathscr{P}$ generated by the mappings $\bar{\phi}$ of Theorem 13.14 as p varies over all prime divisors of n. Then there is a line of $\mathscr{P}$ which is fixed by all the elements of Φ.*

Proof. We first note, although the observation is not essential to the proof, that there is a line of $\mathscr{P}$ fixed by all the elements of Φ if and only if there is a choice of difference set $\mathscr{D}$ such that $\mathscr{D}$ is fixed by all the elements ϕ.

For any integer m let ϕ_m be the mapping given by $\alpha^{\phi_m} = \alpha^m$. Then $\bar{\phi}_n$ is an element of Φ and $\bar{\phi}_n$ fixes those points (α) such that $\alpha^{n-1} = 1$. Since the g.c.d. of $n-1$ and $n^2 + n + 1$ is either 1 or 3, there are either 1 or 3 such points (α).

Case (a). $n \not\equiv 1 \pmod 3$.

In this case the g.c.d. of $n-1$ and $n^2 + n + 1$ is 1 so that $\bar{\phi}_n$ fixes only one point, namely (1), and so, by Theorem 13.3, $\bar{\phi}_n$ has exactly one fixed line. But all the elements of Φ commute with $\bar{\phi}_n$, so they must all fix the same point and line as $\bar{\phi}_n$.

Case (b). $n \equiv 1 \pmod 3$.

In this case the g.c.d. of $n-1$ and $n^2 + n + 1$ is 3 so that $\bar{\phi}_n$ fixes three points, (1), (λ), (λ^2). If these three points were to lie on a line $[\beta]$, then

$1 = \delta_1\beta, \lambda = \delta_2\beta$ and $\lambda^2 = \delta_3\beta$ would imply $\delta_2\delta_1^{-1} = \lambda = \lambda^2\lambda^{-1} = \delta_3\delta_2^{-1}$ which is a contradiction. Thus $\bar{\phi}_n$ fixes the three sides of the triangle with vertices (1), (λ), (λ^2). Any other mapping $\bar{\phi}_m$ say, must fix the point (1) and, since it commutes with $\bar{\phi}_n$, must either permute the points (λ) and (λ^2) or fix them. In either case $\bar{\phi}_m$ must fix the line joining (λ) to (λ^2) and the theorem is proved. ☐

As an illustration of one of its applications we now show how this theorem could have solved the problem of finding a difference set for the plane of order 2. Any group Γ of order 7 is cyclic and, therefore, may be represented as the additive group Z_7. The difference set $\mathcal{D}$ is a subset of three elements and must be such that $\bar{\phi}_2$ leaves it invariant. But, additively, ϕ_2 sends α in Γ onto 2α and thus if α is in $\mathcal{D}$ then so is 2α and 4α (Note that $8\alpha = \alpha$). If we now choose $\alpha = 1$ we have $\mathcal{D} = \{1, 2, 4\}$ which is equivalent to our previous solution.

In a similar way, Theorem 13.14 and 13.15 (plus a little ingenuity), give constructions for difference sets for many prime power orders. But they have an even more powerful use: they permit us to assert that no difference sets can exist for abelian Singer groups of many non-prime power orders. For instance, suppose Γ is an abelian Singer group for a projective plane of order n and that n is divisible by both 2 and 3. If $\mathcal{D}$ is the difference set left invariant by Φ then, since $\bar{\phi}_2$ and $\bar{\phi}_3$ leave $\mathcal{D}$ invariant, if α is in $\mathcal{D}$ then so is $\alpha^{\phi_2} = \alpha^2$ and $\alpha^{\phi_3} = \alpha^3$. But this implies that $\alpha = \alpha^3(\alpha^2)^{-1} = \alpha^2(\alpha)^{-1}$ are two different representations of α as $\delta_1\delta_2^{-1}$. Since this is impossible for any non-identity element of Γ, we must have $\alpha = 1$, i.e. $\mathcal{D} = \{1\}$. But $\mathcal{D}$ must contain more than one element, so we have a contradiction and no such Γ can exist.

Exercise 13.18. Show that there cannot exist an abelian Singer group for a plane whose order is divisible by any of the following pairs of primes: (2, 5), (3, 5), (2, 7), (3, 7).

Exercise 13.19. Using Theorems 13.14 and 13.15 construct a Singer group and difference set for a plane of order 8, and show that they are essentially unique.

We close this section with two final remarks on difference sets. Suppose $\mathcal{P}$ is a projective plane of order n and that it possesses a cyclic Singer group Γ with difference set $\mathcal{D}$. Then it is easy to see that an incidence matrix for $\mathcal{P}$ may be constructed as follows: write down the top row of the matrix with a 0 in every position excepting in position i, where i is in $\mathcal{D}$ (here we must assume that the first position is position 0, the second is position 1 etc.). Then "cycle" the top row one place to the right for the second row and continue in this manner until all the $n^2 + n + 1$ rows have been filled in. The resulting matrix is an incidence matrix for $\mathcal{P}$. Conversely, if a finite projective plane has a cyclic incidence matrix in the sense defined here, it can be shown that it has a cyclic Singer group.

The second observation is that if $\mathcal{P}$ is a projective plane with an abelian Singer group Γ then the mapping $(\alpha) \leftrightarrow [\alpha^{-1}]$ is a polarity of $\mathcal{P}$ whose absolute points are those points (α) such that α^2 is in $\mathcal{D}$. Since Γ has odd order, every element of Γ is the square of some element of Γ. Thus every element of $\mathcal{D}$ is the square of an element of Γ and the polarity has $|\mathcal{D}| = n + 1$ absolute points. Hence by Theorem 12.7, these points all lie on a line (if n is even) or form an oval (if n is odd).

*Exercise 13.20. If $\mathcal{P}$ is a finite projective plane of square order with an abelian Singer group, then show that $\mathcal{P}$ has a unitary polarity (see Chapter XII).

6. Some Non-existence Theorems

In the previous section we proved some theorems of the following type: if a plane of order n has a collineation group with given properties then n must satisfy certain restrictions. A number of other results of this kind exist, some of which have very involved or difficult proofs; but a few are readily accessible. We first mention, without any proof, two of the more difficult theorems. The first concerns Singer groups: as we have already seen, every finite desarguesian projective plane has a cyclic Singer group, and it has been conjectured that any finite plane with a Singer group (cyclic or not) must be desarguesian. This conjecture is, as yet, unresolved. For infinite planes the situation is very different. It is easy to construct non-desarguesian planes with cyclic Singer groups and it was conjectured that any infinite plane with a cyclic Singer group was non-desarguesian. This conjecture was finally proved true by Karzel [3].

Theorem 13.16. *If an infinite projective plane $\mathcal{P}$ possesses a cyclic Singer group, then $\mathcal{P}$ is not desarguesian.*

The next result is due to Hughes and concerns the non-existence of certain collineations in planes of given order n. Its nature is very similar to that of the Bruck-Ryser Theorem (see Chapter III) in the sense that both are concerned with number theoretic formulations of certain combinatorial situations. However the "elementary" proof of the Bruck-Ryser Theorem given in Chapter III seems to be a happy accident, for no such easy proof is known for this.

Theorem 13.17. *Let $\mathcal{P}$ be a finite projective plane of order n, and suppose $\mathcal{P}$ has a collineation α of prime order $p > 2$, fixing N points of $\mathcal{P}$,* *where* $N \equiv 0 \,(\mathrm{mod}\ 2)$. *Then the equation*

$$x^2 = n y^2 + (-1)^{\frac{p-1}{2}} p z^2$$

has a non-trivial solution in integers for x, y, z.

The following theorem has the same basic flavour as Theorem 13.17, but an elementary proof is available.

Theorem 13.18. *Let* $\mathscr{P}$ *be a projective plane of order* $n \equiv 2 \,(\mathrm{mod}\ 4)$, *and suppose* $\mathscr{P}$ *has a collineation of even order. Then* $n = 2$.

Proof. If there is a collineation of even order, then there is one of order two; let it be α. Since n cannot be a square, α is not a Baer collineation, and since n is even, α must be an elation (see Exercise 4.4). So let k be the axis of α and let Q on k be its centre. The n other lines on Q will be denoted by m_i, $1, 2, \ldots, n$, and the n other points on k by P_i, $i = 1, 2, \ldots, n$. The points (other than Q) on a line m_i fall into $n/2$ orbits of length 2, and we can represent them by $X_{i1}, X_{i1}^{\alpha}, X_{i2}, X_{i2}^{\alpha}, \ldots, X_{it}, X_{it}^{\alpha}$, where $t = n/2$. Similarly, the n lines other than k through P_i can be represented by $w_{i1}, w_{i1}^{\alpha}, w_{i2}, w_{i2}^{\alpha}, \ldots, w_{it}, w_{it}^{\alpha}$. Now we construct a square matrix M of order nt, as follows: the rows will be numbered $(1, 1)$, $(1, 2), \ldots, (1, t), (2, 1), (2, 2), \ldots, (n, 1), (n, 2), \ldots, (n, t)$, and the columns numbered similarly. In the intersection of row (i, j) and column (r, s) we put $+1$ if X_{ij} is on w_{rs}, -1 if X_{ij}^{α} is on w_{rs}, and 0 if neither occurs.

We define the "blocks" of M to be the sets of t rows corresponding to the t point orbits on a single line m_i. So the first block is the top t rows of M, the second block the next t rows, and so on. If we consider two rows in the same block, then their inner product, as vectors, must be zero: for no line w_{rs} can contain a point from two different orbits on a single m_i. If we consider two rows from different blocks, then the first row corresponds say, to the points X_{ij} and X_{ij}^{α}, the second to the points X_{ef}, X_{ef}^{α}, where $i \neq e$. Consider the four lines joining one point on m_i to one point on m_e. The line $y = X_{ij}X_{ef}$ or the line $y^{\alpha} = X_{ij}^{\alpha}X_{ef}^{\alpha}$ is a line w_{rs}, so in the (r, s) column, either both rows have $+1$ or both rows have -1. Next the line $z = X_{ij}X_{ef}^{\alpha}$ or the line $z^{\alpha} = X_{ij}^{\alpha}X_{ef}$ is a line w_{uv}, and so in column (u, v) one row has $+1$ and the other has -1. In all other columns at least one of the two rows has a 0. Hence the inner product of the two row is 0.

In order to compute MM' we now have to determine the inner product of a row with itself. Given an orbit on m_i then, for each j, there is a unique s such that one of the points of the given orbit lies on w_{js}. Thus for a given j, there is a unique s such that the entry in the intersection of row (i, j) with column (j, s) is ± 1; for all other values of s the entry is 0. Each of these non-zero entries contributes a 1 to the inner product of the row with itself. Thus, the inner product of any row with itself is n and we have established

$$MM' = nI. \tag{1}$$

We now perform a somewhat peculiar operation on M to give us a new matrix B with only n rows, but still with nt columns. Considering them as vectors we add all the rows of each block. This gives a new vector which, since no two rows of the same block have a non-zero entry in the same column but every line w_{rs} must meet a line m_i in one (and only one) of its orbits, must have ± 1 for each of its entries. The resulting array of vectors gives us our $n \times nt$ matrix B.

Since the inner product of any two rows of B is the sum of inner products of rows of M it is easy to show

$$BB' = ntI. \tag{2}$$

We now investigate Eq. (2). We are given a matrix B, with n rows and nt columns, where $t = n/2$, all of whose entries are ± 1, satisfying (2). If we multiply any column of B by -1, the new matrix still satisfies (2), and if we permute the columns, (2) still remains satisfied. So we may assume that the top row of B consists of $+1$'s only, while the second row must have half its entries $+1$ and half -1; we may assume that the columns are so arranged that the first $nt/2$ columns have $+1$ in the second row, and the last $nt/2$ have -1.

Now suppose that $n > 2$, so that B has at least three rows. The first three rows can be arranged as follows:

$+1$			
$+1$		-1	
$+1$	-1	$+1$	-1
z	$nt/2-z$	y	$nt/2-y$

Taking inner products between the first and third row, we have $2z + 2y - nt = 0$, so $z + y = nt/2$. Taking inner products between the second and third row, we find $z - (nt/2 - z) - y + nt/2 - y = 0$, or $2z - 2y = 0$. Thus $z = y$, and so $z + y = 2z = nt/2$, hence $z = nt/4$. But since $n \equiv 2 \pmod 4$, it is immediate that 4 does not divide nt. This contradiction shows that n must be 2. □

As an interesting "aside", we see that if there is a projective plane of order 10 then it cannot admit any collineations of even order.

Exercise 13.21. Apply Theorem 13.17 to a possible plane of order 10.

7. Appendix (Hall's Multiplier Theorem)

We shall give a proof of Theorem 13.14, but for various reasons, we will change notation. We shall assume that G is an abelian group of order $n^2 + n + 1$, D is a subset of G of $n + 1$ elements, and every g in G, $g \neq 1$, can be represented in exactly one way as $g = d_1 d_2^{-1}$ for some elements d_1, d_2 in D. Let F be the rational field, and let FG be the group algebra of G over F, where we identify the identities of F, G, and FG, and write a typical element of FG as $\Sigma a_x x$. We define some special elements:

$$\delta = \sum_{d \in D} d, \qquad \sigma = \sum_{g \in G} g,$$

Also we define the mapping * of FG onto FG by: $(\Sigma a_x x)^* = \Sigma a_x x^{-1}$. Then it is well known (and easy to prove) that * is an automorphism of FG onto FG. We note that for any element α in FG, $\alpha = \Sigma a_x x$, then $\alpha \sigma = (\Sigma a_x) \sigma$. Now the condition that D is a difference set expresses itself in FG as:

$$\delta \delta^* = \delta^* \delta = (n + 1) + (\sigma - 1) = n + \sigma. \tag{1}$$

Now let $x^\phi = x^p$, where p is a prime divisor of n, so ϕ is an automorphism of G, and hence the mapping ϕ: $\Sigma a_x x \to \Sigma a_x x^p$ is an automorphism of FG. We note that (by using the "binomial theorem"):

$$\delta^\phi = \Sigma d^p = (\Sigma d)^p + p\alpha = \delta^p + p\alpha, \tag{2}$$

where α is an integral element of FG; that is, an element $\Sigma a_x x$, such that all a_x are integers.

Now

$$\delta^\phi \delta^* = \delta^p \delta^* + p\alpha \delta^* = \delta^{p-1}(n + \sigma) + p\alpha \delta^*$$
$$= \delta^{p-1} \sigma + p(\alpha \delta^* + (n/p) \delta^{p-1}) = \delta^{p-1} \sigma + p\alpha_1,$$

where α_1 is also an integral element.

But $\delta \sigma = (n + 1) \sigma$, and so using the fact that n is a multiple of p, the term $\delta^{p-1} \sigma$ can be reduced successively, finally leading to:

$$\delta^\phi \delta^* = \sigma + p\alpha_2, \tag{3}$$

where α_2 is integral.

Now suppose that $\alpha_2 = \Sigma a_x x$. Then for a particular x in G, the coefficient of x on the left hand side of (3) is certainly not negative, and hence we must have $0 \leq 1 + p a_x$, from which we have $a_x \geq -1/p > -1$. Hence if a_x is to be integral, its coefficients are all non-negative.

Now we compute:

$$(\delta^\phi \delta^*)(\delta^\phi \delta^*)^* = \delta^\phi \delta^* \delta \delta^{*\phi} = (\delta^* \delta)(\delta \delta^*)^\phi = (n + \sigma)(n + \sigma)^\phi = (n + \sigma)^2,$$

since $(n + \sigma)^\phi = n + \sigma$. So

$$(\delta^\phi \delta^*)(\delta^\phi \delta^*)^* = n^2 + 2n\sigma + \sigma^2. \tag{4}$$

But on the other hand the left side of (4) is equal to $(\sigma + p\alpha_2)(\sigma + p\alpha_2)^*$ $= \sigma^2 + 2p\alpha_2 \sigma + p^2 \alpha_2 \alpha_2^*$. Hence:

$$n^2 + 2n\sigma = p^2 \alpha_2 \alpha_2^* + 2p\alpha_2 \sigma. \tag{5}$$

So we must compute $p\alpha_2\sigma$, which we do as follows:

$$\delta^\phi\delta^*\sigma = \delta^\phi(n+1)\sigma = (n+1)(\delta\sigma)^\phi = (n+1)^2\sigma.$$

Also from (3):

$$\delta^\phi\delta^*\sigma = \sigma^2 + p\alpha_2\sigma = (n^2+n+1)\sigma + p\alpha_2\sigma.$$

Equating the two expressions, we have $p\alpha_2\sigma = n\sigma$. So (5) yields:

$$n^2 = p^2\alpha_2\alpha_2^*. \tag{6}$$

Now it is easy to see that if as many as two elements of G were actually present in α_2 with non-zero coefficients, then (since all coefficients are non-negative), the expression $\alpha_2\alpha_2^*$ could not equal a multiple of the identity, but would have to have a non-identity group element in it. Since $\alpha_2\alpha_2^* = (n/p)^2$, it thus follows that $\alpha_2 = (n/p)\,g$, for some element g in G.

If $\beta = 1/n - \sigma/[n(n+1)^2]$, we see that $(n+\sigma)\beta = 1$, and so $\delta\beta$ is an inverse for δ^*. Thus (3) becomes:

$$\delta^\phi\delta^* = ng + \sigma = (n+\sigma)g = (\delta g)\delta^*, \tag{7}$$

and hence $\delta^\phi = \delta g$. The equivalent statement in the group G is that $D^\phi = Dg$.

The remaining statements in Theorem 13.14 are obvious. []

References

Brauer's proof of Theorem 13.4 is in [1] and Theorem 13.17 is proved in [2].

1. Brauer, R.: On the connections between the ordinary and the modular characters of groups of finite order. Ann. Math. **42**, 926–935 (1941).
2. Hughes, D. R.: Collineations and generalized incidence matrices. Trans. Am. Math. Soc. **86**, 284–296 (1957).
3. Karzel, H.: Ebene Inzidenzgruppen. Arch. Math. **40**, 10–17 (1964).

XIV. Wagner's Theorem

1. Introduction

In this chapter we prove that a finite affine plane which admits a collineation group acting transitively on its affine lines is a translation plane. As an immediate corollary we show that a finite projective plane with a collineation group acting doubly transitive on its points is desarguesian.

The proof, which is virtually the original one of Wagner [3], uses a number of results on permutation groups. Most of these are standard results which may be found in Wielandt [4] and which are taught in most elementary courses. One result, however, was proved after Wielandt's book was published and so, since it is crucial to the proof, we include a proof of this result in the appendix at the end of the chapter.

2. Some Preliminary Results on Permutation Groups

In this section we list those results on permutation groups which are needed in the proof and have not previously been used in the book.

Result 14.1. *Let Π be a doubly transitive permutation group on a set $\mathcal{S}$ of $n+1$ elements, where n is odd. If, for every element a of $\mathcal{S}$, Π_a contains a subgroup Γ_a of even order which is a Frobenius group on $\mathcal{S} \backslash \{a\}$, then Π_a is primitive on $\mathcal{S} \backslash \{a\}$.*

Result 14.2. *Let Λ be a Frobenius permutation group of order $2n$ on a set $\mathcal{T}$ of n points, where n is odd. Then*

(i) given any pair of distinct points a, b of $\mathcal{T}$ there exists an element of order 2 in Λ which interchanges a and b.

(ii) if Λ' is a subgroup of Λ containing r elements of order 2, $r \geq 1$, then the order of Λ' is $2r$.

Result 14.3. *Let Π be a primitive permutation group on a set $\mathcal{T}$ of n points and let Π contain a normal soluble subgroup Γ. Then Π contains a unique minimal normal subgroup Σ. Moreover Σ is elementary abelian and regular on $\mathcal{T}$, so that n is a prime power. Finally Σ is a subgroup of Γ.*

These three results can be combined to give:

Result 14.4. *Let Π be a doubly transitive permutation group on a set $\mathscr{S}$ of $n+1$ elements, where n is odd. Further let the subgroup Π_a contain a subgroup Γ_a of even order which is a Frobenius group on $\mathscr{S}\backslash\{a\}$. If Σ_a, the kernel of Γ_a, is normal in Π_a then Σ_a is elementary abelian and n is a prime power.*

In order to apply the earlier results to prove Result 14.4, note that, in Result 14.4, since Γ_a contains an element of order 2 not in its kernel, Σ_a is abelian and therefore soluble (see [1]).

Results 14.1 and 14.2 were proved by Wagner with the specific aim of proving the main theorem of this chapter. These results are proved in the appendix.

Finally we list an elementary result which is frequently used in proving Wagner's theorem.

Result 14.5. *Let Γ be a permutation group on a set $\mathscr{S}$. If Σ is a normal subgroup of Γ and if $\mathscr{T}$ is the subset of all elements fixed by Σ then Γ leaves $\mathscr{T}$ invariant.*

3. Two Preliminary Theorems

Theorem 14.6. *Let $\mathscr{A}$ be a finite affine plane and Π be a collineation group of $\mathscr{A}$. Then the following statements are equivalent:*

(i) *Π is transitive on the affine lines of $\mathscr{A}$.*

(ii) *Π is transitive on the affine points of $\mathscr{A}$ and Π is transitive on the special points of $\mathscr{A}$.*

(iii) *Π is transitive on the affine flags of $\mathscr{A}$. (An affine flag is an incident point-line pair (V, l) where V is an affine point and l is an affine line.)*

Proof. Let $\mathscr{A} = \mathscr{P}^{l_\infty}$.

If (i) is true then Π, considered as a collineation group of $\mathscr{P}$, has two line orbits and so, by Theorem 13.4, Π splits the points of $\mathscr{P}$ into two orbits. Since Π fixes l_∞ these orbits must be the affine points of $\mathscr{A}$ and the points of l_∞. Thus we have shown that (i) implies (ii).

Suppose now that (ii) is true. Let C, l and C', l' be any two affine flags of $\mathscr{A}$ and let $D = l \cap l_\infty$, $D' = l' \cap l_\infty$. If the order of $\mathscr{A}$ is n then, by Theorem 3.11, C and C' lie in an orbit of length n^2 while D and D' lie in an orbit of length $n+1$. Since $(n+1, n^2) = 1$, it follows from Result 1.13 that there is a collineation $\pi \in \Pi$ mapping C onto C' and D onto D'. Clearly π maps the flag C, l onto the flag C', l' and we have shown that (ii) implies (iii).

Finally (iii) trivially implies (i) and the theorem is proved. $\quad\square$

If Π is transitive on the affine flags of $\mathscr{A}$ then, since Π_{C^α, l^α} $= \alpha^{-1} \Pi_{C,l} \alpha$, $|\Pi_{C,l}|$ is the same for all affine flags C, l of $\mathscr{A}$. Simple counting now gives

Corollary. *Let $\mathscr{A}$ be a finite affine plane of order n with a collineation group Π transitive on the affine lines. If C, l is any affine flag of $\mathscr{A}$ and D is any point of l_∞ then, putting $|\Pi_{C,l}| = k$, we have: $|\Pi| = n^2(n+1)k$, $|\Pi_D| = n^2 k$, $|\Pi_l| = nk$, $|\Pi_C| = (n+1)k$, $|\Pi_{C,D}| = k$.*

Theorem 14.7. *Let $\mathscr{A}$ be a finite affine plane and Π a collineation group of $\mathscr{A}$. Then Π is transitive on the affine points of $\mathscr{A}$ if, and only if, for every point D on l_∞ the subgroup Π_D is transitive on the affine lines through D.*

Proof. Let Π split the points of l_∞ into s orbits.

(1) Suppose that Π is transitive on the affine points of $\mathscr{A} = \mathscr{P}^{l_\infty}$. Then the total number of point orbits of Π, considered as a collineation group of $\mathscr{P}$, is $s + 1$ so that, by Theorem 13.4, Π has $s + 1$ line orbits in $\mathscr{P}$. As one line orbit is l_∞, Π has s orbits of affine lines. If X, Y are special points in different orbits of Π then, clearly, Π cannot map an affine line through X onto an affine line through Y. Thus, since the number of affine line orbits is the same as the number of special point orbits, all the affine lines which intersect l_∞ in points of the same orbit must themselves lie in one orbit. In particular Π is transitive on the lines through any given special point D. However, any collineation mapping an affine line through D onto another such line must fix D and, hence, Π_D is transitive on the affine lines through D.

Suppose that for all D on l_∞, Π_D is transitive on the affine lines through D. Then, clearly, Π splits the affine lines of $\mathscr{A}$ into s orbits and, thus, has $s + 1$ line orbits in $\mathscr{P}$. Using Theorem 13.4, it follows that Π has $s + 1$ point orbits in $\mathscr{P}$ and so, since the points of l_∞ lie in s orbits, Π is transitive on the affine points of $\mathscr{A}$. $\quad\square$

Corollary. *Let $\mathscr{A}$ be an affine plane and Π a collineation group transitive on the affine lines of $\mathscr{A}$. Further let R, Q be two special points of $\mathscr{A}$ and l, m two affine lines through Q. Then $|\Pi_{R,Q,l}| = |\Pi_{R,Q,m}|$.*

Proof. By Theorem 14.6 Π has two point orbits in $\mathscr{P}$, namely the affine points of $\mathscr{A}$ and the points of l_∞. Since $(n^2, n+1) = 1$, Π_R is transitive on the affine points of $\mathscr{A}$ (by Result 1.13) and hence, by Theorem 14.7, $\Pi_{R,Q}$ is transitive on the affine lines through Q. The corollary now follows immediately. $\quad\square$

4. Some Special Cases

In this section we prove the theorem for certain special cases. More precisely, we show that an affine plane of order n which admits a

4. Some Special Cases

collineation group transitive on its affine lines must be a translation plane if n is either even or a prime power. The rest of the chapter is then devoted to showing that if n is odd then n must be a power of a prime.

Lemma 14.8. *Let $\mathscr{A}$ be a finite affine plane of order n and Π a collineation group transitive on the affine lines of $\mathscr{A}$. Then, if n is either even or a power of an odd prime, $\mathscr{A}$ is a translation plane and Π contains the translation group of $\mathscr{A}$.*

Proof. As a consequence of Theorem 14.6 (ii) and the Corollary to Theorem 4.26 it is sufficient to show that Π contains a non-trivial translation.

For any affine flag C, l let $|\Pi_{C,l}| = k$.

Case (i): n is even.

Let $2^u \| n$ and $2^v \| k$. (Recall $2^u \| n$ implies $2^u | n$ but $2^{u+1} \nmid n$.) By assumption, $u \geq 1$. If Γ is a Sylow 2-subgroup of Π then, by the Corollary to Theorem 14.6, $|\Gamma| = 2^{2u+v}$. Let λ be an element of order 2 in the centre of Γ (such a λ certainly exists since any p-group has a non-trivial centre), let $\mathscr{F}$ be the set of affine points of $\mathscr{A}$ fixed by λ and let $|\mathscr{F}| = s$. By Theorem 4.2 λ is of one of the following types:

(a) λ is a translation. In this case $s = 0$.

(b) λ is an elation with affine axis. In this case $s = n$.

(c) λ fixes a Baer subplane pointwise. In this case $s = n$.

We shall assume that $s = n$ and show that this leads to a contradiction. Since λ is in the centre of Γ, Result 14.5 implies that Γ is a permutation group on the points of $\mathscr{F}$. Let X be any point of $\mathscr{F}$ and let t be the length of the orbit of X under Γ. Since Γ_X is a subgroup of Π_X and $|\Pi_X| = (n+1)k$, $2^v \| |\Pi_X|$ and so $|\Gamma_X||2^v$. However $|\Gamma| = |X\Gamma||\Gamma_X| = 2^{2u+v}$, so that $2^{2u}|t$. Thus we have shown that the number of points in every orbit of Γ is divisible by 2^{2u} and so, since $\mathscr{F}$ is a union of point orbits under Γ, $2^{2u}|s$, i.e. $2^{2u}|n$. But $2^u \| n$ and $u > 0$. This contradiction shows that λ is a translation and, consequently, that $\mathscr{A}$ is a translation plane.

Case (ii): $n = p^s$, with p an odd prime.

Let $p^t \| k$. Let A_1 be any special point of $\mathscr{A}$ and let Γ be a Sylow p-subgroup of Π_{A_1} then, by the Corollary to Theorem 14.6, $|\Gamma| = p^{2s+t}$ (and Γ is a Sylow subgroup of Π). If C is any affine point of $\mathscr{A}$ and if the orbit of C under Γ has length p^u then, by Result 1.13, $|\Gamma_C| = p^{2s+t-u}$. However Γ_C is a subgroup of $\Pi_{A_1,C}$ and, by the Corollary to Theorem 14.6 $|\Pi_{A,C}| = k$, so that $p^{2s+t-u}|k$. Thus $2s+t-u \leq t$ or $2s \leq u$. But, clearly, the orbit of C cannot have length more than p^{2s} so that $2s = u$ and Γ is transitive on the affine points of $\mathscr{A}$.

Let A_2 be any other point of l_∞ such that for all A_k on l_∞, $A_k \neq A_1$, $|\Gamma_{A_2}| \geq |\Gamma_{A_k}|$. Since Γ is transitive on the affine points of $\mathscr{A}$, Γ_{A_2} is transitive on the affine lines through A_2 (by Theorem 14.7). Γ_{A_2} is a p-subgroup of Π_{A_2} so that $\Gamma_{A_2} \subseteq \Gamma'_{A_1}$ where Γ' is a Sylow p-subgroup of

Π_{A_2}. The number of points on l_∞ is $p^s + 1$ which means that any Sylow p-subgroup of Π is a Sylow p-subgroup of Π_{A_k} for some A_k on l_∞. Hence Γ, Γ' are both Sylow p-subgroups of Π and, as such, are conjugate in Π. Thus, as A_2 was chosen to make $|\Gamma_{A_2}|$ maximal amongst all $|\Gamma_{A_k}|$, $k \neq 1$, $|\Gamma_{A_2}| \geq |\Gamma'_{A_j}|$ for all $j \neq 2$ and so, in particular, $|\Gamma_{A_2}| \geq |\Gamma'_{A_1}|$. Hence $\Gamma_{A_2} = \Gamma'_{A_i}$ and it follows from the previous argument that Γ_{A_2} is transitive on the affine lines through A_1.

Let $A_1, A_2, \ldots, A_i$ be all the points of l_∞ left fixed by Γ_{A_2}. Then, since Γ_{A_2} is a subgroup of Γ_{A_j} ($j = 3, \ldots, i$), it follows from the maximal property of Γ_{A_2} that $\Gamma_{A_2} = \Gamma_{A_j}$ ($j = 2, \ldots, i$). Consequently by Theorem 14.3, Γ_{A_2} is transitive on the affine lines through A_j.

If λ is an element of order p in the centre of Γ_{A_2} then, since λ fixes at least two points of l_∞, by Theorem 13.3, λ must fix at least two lines and, consequently, must fix an affine line of $\mathscr{A}$. If λ fixes a line through A_j ($j = 1, \ldots, i$) then the transitivity of Γ_{A_2} on the lines through A_j and Result 14.5 imply that λ fixes A_j linewise: i.e. that λ is a perspectivity with centre A_j. Clearly λ can fix no affine point C as this would imply that λ fixed the lines CA_1, CA_2 and, by the previous argument, λ would be a perspectivity with distinct centres A_1, A_2. Thus, if λ is a perspectivity with centre A_j then λ is a translation.

Suppose λ fixes a line l such that $l \cap l_\infty \neq A_j$ ($j = 1, 2, \ldots, i$). Then the orbit of l under Γ_{A_2} must contain a line m such that $l \cap m = D$ is an affine point and so, by Result 14.4, λ fixes l, m and hence D. But we have just shown that λ cannot fix an affine point and thus λ is a translation with centre one of the A_j ($j = 1, \ldots, i$). $\square$

5. Planes of Type $\mathscr{H}$

A finite projective plane $\mathscr{P}$ is said to be of *type $\mathscr{H}$* if $\mathscr{P}$ contains a point O and a line l, O not on l, such that, for any pair of distinct points A, B on l, $\mathscr{P}$ admits an involutory (B, OA)-honlogy.

Theorem 14.9. *If $\mathscr{P}$ is a finite projective plane of type $\mathscr{H}$ then the order of $\mathscr{P}$ is a power of a prime.*

Proof. Let Π denote the collineation group of $\mathscr{P}$ generated by all involutory homologies which have centre on l and axis through O, and let Γ_A be the subgroup of Π generated by all involutory homologies of $\mathscr{P}$ with centre on l and axis OA, where A is on l. Note that in general $\Gamma_A \neq \Pi_A$.

We now regard Π as a permutation group on the points of l. By the dual of Theorem 4.25 (André's Theorem) $\mathscr{P}$ is (A, AO)-transitive and thus Γ_A is transitive on the points of $l \setminus \{A\}$. Furthermore since every element of Γ_A fixes OA pointwise the only element in Γ_A fixing two distinct points of $\Gamma \backslash \{A\}$ is the identity (i.e. Γ_A is a Frobenius permutation group on the points of $\Gamma \backslash \{A\}$).

Since Π clearly cannot fix A, the transitivity of Γ_A immediately implies that Π acts as a doubly transitive group on the points of l. Finally, since the conjugate of an involutory homology is again an involutory homology, Γ_A is normal in Π_A.

Theorem 14.9 is now proved by appealing to Result 14.4. (In the situation of Theorem 14.9 the kernel of Γ_A is, clearly, the group of elations with axis OA. Since this group is normal in Π_A, the conditions of Result 14.4 are satisfied.) $\square$

6. 2-subplanes

If Π is a collineation group of a projective plane $\mathscr{P}$ a subplane $\mathscr{Q}$ of $\mathscr{P}$ is called a 2-*subplane with respect to* Π if there exists a 2-subgroup of Π whose fixed elements are the points and lines of $\mathscr{Q}$. A *minimal* 2-subplane with respect to Π is a 2-subplane with respect to Π containing no proper subplane with the same property. Since the identity is a 2-group, $\mathscr{P}$ is a 2 subplane of itself so that 2 subplanes exist for any plane with respect to any collineation group. If $\mathscr{P}$ is finite then minimal 2-subplanes also exist for any collineation group.

Theorem 14.10. *Let $\mathscr{P}$ be a finite projective plane and let $\mathscr{Q}$ be a 2-subplane with respect to a collineation group Π. If the order of $\mathscr{P}$ is n and the order of $\mathscr{Q}$ is m, then $n = m^{2^g}$ for some integer g.*

Proof. Let Γ be the 2-subgroup of Π whose fixed elements are the points and lines of $\mathscr{Q}$. If $\Gamma = 1$ then $\mathscr{P} = \mathscr{Q}$ and there is nothing to prove. If $\Gamma \neq 1$ then let γ be a non-trivial involution in the centre of Γ. Since γ fixes the points and lines of $\mathscr{Q}$, γ cannot be a perspectivity and so, by Theorem 4.3, γ fixes a Baer subplane $\mathscr{P}_0$, say, pointwise. Furthermore as $\mathscr{P}_0$ is a Baer subplane the order of $\mathscr{P}_0$ is $\sqrt{n}$. Since γ is in the centre of Γ Result 14.5 shows that Γ leaves $\mathscr{P}_0$ invariant and induces on $\mathscr{P}_0$ a collineation group $\Gamma_0 \cong \Gamma/\Sigma$ where Σ is the subgroup of Γ fixing $\mathscr{P}_0$ pointwise. If $\Gamma_0 = 1$ the result follows and if $\Gamma_0 \neq 1$, we may repeat the argument with $\mathscr{P}_0$ replacing $\mathscr{P}$ and Γ_0 replacing Γ. This process may be repeated until, say at the gth step, $\mathscr{Q} = \mathscr{P}_{g-1}$. Then $n = m^{2^g}$. $\square$

7. Proof of Wagner's Theorem

We now use the preceding results to prove:

Theorem 14.11 (Wagner's theorem). *Let $\mathscr{A}$ be a finite affine plane and Π a collineation group transitive on the affine lines of $\mathscr{A}$. Then $\mathscr{A}$ is a translation plane and Π contains the translation group of $\mathscr{A}$.*

Proof. We shall assume that $\mathscr{A}$ has odd order and show that this implies the order of $\mathscr{A}$ in a prime power. Lemma 14.8 will then prove the theorem.

Throughout the proof we adopt the following notation. If $\mathscr{A} = \mathscr{P}^{l_\infty}$, then $\mathscr{Q}$ is a minimal 2-subplane with respect to Π and $\mathscr{B} = \mathscr{Q}^{l_\infty}$. (Note that since Π is a collineation group of $\mathscr{A}$, Π fixes l_∞ and so l_∞ is in $\mathscr{Q}$.) Further Γ will be a maximal 2-subgroup of Π such that the fixed elements of Γ are the points and lines of $\mathscr{Q}$. Finally Σ is the subgroup of Π leaving $\mathscr{Q}$ invariant.

Wagner's theorem is an immediate consequence of the following lemmas.

Lemma 14.12. *Let $\mathscr{A}$, $\mathscr{B}$, Π, Γ, Σ be defined as above and let $\mathscr{A}$ have odd order n. Regarding Σ as a collineation group of $\mathscr{B}$ we have:*

(i) *for every affine flag in $\mathscr{B}$ there is an involutory homology in Σ fixing that flag.*

(ii) *if A, B are special points of $\mathscr{B}$ with l an affine line of $\mathscr{B}$ through B and if there is an involutory homology in Σ fixing A, B, l then if m is any other affine line of $\mathscr{B}$ through B there exists an involutory homology in Σ fixing A, B, m.*

Proof. Let $2^u \| n+1$, $2^v \| k$ (where $|\Pi_{C,l}| = k$ for any affine flag C, l) and $|\Gamma| = 2^w$.

Since Γ fixes a subplane pointwise, $\Gamma \subseteq \Pi_{C,l}$ for some affine flag C, l and, consequently, $w \leq v$. However, since n is odd, $u \geq 1$ so that Γ is not a Sylow 2-subgroup of Π. Thus there is a 2-subgroup Γ' of Π such that Γ' contains Γ as a normal subgroup of index 2. Since $\Gamma \lhd \Gamma'$, Γ' permutes the fixed points and lines of Γ, i.e. Γ' induces a collineation group on $\mathscr{B}$. Furthermore the choice of Γ as a maximal 2-group whose fixed elements are the points and lines of $\mathscr{B}$ implies that Γ' does not induce the identity on $\mathscr{B}$. The order of the group induced on $\mathscr{B}$ is 2 and so, since the minimality of $\mathscr{B}$ means that Γ' cannot induce a Baer involution on $\mathscr{B}$, Γ' acts on $\mathscr{B}$ as a collineation group generated by an involutory homology. Clearly Γ' must fix an affine flag, D, m say, in $\mathscr{B}$ so that $\Gamma' \subseteq \Pi_{D,m}$. Hence since $|\Gamma'| = 2^{w+1}$ and $|\Pi_{D,m}| = k$, $w+1 \leq v$.

Now let C, l be any affine flag of $\mathscr{B}$. We have shown that Γ is not a Sylow 2-subgroup of $\Pi_{C,l}$ so that there is a 2-subgroup Γ^* in $\Pi_{C,l}$ which contains Γ as a normal subgroup of index 2. The above argument then shows that Γ^* contains an element which acts on $\mathscr{B}$ as an involutory homology fixing C, l and proves (i).

Suppose the conditions of (ii) are satisfied.

The proof is very similar to that of (i); firstly we must show that, if $A, B, l \in \mathscr{B}$, then Γ is not a Sylow 2-subgroup of $\Pi_{A B, l}$. Clearly it is sufficient to show that Γ is not a Sylow 2-subgroup of $\Sigma_{A B, l}$. If $\bar{\Sigma}$ is the subgroup of Σ fixing $\mathscr{B}$ pointwise then $\Gamma \subseteq \bar{\Sigma} \lhd \Sigma_{A,B,l}$ and by assumption, $\Sigma_{A B,l}/\bar{\Sigma}$ contains an element of order 2. Hence Γ is not a Sylow 2-subgroup of $\Sigma_{A B, l}$. By the Corollary of Theorem 14.7 $|\Gamma_{A,B,l}| = |\Gamma_{A,B,m}|$

so that Γ is not a Sylow 2-subgroup of $\Pi_{A\,B,m}$ and, as in the proof of (i), there is a 2-subgroup Γ^* of $\Pi_{A\,B,m}$ which contains Γ as a normal subgroup of index 2. Proceeding as in the proof of (i), Γ^* contains an element which acts on $\mathscr{B}$ as an involutory homology fixing A, B, l. $\square$

Lemma 14.13. *Let $\mathscr{B}$ be a finite affine plane with a collineation group Σ having properties* (i), (ii) *of Lemma 14.12. Then $\mathscr{B}$ is either a translation plane, the dual of a translation plane or is of type $\mathscr{H}$.*

Proof. The proof consists of a detailed case analysis.

Case I. Σ contains no involutory homology with affine centre.

By condition (i) Σ contains an involutory homology α. Let α have centre A, where $A \in l_\infty$, and axis l and let $B = l \cap l_\infty$. If m is any other affine line through B then, since α fixes A, B, l, it follows from condition (ii) that Σ contains an involutory homology β fixing A, B, m. If B is the centre of β then, since $A^\beta = A$, the axis of β passes through A and so, by Lemma 4.22, $\alpha\beta$ is an involutory homology with affine axis. This contradiction shows that β must have centre A and axis m. Hence A is the centre of a homology with axis h for any affine line h through B and so, by the dual of Corollary 1 to Theorem 4.25, $\mathscr{B}$ is (A, l_∞)-transitive.

Case Ia. Σ does not fix A.

In this case Σ maps A onto at least one other point $A' \neq A$ and $\mathscr{B}$ is both (A, l_∞) and (A', l_∞)-transitive which, by Theorem 4.19, implies that $\mathscr{B}$ is a translation plane.

Case Ib. Σ fixes A.

Let P, q be any affine flag such that $C = l_\infty \cap q \neq A$. By condition (i) there exists an involutory homology γ fixing P and q. Since γ also fixes A its centre must be either C or A. If, for every choice of q, γ has centre A and axis q then, by the dual of Corollary 2 to Theorem 4.25, $\mathscr{B}$ is the dual of a translation plane. If, on the other hand, for some line q, γ has centre C then our earlier argument shows that $\mathscr{B}$ is (C, l_∞)-transitive and hence, by Theorem 4.19, a translation plane.

Case II. Only one affine point is the centre of an involutory homology in Σ.

If we denote this point by O then, clearly, $O^\gamma = O$ for all $\gamma \in \Sigma$. Let A, B be any distinct points on l_∞ and let P be any point of OB distinct from O and B. Then, by condition (i), there is an involutory homology $\alpha \in \Sigma$ fixing the flag P, AP. Clearly α is an (A, OB)-homology and so, since such a α exists for all choices of A and B on l_∞, $\mathscr{B}$ has property $\mathscr{H}$.

Case III. There exists at least two affine points which are centres of involutory homologies in Σ.

Let O be one of these centres, let A and B be any two points of l_∞ and let l be any affine line through B. Then since there is an involutory homology fixing A, B and OB, there exists an involutory homology fixing A, B and l.

Case IIIa. Suppose some line m does not contain an affine point which is the centre of an involutory homology.

Let $X = l_\infty \cap m$ and let Y be any point of l_∞ distinct from X. As we have noted above there exists an involutory homology α fixing X, Y and m and, since m does not contain any homology centre, either α has centre X and axis through Y or centre Y and axis m. In either case α has order 2 on the points of $l_\infty \setminus X$ and fixes only Y of this set. Such an α exists for all Y in $l_\infty \setminus X$ and thus, by Lemma 13.5, the group generated by the α's is transitive on $l_\infty \setminus X$. Furthermore, since at least two affine points are homology centres, Σ contains a translation (by Theorem 4.25).

Suppose Σ fixes X. Let C be any affine point of m and B be any point of l_∞ other than X. By condition (i) there is an involutory homology fixing the flag C, CB and, hence, fixing X, B and C. Using the above argument it follows that $\Sigma_{X,C}$ is transitive on $l_\infty \setminus X$.

If γ is the involutory homology fixing X, B, C then γ is either a (X, BC) or a (B, XC)-homology. If γ has centre X then, by taking the conjugates of γ in $\Sigma_{X,C}$ and applying Lemma 4.11, Σ contains an involutory (X, h)-homology for all lines h through C other than XC. (Note that since $\Sigma_{X,C}$ is transitive on the points of $l_\infty \setminus X$ it is transitive on the lines through C other than XC.) From the dual of Corollary 2 of Theorem 4.26 it now follows that $\mathscr{B}$ is (X, XC)-transitive.

If, on the other hand, γ has centre B then, by the dual argument to that used above, every point of $l_\infty \setminus X$ is the centre of an involutory homology with axis XC and $\mathscr{B}$ is (X, XC)-transitive.

By assumption there is an affine point not on XC which is the centre of an involutory homology α. Since $(XC)^\alpha \neq XC$, $\mathscr{B}$ is (X, l) transitive for at least two lines through X and hence, by the dual of Theorem 4.19, is the dual of a translation plane.

Case IIIb. Every affine line contains an affine point which is the centre of an involutory homology.

If A is any point of l_∞ and l is any affine line through A then any involutory homology with affine centre on l fixes l and no other affine line through A. Hence, by Lemma 13.5, Σ_A is transitive on the affine lines through A for all $A \in l_\infty$. Hence, by Theorem 14.7, Σ is transitive on the affine points of $\mathscr{B}$ and thus, by Corollary 2 to Theorem 4.25, $\mathscr{B}$ is a translation plane.

We are now in a position to complete the proof of Wagner's theorem. By Lemmas 14.12, 14.13 any minimal 2-subplane $\mathscr{B}$ of $\mathscr{A}$ is either a translation plane, the dual of a translation plane or of type $\mathscr{H}$. But translation planes and planes of type $\mathscr{H}$ all have prime power order and so, since, by Theorem 14.10, the order of $\mathscr{A}$ is a power of the order of $\mathscr{B}$, the order of $\mathscr{A}$ is a power of a prime. Lemma 14.8 now completes the proof. $\square$

As a corollary to Wagner's theorem we prove the celebrated Ostrom-Wagner theorem.

Theorem 14.13 (The Ostrom-Wagner Theorem). *Let $\mathscr{P}$ be a finite projective plane with a collineation group Π which is doubly transitive on the points of $\mathscr{P}$. Then $\mathscr{P}$ is desarguesian and Π contains the little projective group of $\mathscr{P}$.*

Proof. By Exercise 13.5 Π is doubly transitive on the lines of $\mathscr{P}$ so that Π_l is transitive on the affine lines of $\mathscr{P}^l$. By Wagner's theorem $\mathscr{P}^l$ is a translation plane with respect to l and hence, by the transitivity on lines, $\mathscr{P}$ is a Moufang plane. But, by the Corollary to Theorem 6.20 a finite Moufang plane is desarguesian.

That Π contains the little projective group follows from the fact that Π contains all elations with axis l for any l in $\mathscr{P}$. □

Wagner's theorem poses two interesting questions. What are the translation planes $\mathscr{A}$ satisfying the theorem? What are the groups satisfying the theorem? The second question has been answered when $\mathscr{A}$ is desarguesian and Π is soluble in [2]. (In the same paper two non-isomorphic planes of order 25 which satisfy the theorem are found.)

Exercise 14.1. Show that the full collineation group of any Hall plane of order greater than 9 is not transitive on the affine lines.

**Exercise 14.2. Show that the full collineation group of the Hall plane $\mathscr{A}$ of order 9 is transitive on the affine lines of $\mathscr{A}$.

8. Appendix (Proof of Some Permutation Group Theorems)

We now give proofs of Results 14.1 and 14.2.

Result 14.2. *Let Λ be a Frobenius permutation group of order $2n$ on a set $\mathscr{S}$ of n points, where n is odd. Then*

(i) *given any two distinct points a, b of $\mathscr{S}$ there exists an element of order 2 in Λ which interchanges a and b*

(ii) *if Λ' is a subgroup of Λ containing exactly r elements of order 2, $r \geqq 1$, then the order of Λ' is $2r$.*

Proof. Since n is odd every element of order 2 in Λ fixes a point of $\mathscr{S}$. Furthermore, since $|\Lambda| = 2n$, Λ_a has order 2, for any $a \in \mathscr{S}$. Thus there is a one-to-one correspondence between the elements of order 2 in Λ and the points of $\mathscr{S}$. If α is the element of order 2 fixing the point a, we shall denote it by α_a.

(i) Let a, b be any two distinct points of $\mathscr{S}$. If c, d are such that $a^{\alpha_c} = a^{\alpha_d} \neq a$ then $\alpha_c \alpha_d^{-1}$ fixes both a and a^{α_d} which, since Λ is a Frobenius group, implies $\alpha_c = \alpha_d$ and $c = d$. Consequently a has $n-1$ distinct images a^{α_c} as c varies over $\mathscr{S}\setminus\{a\}$. One of these must be b, which proves (i).

(ii) Let α_a be an element of order 2 in Λ', let $\mathscr{T}$ be the orbit of Λ' containing a and let $|\mathscr{T}| = t$. Since α_a fixes only one point of $\mathscr{T}$, t is odd. Furthermore since $\Lambda'_a \subseteq \Lambda_a$ and $|\Lambda_a| = 2 \geq |\Lambda'_a|$, $\Lambda'_a = \Lambda_a$ and hence, the order of Λ' is $2t$. Since Λ' is a subgroup of the Frobenius group Λ, only the identity of Λ' fixes 2 points of $\mathscr{T}$ so that Λ' is a Frobenius group on $\mathscr{T}$. The argument used at the beginning of the proof now establishes a one-to-one correspondence between the elements of order 2 in Λ' and the points of $\mathscr{T}$. Thus $r = t$ and $|\Lambda'| = 2r$. ☐

Result 14.1. *Let Π be a doubly transitive group on a set $\mathscr{S}$ consisting of $n+1$ points, where n is odd. If a is a point of $\mathscr{S}$ let Π_a contain a subgroup Γ_a of even order which is a Frobenius group on $\mathscr{S}\setminus\{a\}$. Then Π_a is primitive on $\mathscr{S}\setminus\{a\}$.*

Proof. Since Π is transitive on $\mathscr{S}$ if follows at once by taking conjugates of Π_a that for every $s \in \mathscr{S}$ there is a subgroup Γ_s which fixes s and acts as a Frobenius group on $\mathscr{S}\setminus\{s\}$. Let Σ_s be the kernel of Γ_s and let Γ'_s be the group generated by Σ_s and an element of order 2 in Γ_s. Since the element of order 2 in Γ_s fixes exactly one element of $\mathscr{S}\setminus\{s\}$ it does not belong to Σ_s so that, by the normality of Σ_s in Γ_s, $|\Gamma'_s| = 2|\Sigma_s| = 2n$. Clearly Γ'_s is a Frobenius group on $\mathscr{S}\setminus\{s\}$.

The proof of the result will now be by contradiction. We shall assume that Π_a is imprimitive on $\mathscr{S}\setminus\{a\}$ and let $\mathscr{T}_1, \mathscr{T}_2, \ldots, \mathscr{T}_r$ denote the sets of imprimitivity on $\mathscr{S}\setminus\{a\}$, where $r \geq 2$. The number of points in each of these sets of imprimitivity is a constant, t say. Clearly $t \mid n$ so that t is odd.

Let $b_1, b_2, \ldots, b_t$ denote the points of $\mathscr{T}_1$, $t \geq 2$, and let $c \in \mathscr{T}_r$. For any i, $(i = 1, \ldots, t)$, we define γ_i to be an involution in Γ'_c which interchanges a and b_i. (That the γ_i exist follows from (i) of Result 14.2.)

If we let $\mathscr{T}$ be the set $\{a\} \cup \mathscr{T}_1$, then we now show that $\mathscr{T}^{\gamma_i} = \mathscr{T}$ for $i = 1, 2, \ldots, t$. Certainly, by definition, $a^{\gamma_i} \in \mathscr{T}$, so suppose that for some j and k, $b_k^{\gamma_j} = d \notin \mathscr{T}$ and let δ be an element of order 2 in Γ'_d which interchanges a with b_j. Then $\gamma_j \delta$ fixes a, b_j and maps b_k onto d. Hence $\gamma_j \delta \in \Pi_a$ but does not preserve the system of imprimitivity $\mathscr{T}_1$. This contradiction proves that $\mathscr{T}^{\gamma_i} = \mathscr{T}$.

Now let Δ be the group generated by $\gamma_1, \gamma_2, \ldots, \gamma_t$. Clearly Δ is transitive on $\mathscr{T}$ and, since Δ is a subgroup of Γ'_c which is a Frobenius group on $\mathscr{S}\setminus\{c\}$, every non-identity element of Δ either fixes no point of $\mathscr{S}$ except c or has order 2 and fixes exactly one other point. However $|\mathscr{T}| = t+1$ which is even, so that no element of order 2 in Δ can fix any point of $\mathscr{T}$. Thus Δ is regular on $\mathscr{T}$ and $|\Delta| = t+1$. But, since Δ contains at least t elements of order 2, namely the γ_i, it follows from part (ii) of Result 14.2 that the order of Δ is at least $2t$. This leads to the contradiction $t+1 \geq 2t$ (impossible since $t > 1$), and proves Result 14.1. ☐

References

The soluble collineation groups of the affine desarguesian plane are discussed in [2].

1. Burnside, W.: Theory of groups of finite order. New York: Dover 1955.
2. Foulser, D. A.: The flag transitive collineation groups of the finite desarguesian affine planes. Can. J. Math. 16, 443–472 (1964).
3. Wagner, A.: On finite affine line transitive planes. Math. Z. 87, 1–11 (1965).
4. Wielandt, H.: Finite permutation groups. New York: Academic Press 1964.

XV. Appendix

In this brief chapter we discuss some topics which are either related to the material in this book, or follow logically upon it.

There are a number of the following type of "configurational theorems" in projective planes: a certain configurational condition is satisfied in a finite projective plane if and only if the plane is desarguesian. Many of these results can be found in [6] but a considerable amount of work has been done in this area since that book was published. Gleason's proof that a finite Fano plane (i.e. one in which every quadrangle generates a subplane of order 2), is desarguesian is perhaps the most spectacular result in this field. This theorem is connected with many other beautiful and subtle results, as well as many unsolved problems. A brief discussion of these results and the references for further reading are in Section 3.4 of [3].

Another large area of research is that of the theory of collineation groups. The work on group theory and its connections with projective planes, especially finite planes, has been extensive and, on occasions, has involved very deep group theory indeed. Considerable work has been done on questions of the sort: if the plane $\mathscr{P}$ has a collineation group which acts transitively (or two-transitively or three-transitively) on a certain part of $\mathscr{P}$, then what can we say about $\mathscr{P}$ and/or the group? Wagner's theorem and the Ostrom-Wagner theorem (see Chapter XIV), are the two classic examples and have paved the way to many other results using similar techniques. Lüneburg and others have investigated a rather different type of problem: if a certain group is known to act on a finite projective plane $\mathscr{P}$ then what can be said about $\mathscr{P}$? Often problems of this type require a detailed knowledge of the structure of the given group. Once again references to these results are in [3] (Section 4.2).

The Lenz-Barlotti Classification (see Chapter IV) has led to many attractive results in the subject, although a number of questions remain unsolved. Besides Section 3.1 of [3] an excellent survey article on this topic is in [8].

As well as further problems and generalizations which are really concerned with the incidence properties of projective planes, there is also the topological side to these planes. Indeed this is one of the most intuitively evident aspects of geometry. Considerable work has been done on topological projective planes, above all by a German school including Pickert and Salzmann (see, for example, [7]). Similar considerations have led to Hjelmslev planes which are defined in [3].

The area of *inversive* (or *Möbius*) *planes* is closely related to projective and affine planes, and cannot be studied without a good background in projective planes, plus a little in projective geometries of g-dimension 3. The beautiful theorem of Dembowski, proving that finite inversive planes of even order are "egglike", was followed by a wave of activity in this field, most of which is in [3]. The central problem at the moment appears to be the classification of inversive planes of odd order in a somewhat similar manner; but there is also the very interesting question of whether there are any more inversive planes of even order (besides the miquelian, or "standard" ones, and the examples constructed from the Suzuki groups).

These considerations would lead naturally to considering "designs", that is, sets of points with certain distinguished subsets to be called *blocks*, with certain incidence properties. Usually we want to insist that each block contain the same number k of points, and that for some integer t, all subsets of t distinct points are contained in the same number λ of blocks. Then the object is called a *t-design for* (v, k, λ), where v is the number of points; it is easy to see that a t-design is also an s-design, for any positive integer $s < t$. The exercises at the end of Chapter IV show that finite projective planes are exactly the 2-designs for $(n^2 + n + 1, n + 1, 1)$, while finite affine planes are exactly the 2-designs for $(n^2, n, 1)$; finite inversive planes are exactly the 3-designs for $(n^2 + 1, n + 1, 1)$. But many other examples of t-designs are known, and they fall into a large number of families of various sorts. It is interesting that except for trivial examples (i.e., every subset of k points is a block), no 6-designs (and hence no t-designs with $t \geq 6$) are known. Much of [3] is actually concerned with these t-designs, and slight generalizations of them.

As is evident, Dembowski [3] is the most important single reference in connection with additional material; no serious student of the field can be without it. Pickert's book [6] covers a much wider area, including infinite problems (which Dembowski rigourously excludes) and problems of a more obviously "geometric" nature. A number of other books touch on projective planes in one way or another. Almost all modern books on projective geometry have at least some reference to the "pathological" situation in g-dimension 2, and some of these books approach planes in a manner somewhat like ours in Chapters III to VI; although their treatment is usually more compressed. Also, it is becoming more popular and usual to pay some attention to geometry as a kind of applied linear algebra, or as the proper setting for many problems of linear algebra; this is especially true of the influential *Geometric Algebra* of Artin [1], and the excellent *Linear Geometry* of Gruenberg and Weir [4]. At least two group theory books deal with projective planes and designs in some detail:

Carmichael [2], and Hall [5]. The treatment in Carmichael is not always as rigourous or careful as one might wish, but many examples (especially of designs and so on) are listed there; the last chapter of Hall has a very extensive treatment of selected topics from projective planes, usually with a group theory slant, and some interesting material which we have excluded is to be found there.

References

1. Artin, E.: Geometric Algebra. New York-London: Interscience 1957.
2. Carmichael, R. D.: Introduction to the theory of groups of finite order. New York: Dover 1956 (Reprint).
3. Dembowski, H. P.: Finite Geometries. Berlin-Heidelberg-New York: Springer 1968.
4. Gruenburg, K. W., Weir, A. J.: Linear Geometry. New York-London: Van Nostrand 1967.
5. Hall, M.: The Theory of Groups. New York: Macmillan 1959.
6. Pickert, G.: Projektive Ebenen. Berlin-Göttingen-Heidelberg: Springer 1955.
7. Salzmann, H.: Topological planes. Advan. Math. 2, 1–60 (1967).
8. Yaqub, J. C. D. S.: The Lenz-Barlotti classification. Proc. Proj. Geometry Conference, Univ. of Illinois Chicago, 1967.

Index

Index

Graduate Texts in Mathematics

Soft and hard cover editions are available for each volume.

For information

A student approaching mathematical research is often discouraged by the sheer volume of the literature and the long history of the subject, even when the actual problems are readily understandable. The new series, Graduate Texts in Mathematics, is intended to bridge the gap between passive study and creative understanding; it offers introductions on a suitably advanced level to areas of current research. These introductions are neither complete surveys, nor brief accounts of the latest results only. They are textbooks carefully designed as teaching aids; the purpose of the authors is, in every case, to highlight the characteristic features of the theory.

Graduate Texts in Mathematics can serve as the basis for advanced courses. They can be either the main or subsidiary sources for seminars, and they can be used for private study. Their guiding principle is to convince the student that mathematics is a living science.